Theodore E. Simos
Editor

Recent Advances in Computational and Applied Mathematics

Springer

Editor
Theodore E. Simos
Department of Mathematics
College of Sciences
King Saud University
P.O. Box 2455
Riyadh 11451
Saudi Arabia
and
Laboratory of Computational Sciences
Department of Computer Science
and Technology
University of Peloponnese
22100 Tripolis
Greece
tsimos.conf@gmail.com

ISBN 978-90-481-9980-8 e-ISBN 978-90-481-9981-5
DOI 10.1007/978-90-481-9981-5
Springer Dordrecht Heidelberg London New York

Library of Congress Control Number: 2010938640

Mathematics Subject Classification (2000): 65, 65D20, 65Rxx, 65Lxx, 30, 30G35, 31B05, 31B35, 31, 33C05, 33C10, 33C15, 35A08, 35G15, 35Q40, 35R01, 41A60, 49Jxx, 49Nxx, 76-xx, 76A05, 81Q05, 81Qxx, 93Exx

Cover design: WMXDesign

Printed on acid-free paper

Springer is part of Springer Science+Business Media (www.springer.com)

Preface

This volume includes an exciting collection of papers on computational and applied mathematics presenting the recent advances in several areas of this field. All the papers have been peer reviewed by at least three reviewers.

In the paper entitled: "Fifty Years of Stiffness" by Luigi Brugnano, Francesca Mazzia and Donato Trigiante a review on the evolution of stiffness is presented. The authors also given a precise definition of stiffness which encompasses all the previous ones.

In the paper entitled: "Efficient Global Methods for the Numerical Solution of Nonlinear Systems of Two point Boundary Value Problems" by Jeff R. Cash and Francesca Mazzia, the authors investigated the numerical methods for the solution of nonlinear systems of two point boundary value problems in ordinary differential equations. More specifically they answer to the question: "which codes are currently available for solving these problems and which of these codes might we consider as being state of the art". Finally the authors included some new codes for BVP's which are written in MATLAB. These codes was not available before and allow us for the first time in the literature the possibility of comparing some important MATLAB codes for solving boundary value problems.

In the paper entitled: "Advances on collocation based numerical methods for Ordinary Differential Equations and Volterra Integral Equations" by D. Conte, R. D'Ambrosio, B. Paternoster a survey on collocation based numerical methods for the numerical integration of Ordinary Differential Equations and Volterra Integral Equations (VIEs) is presented. This survey starts from the classical collocation methods and arrive to the important modifications appeared in the literature. The authors consider also the multistep case and the usage of basis of functions other than polynomials.

In the paper entitled: "Basic Methods for Computing Special Functions" by Amparo Gil, Javier Segura and Nico M. Temme, the authors given a survey of methods for the numerical evaluation of special functions, that is, the functions that arise in many problems in the applied sciences. They considered a selection of basic methods which are used frequently in the numerical evaluation of special functions. They discussed also several other methods which are available. Finally, they given examples of recent software for special functions which use the above mentioned methods and they mentioned a list of new bibliography on computational aspects of special functions available on our website.

In the paper entitled: "Melt Spinning: Optimal Control and Stability Issue" by Thomas Gotz and Shyam S.N. Perera, the authors studied a mathematical model which describe the melt spinning process of polymer fibers. The authors used Newtonian and non-Newtonian models in order to describe the rheology of the polymeric material. They also investigated two important properties, the optimization and the stability of the process.

In the paper entitled: "On orthonormal polynomial solutions of the Riesz system in $\mathbb{R}^3$" by K. Gürlebeck and J. Morais, a special orthogonal system of polynomial solutions of the Riesz system in $\mathbb{R}^3$ is studied. This system presents a proportion with the complex case of the Fourier exponential functions $\{e^{in\theta}\}_{n\geq 0}$ on the unit circle and has the additional property that also the scalar parts of the polynomials form an orthogonal system. An application of the properties of the above system to the explicit calculation of conjugate harmonic functions with a certain regularity is also presented.

In the paper entitled: "Brief survey on the CP methods for the Schrödinger equation" by L.Gr. Ixaru, a review of the CP methods is presented. The authors investigated, after years of research in the subject all the advantages over other methods.

In the paper entitled: "Symplectic Partitioned Runge–Kutta methods for the numerical integration of periodic and oscillatory problems" by Z. Kalogiratou, Th. Monovasilis and T.E. Simos an investigation on Symplectic Partitioned Runge–Kutta methods (SPRK) is presented. More specifically they present the methodology for the construction of the exponentially/trigonometrically fitted SPRK. They applied the above methodology to methods with corresponding order up to fifth. The trigonometrically-fitted approach is based on two different types of construction: (i) fitting at each stage and (ii) Simos's approach. The authors also derived SPRK methods with minimal phase-lag as well as phase-fitted SPRK methods. Finally, they applied the methods to several problems.

In the paper entitled: "On the Klein-Gordon equation on some examples of conformally flat spin 3-manifolds" by Rolf Sören Kraußhar a review about recent results on the analytic treatment of the Klein-Gordon equation on some conformally flat 3-tori and on 3-spheres is presented. The paper has two parts. In the first part the time independent Klein-Gordon equation $(\Delta - \alpha^2)u = 0$ $(\alpha \in \mathbb{R})$ on some conformally flat 3-tori associated with a representative system of conformally inequivalent spinor bundles is considered. In the second part a unified approach to represent the solutions to the Klein-Gordon equation on 3-spheres is described.

The *hp* version of the finite element method (*hp*-FEM) combined with adaptive mesh refinement is a particularly efficient method. For this method a single error estimate can not simultaneously determine whether it is better to do the refinement by h or by p. Several strategies for making this determination have been proposed over the years. In the paper entitled: "A Survey of *hp*-Adaptive Strategies for Elliptic Partial Differential Equations" by William F. Mitchell and Marjorie A. McClain, the authors studied these strategies and demonstrate the exponential convergence rates with two classic test problems.

In the paper entitled: "Vectorized Solution of ODEs in MATLAB with Control of Residual and Error" by L.F. Shampine a study on vectorization which is very important to the efficient computation in the popular problem-solving environment MATLAB is presented. More specifically, the author derived a new error control procedure which is based on vectorization. An explicit Runge—Kutta (7,8) pair of formulas that exploits vectorization is obtained. The new proposed method controls the local error at 8 points equally spaced in the span of a step. A new solver which is based on the above mentiobed pair and it is called `odevr7` is developed. This solver is much more efficient than the solver `ode45` which is recommended by MATLAB.

In the paper entitled: "Forecasting equations in complex-quaternionic setting" by W. Sprössig, the author considered classes of fluid flow problems under given initial value and boundary value conditions on the sphere and on ball shells in $\mathbb{R}^3$. The author interest is emphasized to the forecasting equations and the deduction of a suitable quaternionic operator calculus.

In the paper entitled: "Symplectic exponentially-fitted modified Runge–Kutta methods of the Gauss type: revisited" by G. Vanden Berghe and M. Van Daele, the development of symmetric and symplectic exponentially-fitted Runge–Kutta methods for the numerical integration of Hamiltonian systems with oscillatory solutions is studied. New integrators are obtained following the six-step procedure of Ixaru and Vanden Berghe (*Exponential Fitting*, Kluwer Academic, 2004).

We would like to express our gratitude to the numerous (anonymous) referees, to Prof. Hélène de Rode, the President of the European Academy of Sciences for giving us the opportunity to come up with this guest editorial work.

University of Peloponnese T.E. Simos

Contents

Contributors

Luigi Brugnano Dipartimento di Matematica, Università di Firenze, Viale Morgagni 67/A, 50134 Firenze, Italy, luigi.brugnano@unifi.it

Jeff R. Cash Department of Mathematics, Imperial College, South Kensington, London SW7, UK, j.cash@imperial.ac.uk

Dajana Conte Dipartimento di Matematica e Informatica, Università di Salerno, Fisciano, Italy, dajconte@unisa.it

Raffaele D'Ambrosio Dipartimento di Matematica e Informatica, Università di Salerno, Fisciano, Italy, rdambrosio@unisa.it

M. Daele Vakgroep Toegepaste Wiskunde en Informatica, Universiteit Gent, Krijgslaan 281-S9, 9000 Gent, Belgium

Thomas Götz Department of Mathematics, TU Kaiserslautern, 67663 Kaiserslautern, Germany, goetz@mathematik.uni-kl.de

K. Gürlebeck Institut für Mathematik/Physik, Bauhaus-Universität Weimar, Coudraystr. 13B, 99421 Weimar, Germany, klaus.guerlebeck@uni-weimar.de

Amparo Gil Departamento de Matemática Aplicada y CC. de la Computación, ETSI Caminos, Universidad de Cantabria, 39005 Santander, Spain amparo.gil@unican.es

L.Gr. Ixaru Department of Theoretical Physics, "Horia Hulubei" National Institute of Physics and Nuclear Engineering, P.O. Box MG-6, Bucharest, Romania, ixaru@theory.nipne.ro; Academy of Romanian Scientists, 54 Splaiul Independentei, 050094, Bucharest, Romania

Z. Kalogiratou Department of Informatics and Computer Technology, Technological Educational Institute of Western Macedonia at Kastoria, P.O. Box 30, 521 00, Kastoria, Greece

Francesca Mazzia Dipartimento di Matematica, Università di Bari, Via Orabona 4, 70125 Bari, Italy, mazzia@dm.uniba.it

Francesca Mazzia Dipartimento di Matematica, Università di Bari, Via Orabona 4, 70125 Bari, Italy

Marjorie A. McClain Mathematical and Computational Sciences Division, National Institute of Standards and Technology, Gaithersburg, MD 20899-8910, USA

William F. Mitchell Mathematical and Computational Sciences Division, National Institute of Standards and Technology, Gaithersburg, MD 20899-8910, USA, william.mitchell@nist.gov

Th. Monovasilis Department of International Trade, Technological Educational Institute of Western Macedonia at Kastoria, P.O. Box 30, 521 00, Kastoria, Greece

J. Morais Institut für Mathematik/Physik, Bauhaus-Universität Weimar, Coudraystr. 13B, 99421 Weimar, Germany, jmorais@mat.ua.pt

Beatrice Paternoster Dipartimento di Matematica e Informatica, Università di Salerno, Fisciano, Italy, beapat@unisa.it

Shyam S.N. Perera Department of Mathematics, University of Colombo, Colombo 03, Sri Lanka, ssnp@maths.cmb.ac.lk

Rolf Sören Kraußhar Fachbereich Mathematik, Technische Universität Darmstadt, Schloßgartenstraße 7, 64289 Darmstadt, Germany
krausshar@mathematik.tu-darmstadt.de

Javier Segura Departamento de Matemáticas, Estadística y Computación, Universidad de Cantabria, 39005 Santander, Spain, javier.segura@unican.es

L.F. Shampine 1204 Chesterton Dr., Richardson, TX 75080, USA
lfshampine@aol.com

T.E. Simos Laboratory of Computational Sciences, Department of Computer Science and Technology, Faculty of Science and Technology, University of Peloponnessos, 22100, Tripolis, Greece, tsimos@mail.ariadne-t.gr; Department of Mathematics, College of Sciences, King Saud University, P.O. Box 2455, Riyadh 11451, Saudi Arabia

W. Sprössig Fakultaet fuer Mathematik und Informatik, TU Bergakademie Freiberg, Prueferstraße 9, 09596 Freiberg, Germany, sproessig@math.tu-freiberg.de

Nico M. Temme CWI, Science Park 123, 1098 XG Amsterdam, The Netherlands, Nico.Temme@cwi.nl

Donato Trigiante Dipartimento di Energetica, Università di Firenze, Via Lombroso 6/17, 50134 Firenze, Italy, trigiant@unifi.it

G. Vanden Berghe Vakgroep Toegepaste Wiskunde en Informatica, Universiteit Gent, Krijgslaan 281-S9, 9000 Gent, Belgium, guido.vandenberghe@ugent.be

Chapter 1
Fifty Years of Stiffness

**Luigi Brugnano, Francesca Mazzia,
and Donato Trigiante**

Abstract The notion of *stiffness*, which originated in several applications of a different nature, has dominated the activities related to the numerical treatment of differential problems for the last fifty years. Contrary to what usually happens in Mathematics, its definition has been, for a long time, not formally precise (actually, there are too many of them). Again, the needs of applications, especially those arising in the construction of robust and general purpose codes, require nowadays a formally precise definition. In this paper, we review the evolution of such a notion and we also provide a precise definition which encompasses all the previous ones.

Keywords Stiffness · ODE problems · Discrete problems · Initial value problems · Boundary value problems · Boundary value methods

Mathematics Subject Classification (2000) 65L05 · 65L10 · 65L99

Frustra fit per plura quod potest per pauciora.
Razor of W. of Ockham, doctor invincibilis.

Work developed within the project "Numerical methods and software for differential equations".

L. Brugnano (✉)
Dipartimento di Matematica, Università di Firenze, Viale Morgagni 67/A, 50134 Firenze, Italy
e-mail: luigi.brugnano@unifi.it

F. Mazzia
Dipartimento di Matematica, Università di Bari, Via Orabona 4, 70125 Bari, Italy
e-mail: mazzia@dm.uniba.it

D. Trigiante
Dipartimento di Energetica, Università di Firenze, Via Lombroso 6/17, 50134 Firenze, Italy
e-mail: trigiant@unifi.it

T.E. Simos (ed.), *Recent Advances in Computational and Applied Mathematics*,
DOI 10.1007/978-90-481-9981-5_1,

1.1 Introduction

The struggle generated by the duality short times–long times is at the heart of human culture in almost all its aspects. Here are just a few examples to fix the idea:

- in historiography: Braudel's distinction among the geographic, social and individual times;[1]
- in the social sphere: Societies are organized according to three kinds of laws, i.e., codes (regulating short term relations), constitutions (regulating medium terms relations), and ethical laws (long term rules) often not explicitly stated but religiously accepted;
- in the economy sphere: the laws of this part of human activities are partially unknown at the moment. Some models (e.g., the Goodwin model [19]), permits us to say, by taking into account only a few variables, that the main evolution is periodic in time (and then predictable), although we are experiencing an excess of periodicity (chaotic behavior). Nevertheless, some experts claim (see, e.g., [18]) that the problems in the predictability of the economy are mainly due to a sort of gap in passing information from a generation to the next ones, i.e. to the conflict between short time and long time behaviors.[2]

Considering the importance of this concept, it would have been surprising if the duality "short times–long times" did not appear somewhere in Mathematics. As a matter of fact, this struggle not only appears in our field but it also has a name: *stiffness*.

Apart from a few early papers [10, 11], there is a general agreement in placing the date of the introduction of such problems in Mathematics to around 1960 [17]. They were the necessities of the applications to draw the attention of the mathematical community towards such problems, as the name itself testifies: "*they have been termed stiff since they correspond to tight coupling between the driver and the driven components in servo-mechanism*" ([12] quoting from [11]).

Both the number and the type of applications proposing difficult differential problems has increased exponentially in the last fifty years. In the early times, the problems proposed by applications were essentially initial value problems and, consequently, the definition of stiffness was clear enough and shared among the few experts, as the following three examples evidently show:

D1: *Systems containing very fast components as well as very slow components* (Dahlquist [12]).

D2: *They represent coupled physical systems having components varying with very different times scales: that is they are systems having some components varying much more rapidly than the others* (Liniger [31], translated from French).

---

[1]Moreover, his concept of *structure*, i.e. events which are able to accelerate the normal flow of time, is also interesting from our point of view, because it somehow recalls the mathematical concept of large variation in small intervals of time (see later).

[2]Even Finance makes the distinction between short time and long time traders.

D3: *A stiff system is one for which λ_{max} is enormous so that either the stability or the error bound or both can only be assured by unreasonable restrictions on h... Enormous means enormous relative to the scale which here is $\bar{t}$ (the integration interval)*... (Miranker [34]).

The above definitions are rather informal, certainly very far from the precise definitions we are accustomed to in Mathematics, but, at least, they agree on a crucial point: the relation among stiffness and the appearance of different time-scales in the solutions (see also [24]).

Later on, the necessity to encompass new classes of difficult problems, such as Boundary Value Problems, Oscillating Problems, etc., has led either to weaken the definition or, more often, to define some consequence of the phenomenon instead of defining the phenomenon itself. In Lambert's book [29] five propositions about stiffness, each of them capturing some important aspects of it, are given. As matter of fact, it has been also stated that no universally accepted definition of stiffness exists [36].

There are, in the literature, other definitions based on other numerical difficulties, such as, for example, large Lipschitz constants or logarithmic norms [37], or non-normality of matrices [23]. Often is not even clear if stiffness refers to particular solutions (see, e.g. [25]) or to problems as a whole.

Sometimes one has the feeling that stiffness is becoming so broad to be nearly synonymous of difficult.

At the moment, even if the old intuitive definition relating stiffness to multiscale problems survives in most of the authors, the most successful definition seems to be the one based on particular effects of the phenomenon rather than on the phenomenon itself, such as, for example, the following almost equivalent items:

D4: *Stiff equations are equations where certain implicit methods ... perform better, usually tremendous better, than explicit ones* [11].

D5: *Stiff equations are problems for which explicit methods don't work* [21].

D6: *If a numerical method with a finite region of absolute stability, applied to a system with any initial condition, is forced to use in a certain interval of integration a step length which is excessively small in relation to the smoothness of the exact solution in that interval, then the system is said to be stiff in that interval* [29].

As usually happens, describing a phenomenon by means of its effects may not be enough to fully characterize the phenomenon itself. For example, saying that fire is what produces ash, would oblige firemen to wait for the end of a fire to see if the ash has been produced. In the same way, in order to recognize stiffness according to the previous definitions, it would be necessary to apply first one[3] explicit method and see if it works or not. Some authors, probably discouraged by the above defeats in giving a rigorous definition, have also affirmed that a rigorous mathematical definition of stiffness is not possible [20].

It is clear that this situation is unacceptable for at least two reasons:

[3]It is not clear if one is enough: in principle the definition may require to apply all of them.

- it is against the tradition of Mathematics, where objects under study have to be *precisely* defined;
- it is necessary to have the possibility to recognize *operatively* this class of problems, in order to increase the efficiency of the numerical codes to be used in applications.

Concerning the first item, our opinion is that, in order to gain in precision, it would be necessary to revise the concept of *stability* used in Numerical Analysis, which is somehow different from the homonym concept used in all the other fields of Mathematics, where stable are equilibrium points, equilibrium sets, reference solutions, etc., but not equations or problems[4] (see also [17] and [30]).

Concerning the second item, *operatively* is intended in the sense that the definition must be stated in terms of *numerically observable* quantities such as, for example, norms of vectors or matrices. It was believed that, seen from the applicative point of view, a formal definition of stiffness would not be strictly necessary: *Complete formality here is of little value to the scientist or engineer with a real problem to solve* [24].

Nowadays, after the great advance in the quality of numerical codes,[5] the usefulness of a formal definition is strongly recognized, also from the point of view of applications: *One of the major difficulties associated with the study of stiff differential systems is that a good mathematical definition of the concept of stiffness does not exist* [6].

In this paper, starting from ideas already partially exposed elsewhere [2, 4, 26], we will try to unravel the question of the definition of stiffness and show that a precise and operative definition of it, which encompasses all the known facets, is possible.

In order to be as clear as possible, we shall start with the simpler case of initial value for a single linear equation and gradually we shall consider more general cases and, eventually, we shall synthesize the results.

1.2 The Asymptotic Stability Case

For initial value problems for ODEs, the concept of stability concerns the behavior of a generic solution $y(t)$, in the neighborhood of a reference solution $\bar{y}(t)$, when the initial value is perturbed. When the problem is linear and homogeneous, the difference, $e(t) = y(t) - \bar{y}(t)$, satisfies the same equation as $\bar{y}(t)$. For nonlinear problems, one resorts to the linearized problem, described by the variational equation, which, essentially, provides valuable information only when $\bar{y}(t)$ is asymptotically stable. Such a variational equation can be used to generalize to nonlinear problems the arguments below which, for sake of simplicity, concerns only the linear case.

[4]Only in particular circumstances, for example in the linear case, it is sometimes allowed the language abuse: the nonlinear case may contain simultaneously stable and unstable solutions.

[5]A great deal of this improvement is due to the author of the previous sentence.

Originally, stiffness was almost always associated with initial value problems having asymptotically stable equilibrium points (dissipative problems) (see, e.g., Dahlquist [13]). We then start from this case, which is a very special one. Its peculiarities arise from the following two facts:[6]

- it is the most common in applications;
- there exists a powerful and fundamental theorem, usually called *Stability in the first approximation Theorem* or *Poincaré-Liapunov Theorem*, along with its corollary due to Perron[7], which allows us to reduce the study of stability of critical points, of a very large class of nonlinearities, to the study of the stability of the corresponding linearized problems (see, e.g., [9, 27, 35, 38]).

The former fact explains the pressure of applications for the treatment of such problems even before the computer age. The latter one provides, although not always explicitly recognized, the mathematical solid bases for the profitable and extensive use, in Numerical Analysis, of the linear test equation to study the fixed-h stability of numerical methods.

We shall consider explicitly the case where the linearized problem is autonomous, although the following definitions will take into account the more general case.

Our starting case will then be that of an initial value problem having an asymptotically stable reference solution, whose representative is, in the scalar case,

$$\begin{aligned} y' &= \lambda y, \quad t \in [0, T], \quad \mathrm{Re}\,\lambda < 0, \\ y(0) &= \eta, \end{aligned} \tag{1.2.1}$$

where the reference solution (an equilibrium point, in this case) has been placed at the origin. From what is said above, it turns out that it is not by chance that it coincides with the famous test equation.

Remark 1.1 It is worth observing that the above test equation is not less general than $y' = \lambda y + g(t)$, which very often appears in the definitions of stiffness: the only difference is the reference solution, which becomes $\bar{y}(t) = \int_0^t e^{\lambda(t-s)} g(s)ds$, but not the topology of solutions around it. This can be easily seen by introducing the new variable $z(t) = y(t) - \bar{y}(t)$ which satisfies exactly equation (1.2.1) and then, trivially, must share the same stiffness. Once the solution $z(t)$ of the homogeneous equation has been obtained, the solution $y(t)$ is obtained by adding to it $\bar{y}(t)$ which, in principle, could be obtained by means of a quadrature formula. This allows us to conclude that if any stiffness is in the problem, this must reside in the homogeneous part of it, i.e., in problem (1.2.1).

[6] We omit, for simplicity, the other fact which could affect new definitions, i.e., the fact that the solutions of the linear equation can be integrated over any large interval because of the equivalence, in this case, between asymptotic and exponential stability.

[7] It is interesting to observe that the same theorem is known as the *Ostrowsky's Theorem*, in the theory of iterative methods.

Remark 1.2 We call attention to the interval of integration $[0, T]$, which depends on our need for information about the solution, even if the latter exists for all values of t. This interval must be considered as datum of the problem. This has been sometimes overlooked, thus creating some confusion.

Having fixed problem (1.2.1), we now look for a mathematical tool which allows us to state formally the intuitive concept, shared by almost all the definitions of stiffness: i.e., we look for one or two parameters which tell us if in $[0, T]$ the solution varies rapidly or not. This can be done easily by introducing the following two measures for the solution of problem (1.2.1):

$$\kappa_c = \frac{1}{|\eta|} \max_{t\in[0,T]} |y(t)|, \qquad \gamma_c = \frac{1}{|\eta|}\frac{1}{T}\int_0^T |y(t)|dt, \tag{1.2.2}$$

which, in the present case, assume the values:

$$\kappa_c = 1, \qquad \gamma_c = \frac{1}{|\mathrm{Re}\,\lambda|T}(1 - e^{\mathrm{Re}\,\lambda T}) \approx \frac{1}{|\mathrm{Re}\,\lambda|T} = \frac{T^*}{T},$$

where $T^* = |\mathrm{Re}\,\lambda|^{-1}$ is the transient time. The two measures κ_c, γ_c are called *conditioning parameters* because they measure the sensitivity of the solution subject to a perturbation of the initial conditions in the infinity and in the l_1 norm.

Sometimes, it would be preferable to use a lower value of γ_c, i.e.,

$$\gamma_c = \frac{1}{|\lambda|T}. \tag{1.2.3}$$

This amounts to consider also the oscillating part of the solution (see also Remark 1.5 below).

By looking at Fig. 1.1, one realizes at once that a rapid variation of the solution in $[0, T]$ occurs when $k_c \gg \gamma_c$. It follows then that the parameter

$$\sigma_c = \frac{k_c}{\gamma_c} \equiv \frac{T}{T^*}, \tag{1.2.4}$$

which is the ratio between the two characteristic times of the problem, is more significant. Consequently, the definition of stiffness follows now trivially:

Definition 1.3 The initial value problem (1.2.1) is *stiff* if $\sigma_c \gg 1$.

The parameter σ_c is called *stiffness ratio*.

Remark 1.4 The width of the integration interval T plays a fundamental role in the definition. This is an important point: some authors, in fact, believe that stiffness should concern equations; some others believe that stiffness should concern problems, i.e., equations and data. We believe that both statements are partially correct: stiffness concerns equations, integration time, and a set of initial data (not a specific one of them). Since this point is more important in the non scalar case, it will be discussed in more detail later.

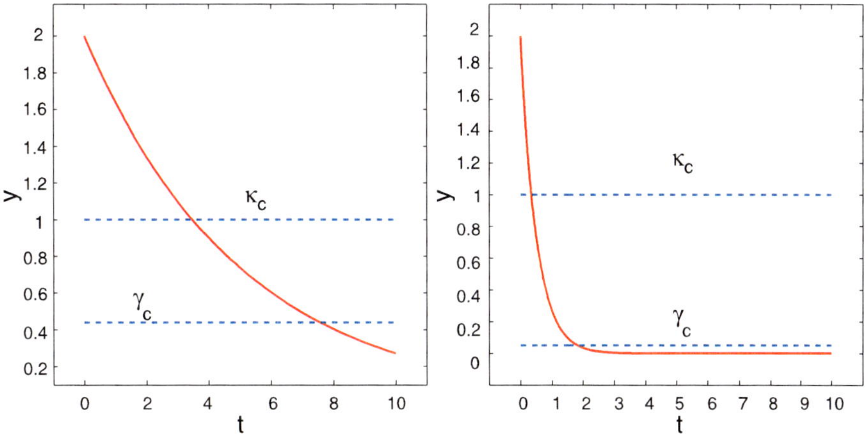

Fig. 1.1 Solutions and values of k_c and γ_c in the cases $\lambda = -0.2$ (*left plot*) and $\lambda = -2$ (*right plot*)

Remark 1.5 When γ_c is defined according to (1.2.3), the definition of stiffness continues to be also meaningful in the case $\mathrm{Re}\,\lambda = 0$, i.e., when the critical point is only marginally stable. In fact, let

$$\lambda = i\omega \equiv i\frac{2\pi}{T^*}.$$

Then,

$$\sigma_c = 2\pi\frac{T}{T^*},$$

and the definition encompasses also the case of *oscillating stiffness* introduced by some authors (e.g., [34]). Once again the stiffness is the ratio of two times. If information about the solution on the smaller time scale is needed, an adequately small stepsize should be used. It is worth noting that high oscillating systems (with respect to T) fall in the class of problems for which explicit methods do not work, and then are stiff according to definitions D4–D6.

When $\lambda = 0$, then $k_c = \gamma_c = \sigma_c = 1$.

In the case $\mathrm{Re}\,\lambda > 0$ (i.e., the case of an unstable critical point), both parameters k_c and γ_c grow exponentially with time. This implies that small variations in the initial conditions will imply exponentially large variations in the solutions, both pointwise and on average: i.e., the problem is *ill conditioned*.

Of course, the case $\mathrm{Re}\,\lambda = 0$ considered above cannot be considered as representative of more difficult nonlinear equations, since linearization is in general not allowed in such a case.

The linearization is not the only way to study nonlinear differential (or difference) equations. The so called *Liapunov second method* can be used as well (see, e.g., [22, 27, 38]). It has been used, in connection with stiffness in [5, 13–17], although not always explicitly named.[8] Anyway, no matter how the asymptotic stability of a reference solution is detected, the parameters (1.2.2) and Definition 1.3 continue to be valid. Later on, the problem of effectively estimating such parameters will also be discussed.

1.2.1 The Discrete Case

Before passing to the non scalar case, let us now consider the discrete case, where some interesting additional considerations can be made. Here, almost all we have said for the continuous case can be repeated. The first approximation theorem can be stated almost in the same terms as in the continuous case (see e.g. [28]).

Let the interval $[0, T]$ be partitioned into N subintervals of length $h_n > 0$, thus defining the mesh points: $t_n = \sum_{j=1}^{n} h_j$, $n = 0, 1, \dots, N$.

The linearized autonomous problem is now:

$$y_{n+1} = \mu_n y_n, \qquad n = 0, \dots, N-1, \qquad y_0 = \eta, \tag{1.2.5}$$

where the $\{\mu_n\}$ are complex parameters. The conditioning parameters for (1.2.5), along with the stiffness ratio, are defined as:

$$\kappa_d = \frac{1}{|\eta|} \max_{i=0,\dots,N} |y_i|, \qquad \gamma_d = \frac{1}{|\eta|}\frac{1}{T}\sum_{i=1}^{N} h_i \max(|y_i|, |y_{i-1}|),$$
$$\sigma_d = \frac{k_d}{\gamma_d}. \tag{1.2.6}$$

This permits us to define the notion of *well representation* of a continuous problem by means of a discrete one.

Definition 1.6 The problem (1.2.1) is *well represented* by (1.2.5) if

$$k_c \approx k_d, \tag{1.2.7}$$

$$\gamma_c \approx \gamma_d. \tag{1.2.8}$$

In the case of a constant mesh-size h, $\mu_n \equiv \mu$ and it easily follows that the condition (1.2.7) requires $|\mu| < 1$. It is not difficult to recognize the usual A-stability conditions for one-step methods (see Table 1.1). Furthermore, it is easily recognized that the request that condition (1.2.7) holds uniformly with respect to $h\lambda \in \mathbb{C}^-$ implies that the numerical method producing (1.2.5) must be implicit.

What does condition (1.2.8) require more? Of course, it measures how faithfully the integral $\int_0^T |y(t)|dt$ is approximated by the quadrature formula $\sum_{i=1}^{N} h_i \cdot \max(|y_i|, |y_{i-1}|)$, thus giving a sort of global information about the behavior of the

[8] Often, it appears under the name of one-sided Lipschitz condition.

Table 1.1 Condition (1.2.7) for some popular methods

Method	μ	Condition
Explicit Euler	$1+h\lambda$	$\|1+h\lambda\|<1$
Implicit Euler	$\frac{1}{1-h\lambda}$	$\left\|\frac{1}{1-h\lambda}\right\|<1$
Trapezoidal rule	$\frac{1+h\lambda/2}{1-h\lambda/2}$	$\left\|\frac{1+h\lambda/2}{1-h\lambda/2}\right\|<1$

method producing the approximations $\{y_i\}$. One of the most efficient global strategies for changing the stepsize is based on monitoring this parameter [3, 4, 7, 8, 32, 33]. In addition to this, when finite precision arithmetic is used, then an interesting property of the parameter γ_d occurs [26]: if it is smaller than a suitably small threshold, this suggests that we are doing useless computations, since the machine precision has already been reached.

1.2.2 The non Scalar Case

In this case, the linearized problem to be considered is

$$y' = Ay, \quad t \in [0, T], \quad y(0) = \eta, \tag{1.2.9}$$

with $A \in \mathbb{R}^{m\times m}$ and having all its eigenvalues with negative real part. It is clear from what was said in the scalar case that, denoting by $\Phi(t) = e^{At}$ the fundamental matrix of the above equation, the straightforward generalization of the definition of the conditioning parameters (1.2.2) would lead to:

$$\kappa_c = \max_{t\in[0,T]} \|\Phi(t)\|, \qquad \gamma_c = \frac{1}{T}\int_0^T \|\Phi(t)\| dt, \qquad \sigma_c = \frac{\kappa_c}{\gamma_c}. \tag{1.2.10}$$

Indeed, these straight definitions *work most of the time*, as is confirmed by the following example, although, as we shall explain soon, not always.

Example 1.7 Let us consider the well-known Van der Pol's problem,

$$\begin{aligned} y_1' &= y_2, \\ y_2' &= -y_1 + \mu\, y_2(1-y_1^2), \quad t \in [0, 2\mu], \\ y(0) &= (2,\ 0)^T, \end{aligned} \tag{1.2.11}$$

whose solution approaches a limit cycle of period $T \approx 2\mu$. It is also very well-known that, the larger the parameter μ, the more difficult the problem is. In Fig. 1.2 we plot the parameter $\sigma_c(\mu)$ (as defined in (1.2.10)) for μ ranging from 0 to 10^3. Clearly, stiffness increases with μ.

Even though (1.2.10) works for this problem, this is not true in general. The problem is that the definition of stiffness as the ratio of two quantities may require a lower bound for the denominator. While the definition of κ_c remains unchanged, the definition of γ_c is more entangled. Actually, we need two different estimates of such a parameter:

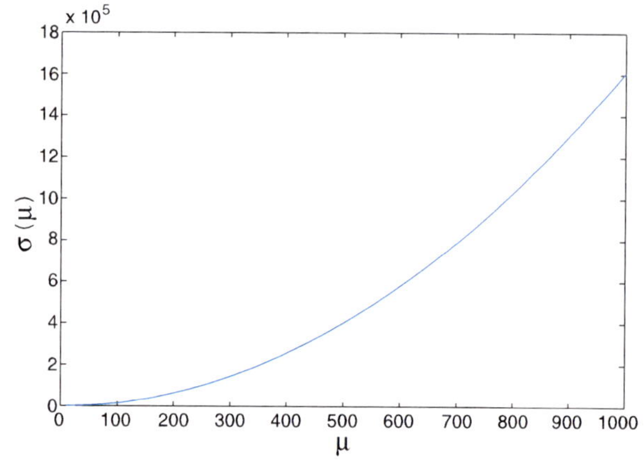

Fig. 1.2 Estimated stiffness ratio of Van der Pol's problem (1.2.11)

- an upper bound, to be used for estimating the conditioning of the problem in l_1 norm;
- a lower bound, to be used in defining σ_c and, then, the stiffness.

In the definition given in [2, 4], this distinction was not made, even though the definition was (qualitatively) completed by adding

$$\text{“\textit{for at least one of the modes}”}. \tag{1.2.12}$$

We shall be more precise in a moment. In the meanwhile, it is interesting to note that the clarification contained in (1.2.12) is already in one of the two definitions given by Miranker [34]:

A system of differential equations is said to be stiff on the interval $(0,\bar{t})$ *if there exists a solution of that system a component of which has a variation on that interval which is large compared to* $\frac{1}{\bar{t}}$,

where it should be stressed that the definition considers equations and not problems: this implies that the existence of largely variable components may appear for at least one choice of the initial conditions, not necessary for a specific one.

Later on, the definition was modified so as to translate into formulas the above quoted sentence (1.2.12). The following definitions were then given (see, e.g., [26]):

$$\begin{aligned} \kappa_c(T,\eta) &= \frac{1}{\|\eta\|}\max_{0\le t\le T}\|y(t)\|, \qquad & \kappa_c(T) &= \max_{\eta}\kappa_c(T,\eta), \\ \gamma_c(T,\eta) &= \frac{1}{T\|\eta\|}\int_0^T\|y(t)\|dt, \qquad & \gamma_c(T) &= \max_{\eta}\gamma_c(T,\eta) \end{aligned} \tag{1.2.13}$$

and

$$\sigma_c(T) = \max_{\eta}\frac{\kappa_c(T,\eta)}{\gamma_c(T,\eta)}. \tag{1.2.14}$$

The only major change regards the definition of σ_c. Let us be more clear on this point with an example, since it leads to a controversial question in the literature: i.e., the dependence of stiffness from the initial condition. Let $A = \mathrm{diag}(\lambda_1, \lambda_2, \ldots, \lambda_m)$ with $\lambda_i < 0$ and $|\lambda_1| > |\lambda_2| > \cdots > |\lambda_m|$. The solution of problem (1.2.9) is $y(t) = e^{At}\eta$.

If σ_c is defined according to (1.2.10), it turns out that $\|e^{At}\| = e^{\lambda_m t}$ and, then, $\gamma_c(T) \approx \frac{1}{T|\lambda_m|}$. If, however, we take $\eta = (1, 0, \ldots, 0)^T$, then $y(t) = e^{\lambda_1 t}$ and $\gamma_c(T)$ becomes $\gamma_c(T) \approx \frac{1}{T|\lambda_1|}$. Of course, by changing the initial point, one may activate each one of the *modes*, i.e. the functions $e^{\lambda_i t}$ on the diagonal of the matrix e^{At}, leaving silent the others. This is the reason for specifying, in the older definition, the quoted sentence (1.2.12). The new definition (1.2.14), which essentially poses as the denominator of the ratio σ_c the smallest value among the possible values of $\gamma_c(T, \eta)$, is more compact and complies with the needs of people working on the construction of codes, who like more operative definitions. For the previous diagonal example, we have that k_c continues to be equal to 1, while $\gamma_c(T) = \frac{1}{T|\lambda_1|}$.

Having got the new definition (1.2.14) of $\sigma_c(T)$, the definition of stiffness continues to be given by Definition 1.3 given in the scalar case, i.e., the problem (1.2.9) is *stiff* if $\sigma_c(T) \gg 1$.

How does this definition reconcile with the most used definition of stiffness for the linear case, which considers the "smallest" eigenvalue λ_m as well? The answer is already in Miranker's definition D3. In fact, usually the integration interval is chosen large enough to provide complete information on the behavior of the solution. In this case, until the slowest mode has decayed enough, i.e. $T = 1/|\lambda_m|$, which implies

$$\sigma_c\left(T = \frac{1}{|\lambda_m|}\right) = \left|\frac{\lambda_1}{\lambda_m}\right|, \tag{1.2.15}$$

which, when much larger than 1, coincides with the most common definition of stiffness in the linear case. However, let us insist on saying that if the interval of integration is much smaller than $1/|\lambda_m|$, *the problem may be not stiff* even if $|\frac{\lambda_1}{\lambda_m}| \gg 1$.

The controversy about the dependence of the definition of stiffness on the initial data is better understood by considering the following equation given in [29, pp. 217–218]:

$$\frac{d}{dt}\begin{pmatrix} y_1 \\ y_2 \end{pmatrix} = \begin{pmatrix} -2 & 1 \\ -1.999 & 0.999 \end{pmatrix}\begin{pmatrix} y_1 \\ y_2 \end{pmatrix} + \begin{pmatrix} 2\sin t \\ 0.999(\sin t - \cos t) \end{pmatrix},$$

whose general solution is

$$\begin{pmatrix} y_1 \\ y_2 \end{pmatrix} = c_1 e^{-t}\begin{pmatrix} 1 \\ 1 \end{pmatrix} + c_2 e^{-0.001t}\begin{pmatrix} 1 \\ 1.999 \end{pmatrix} + \begin{pmatrix} \sin t \\ \cos t \end{pmatrix}.$$

The initial condition $y(0) = (2, 3)^T$ requires $c_2 = 0$ and, then, the slowest mode is not activated: the solution rapidly reaches the reference solution. If this information was known beforehand, one could, in principle, choose the interval of integration T much smaller than $\frac{1}{0.001}$. This, however, does not take into account the fact that the computer uses finite precision arithmetic, which may not represent exactly the initial condition η. To be more precise, let us point out that the slowest mode is not activated only if the initial condition is on the line $y_2(0) - y_1(0) - 1 = 0$. Any irrational value of $y_1(0)$ will not be well represented on the computer. This is enough

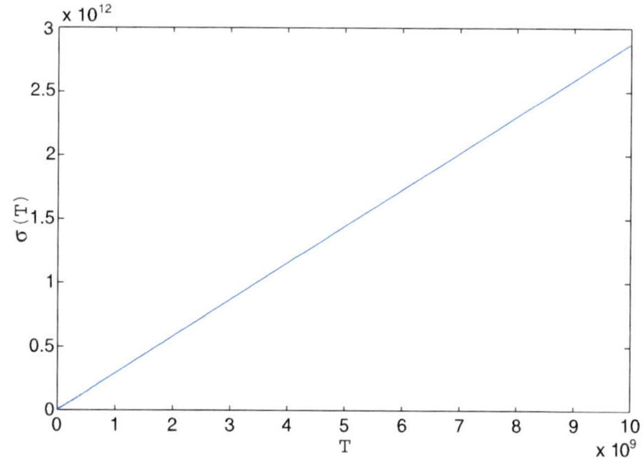

Fig. 1.3 Estimated stiffness ratio of Robertson's problem (1.2.16)

to activate the silent mode. Of course, if one is sure that the long term contribution to the solution obtained on the computer is due to this kind of error, a small value of T can always be used. But it is rare that this information is known in advance. For this reason, we consider the problem to be stiff, since we believe that the definition of stiffness cannot distinguish, for example, between rational and irrational values of the initial conditions. Put differently, initial conditions are like a fuse that may activate stiffness.

We conclude this section by providing a few examples, which show that Definition 1.3, when σ_c is defined according to (1.2.14), is able to adequately describe the stiffness of nonlinear and/or non autonomous problems as well.

Example 1.8 Let us consider the well-known Robertson's problem:

$$\begin{aligned} y_1' &= -0.04y_1 + 10^4 y_2 y_3, \\ y_2' &= 0.04y_1 - 10^4 y_2 y_3 - 3\times 10^7 y_2^2, \quad t\in[0,T], \\ y_3' &= 3\times 10^7 y_2^2, \\ y(0) &= (1,0,0)^T. \end{aligned} \tag{1.2.16}$$

Its stiffness ratio with respect to the length T of the integration interval, obtained through the linearized problem and considering a perturbation of the initial condition of the form $(0, \varepsilon, -\varepsilon)^T$, is plotted in Fig. 1.3. As it is well-known, the figure confirms that for this problem stiffness increases with T.

Example 1.9 Let us consider the so-called Kreiss problem [21, p. 542], a linear and non autonomous problem:

$$y' = A(t)y, \quad t\in[0,4\pi], \quad y(0) \quad \text{fixed}, \tag{1.2.17}$$

where

$$A(t) = Q^T(t)\Lambda_\varepsilon Q(t), \tag{1.2.18}$$

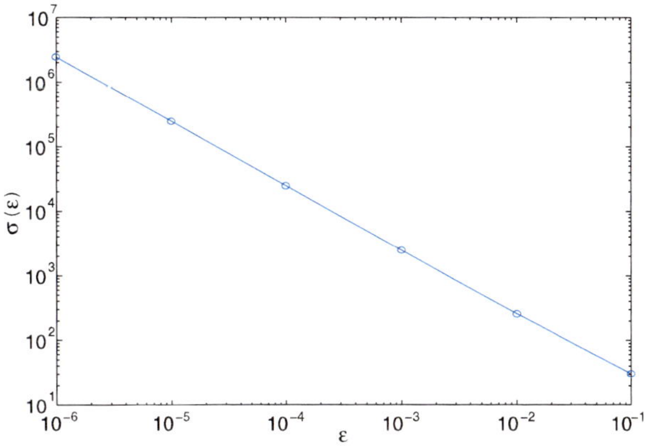

Fig. 1.4 Estimated stiffness ratio of the Kreiss problem (1.2.17)–(1.2.19)

and

$$Q(t)=\begin{pmatrix}\cos t & \sin t\\ -\sin t & \cos t\end{pmatrix},\qquad \Lambda_\varepsilon=\begin{pmatrix}-1 & \\ & -\varepsilon^{-1}\end{pmatrix}. \tag{1.2.19}$$

Its stiffness ratio with respect to the small positive parameter ε, obtained by considering a perturbation of the initial condition of the form $(-\varepsilon,\,1)^T$, is plotted in Fig. 1.4. As one expects, the figure confirms that the stiffness of the problem behaves as ε^{-1}, as ε tends to 0.

Example 1.10 Let us consider the following linear and non autonomous problem, a modification of problem (1.2.17), that we call "modified Kreiss problem":[9]

$$y'=A(t)y,\qquad t\in[0,4\pi],\qquad y(0)\quad \text{fixed}, \tag{1.2.20}$$

where

$$A(t)=Q_\varepsilon^{-1}(t)P^{-1}\Lambda_\varepsilon P Q_\varepsilon(t), \tag{1.2.21}$$

and

$$P=\begin{pmatrix}-1 & 0\\ 1 & 1\end{pmatrix},\qquad Q_\varepsilon(t)=\begin{pmatrix}1 & \varepsilon\\ e^{\sin t} & e^{\sin t}\end{pmatrix},\qquad \Lambda_\varepsilon=\begin{pmatrix}-1 & \\ & -\varepsilon^{-1}\end{pmatrix}. \tag{1.2.22}$$

Its stiffness ratio with respect to the small positive parameter ε, obtained by considering a perturbation of the initial condition of the form $(-\varepsilon,\,1)^T$, is shown in Fig. 1.5. Also in this case the stiffness of the problem behaves as ε^{-1}, as ε tends to 0.

Remark 1.11 It is worth mentioning that, in the examples considered above, we numerically found that

$$\max_\eta \frac{\kappa_c(T,\eta)}{\gamma_c(T,\eta)}$$

[9]This problem has been suggested by J.I. Montijano.

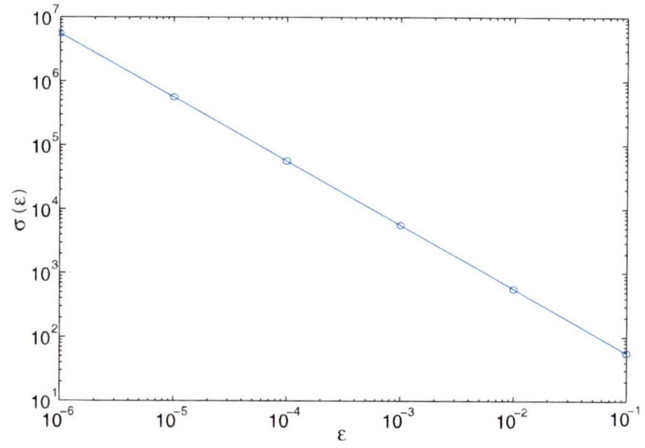

Fig. 1.5 Estimated stiffness ratio of the modified Kreiss problem (1.2.20)–(1.2.22)

is obtained by considering an initial condition η in the direction of the eigenvector of the Jacobian matrix (computed for $t \approx t_0$) associated to the dominant eigenvalue. We note that, for an autonomous linear problem, if A is diagonalizable, this choice activates the *mode* associated with λ_1, i.e., the eigenvalue of maximum modulus of A.

1.2.3 The Non Scalar Discrete Case

As for the scalar case, what we said for the continuous problems can be repeated, *mutatis mutandis*, for the discrete ones. For brevity, we shall skip here the details for this case, also because they can be deduced from those described in the more general case discussed in the next section.

1.3 Boundary Value Problems (BVPs)

The literature about BVPs is far less abundant than that about IVPs, both in the continuous and in the discrete case. While there are countless books on the latter subject presenting it from many points of view (e.g., stability of motion, dynamical systems, bifurcation theory, etc.), there are many less books about the former. More importantly, the subject is usually presented as a by product of the theory of IVPs. This is not necessarily the best way to look at the question, even though many important results can be obtained this way. However, it may sometimes be more useful to look at the subject the other way around. Actually, the question is that IVPs are naturally a subclass of BVPs. Let us informally clarify this point without many technical details which can be found, for example, in [4].

IVPs transmit the initial information "from left to right". Well conditioned IVPs are those for which the initial value, along with the possible initial errors, decay moving from left to right. FVPs (Final Value problems) are those transmitting information "from right to left" and, of course, well conditioning should hold when the time, or the corresponding independent variable, varies towards $-\infty$. More precisely, considering the scalar test equation (1.2.1), the asymptotically stability for IVPs and FVPs requires $\mathrm{Re}\,\lambda < 0$ and $\mathrm{Re}\,\lambda > 0$, respectively. BVPs transmit information both ways. Consequently, they cannot be scalar problems but vectorial of dimension at least two. We need then to refer to the test equation (1.2.9). It can be affirmed that a well conditioned linear BVP needs to have eigenvalues with both negative and positive real parts (*dichotomy*, see, e.g., [1, 4]). More precisely: the number of eigenvalues with negative real part has to match the amount of information transmitted "from left to right", and the number of eigenvalues with positive real part has to match the amount of information traveling "from right to left". For brevity, we shall call the above statement *continuous matching rule*. Of course, if there are no final conditions, then the problem becomes an IVP and, as we have seen, in order to be well conditioned, it must have all the eigenvalues with negative real part. In other words, the generalization of the case of asymptotically stable IVPs is the class of well conditioned BVPs *because both satisfy the continuous matching rule*. This is exactly what we shall assume hereafter.

Similar considerations apply to the discrete problems, where the role of the imaginary axis is played by the unit circumference in the complex plane. It is not surprising that a numerical method will *well represent* a continuous autonomous linear BVP if the corresponding matrix has as many eigenvalues inside the unit circle as the number of initial conditions and as many eigenvalues outside the unit circle as the number of final conditions (*discrete matching rule*).

Remark 1.12 The idea that IVPs are a subset of BVPs is at the root of the class of methods called *Boundary Value Methods (BVMs)* which permits us, thanks to the discrete matching rule, to define high order and perfectly A-stable methods (i.e., methods having the imaginary axis separating the stable and unstable domains), which *overcome the Dahlquist's barriers*, and are able to solve both IVPs and BVPs (see, e.g., [4]).

Remark 1.13 From this point of view, the popular *shooting method*, consisting of transforming a BVP into an IVP and then applying a good method *designed for IVPs*, does not appear to be such a good idea. As matter of fact, even a very well conditioned linear BVP, i.e. one which satisfies the continuous matching rule, will be transformed in a badly conditioned IVP, since the matrix of the continuous IVP shall, of course, contain eigenvalues with positive real part. This will prevent the discrete matching rule to hold.

1.3.1 Stiffness for BVPs

Coming back to our main question, stiffness for BVPs is now defined by generalizing the idea already discussed in the previous sections.

As in the previous cases, we shall refer to linear problems, but the definitions will also be applicable to nonlinear problems as well. Moreover, according to what is stated above, we shall only consider the case where the problems are well conditioned (for the case of ill conditioned problems, the arguments are slightly more entangled, see e.g. [7]). Then, let us consider the linear and non autonomous BVP:

$$y' = A(t)y, \quad t \in [0, T], \quad B_0 y(0) + B_1 y(T) = \eta, \tag{1.3.1}$$

where $y(t), \eta \in \mathbb{R}^m$ and $A(t), B_0, B_1 \in \mathbb{R}^{m\times m}$. The solution of the problem (1.3.1) is

$$y(t) = \Phi(t) Q^{-1} \eta,$$

where $\Phi(t)$ is the fundamental matrix of the problem such that $\Phi(0) = I$, and $Q = B_a + B_b \Phi(T)$, which has to be nonsingular, in order for (1.3.1) to be solvable.[10]

As in the continuous IVP case, the conditioning parameters are defined (see (1.2.13)) as:

$$\begin{aligned} \kappa_c(T,\eta) &= \frac{1}{\|\eta\|} \max_{0\le t\le T} \|y(t)\|, \qquad & \kappa_c(T) &= \max_\eta \kappa_c(T,\eta), \\ \gamma_c(T,\eta) &= \frac{1}{T\|\eta\|} \int_0^T \|y(t)\| dt, \qquad & \gamma_c(T) &= \max_\eta \gamma_c(T,\eta). \end{aligned} \tag{1.3.2}$$

Consequently, the stiffness ratio is defined as (see (1.2.14)):

$$\sigma_c(T) = \max_\eta \frac{\kappa_c(T,\eta)}{\gamma_c(T,\eta)},$$

and the problem is stiff if $\sigma_c(T) \gg 1$. Moreover, upper bounds of $\kappa_c(T)$ and $\gamma_c(T)$ are respectively given by:

$$\kappa_c(T) \le \max_{0\le t\le T} \|\Phi(t) Q^{-1}\|, \qquad \gamma_c(T) \le \frac{1}{T}\int_0^T \|\Phi(t)Q^{-1}\| dt. \tag{1.3.3}$$

Thus, the previous definitions naturally extend to BVPs the results stated for IVPs. In a similar way, when considering the discrete approximation of (1.3.1), for the sake of brevity provided by a suitable one-step method over a partition π of the interval $[0, T]$, with subintervals of length h_i, $i = 1, \dots, N$, the discrete problem will be given by

$$y_{n+1} = R_n y_n, \quad n = 0, \dots, N-1, \quad B_0 y_0 + B_1 y_N = \eta, \tag{1.3.4}$$

whose solution is given by

$$y_n = \left(\prod_{i=0}^{n-1} R_i\right) Q_N^{-1} \eta, \quad Q_N = B_0 + B_1 \prod_{i=0}^{N-1} R_i.$$

[10]Observe that, in the case of IVPs, $B_0 = I$ and $B_1 = O$, so that $Q = I$.

The corresponding discrete conditioning parameters are then defined by:

$$\kappa_d(\pi,\eta) = \frac{1}{\|\eta\|}\max_{0\le n\le N}\|y_n\|, \qquad \kappa_d(\pi) = \max_\eta \kappa_d(\pi,\eta),$$
$$\gamma_d(\pi,\eta) = \frac{1}{T\|\eta\|}\sum_{i=1}^{N} h_i \max(\|y_i\|,\|y_{i-1}\|), \qquad \gamma_d(\pi) = \max_\eta \gamma_d(\pi,\eta), \tag{1.3.5}$$

and

$$\sigma_d(\pi) = \max_\eta \frac{\kappa_d(\pi,\eta)}{\gamma_d(\pi,\eta)}.$$

According to Definition 1.6, we say that the discrete problem[11] (1.3.4) *well represents* the continuous problem (1.3.1) if

$$\kappa_d(\pi) \approx \kappa_c(T), \qquad \gamma_d(\pi) \approx \gamma_c(T). \tag{1.3.6}$$

Remark 1.14 It is worth mentioning that innovative mesh-selection strategies for the efficient numerical solution of stiff BVPs have been defined by requiring the match (1.3.6) (see, e.g., [3, 4, 7, 8, 26]).

1.3.2 Singular Perturbation Problems

The numerical solution of singular perturbation problems can be very difficult because they can have solutions with very narrow regions of rapid variation characterized by boundary layers, shocks, and interior layers. Usually, the equations depend on a small parameter, say ε, and the problems become more difficult as ε tends to 0. It is not always clear, however, how the width of the region of rapid variation is related to the parameter ε. By computing the stiffness ratio $\sigma_c(T)$, we observe that singularly perturbed problems are stiff problems. Moreover, as the following examples show, the parameter $\sigma_c(T)$ provides us also with information about the width of the region of rapid variation.

The examples are formulated as second order equations: of course, they have to be transformed into corresponding first order systems, in order to apply the results of the previous statements.

Example 1.15 Let us consider the linear singularly perturbed problem:

$$\varepsilon y'' + ty' = -\varepsilon\pi^2\cos(\pi t) - \pi t\sin(\pi t), \quad y(-1) = -2, \quad y(1) = 0, \tag{1.3.7}$$

whose solution has, for $0 < \varepsilon \ll 1$, a turning point at $t = 0$ (see Fig. 1.6). The exact solution is $y(t) = \cos(\pi t) + \exp((t-1)/\sqrt{\varepsilon}) + \exp(-(t+1)/\sqrt{\varepsilon})$.

In Fig. 1.7 we plot an estimate of the stiffness ratio obtained by considering two different perturbations of the boundary conditions of the form $(1,0)^T$ and $(0,1)^T$.

[11] It is both defined by the used method and by the considered mesh.

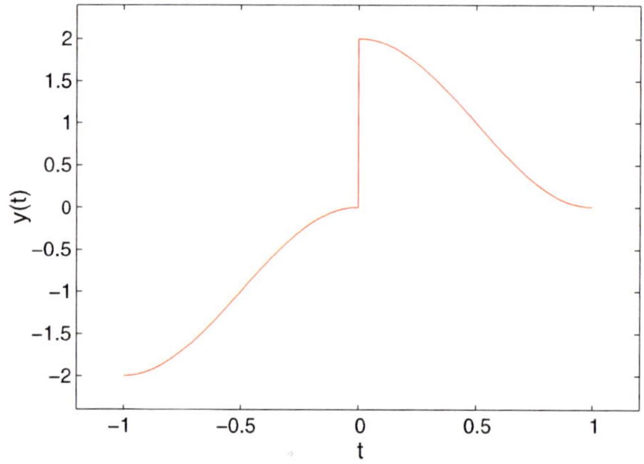

Fig. 1.6 Problem (1.3.7), $\varepsilon = 10^{-8}$

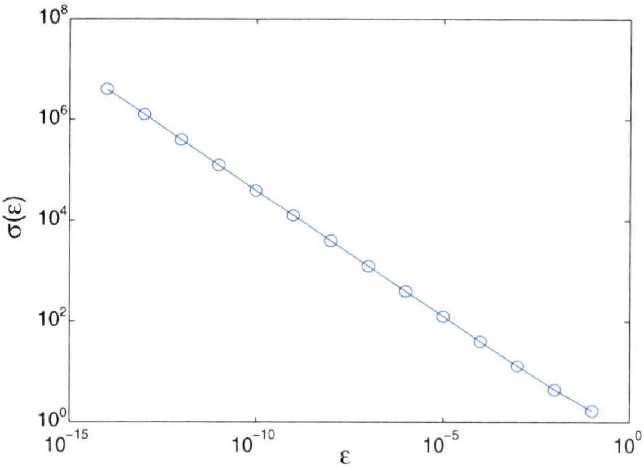

Fig. 1.7 Estimated stiffness ratio of problem (1.3.7)

The parameter ε varies from 10^{-1} to 10^{-14}. We see that the (estimated) stiffness parameter grows like $\sqrt{\varepsilon^{-1}}$.

Example 1.16 Let us consider the following nonlinear problem:

$$\varepsilon y'' + \exp(y)y' - \frac{\pi}{2}\sin\left(\frac{\pi t}{2}\right)\exp(2y) = 0, \quad y(0) = 0, \quad y(1) = 0. \tag{1.3.8}$$

This problem has a boundary layer at $t = 0$ (see Fig. 1.8). In Fig. 1.9 we plot an estimate of the stiffness ratio obtained by considering two different perturbations of the boundary conditions of the form $(1, 0)^T$ and $(0, 1)^T$. The parameter ε varies from 1 to 10^{-8}. We see that the (estimated) stiffness parameter grows like ε^{-1}, as ε tends to 0.

Example 1.17 Let us consider the nonlinear Troesch problem:

$$y'' = \mu \sinh(\mu y), \quad y(0) = 0, \quad y(1) = 1. \tag{1.3.9}$$

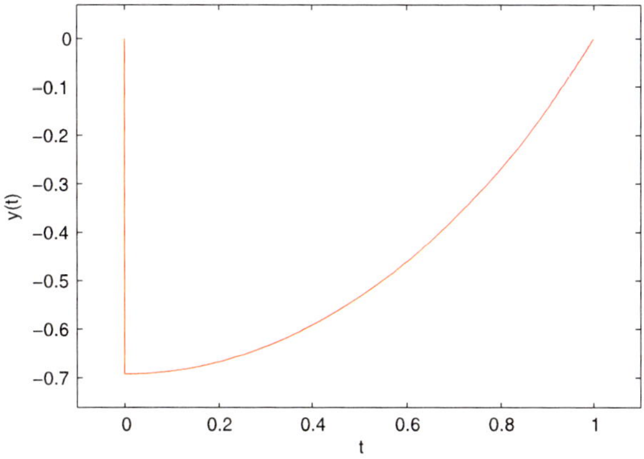

Fig. 1.8 Problem (1.3.8), $\varepsilon = 10^{-6}$

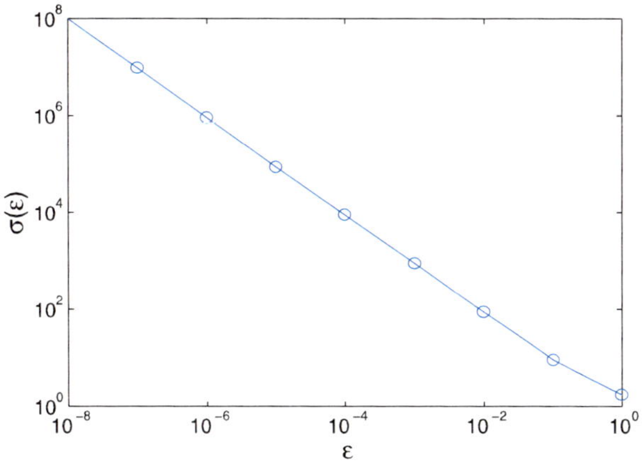

Fig. 1.9 Estimated stiffness ratio of problem (1.3.8)

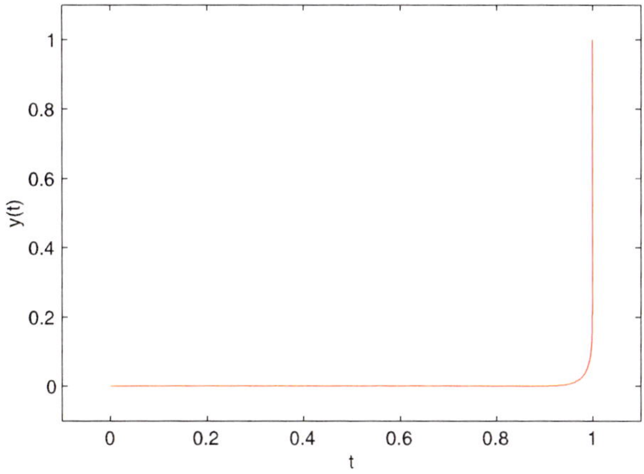

Fig. 1.10 Troesch's problem (1.3.9), $\mu = 50$

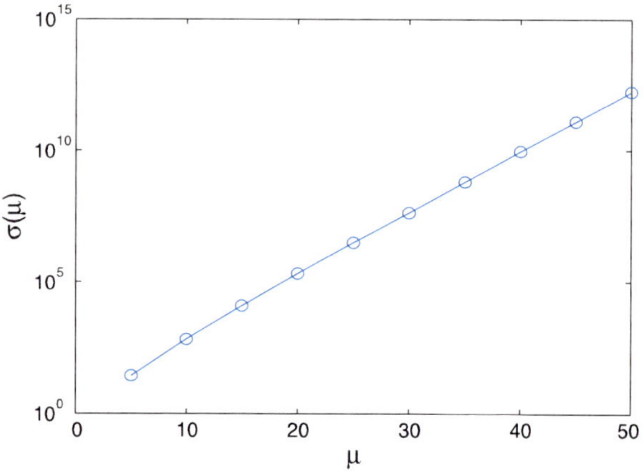

Fig. 1.11 Estimated stiffness ratio of Troesch's problem (1.3.9)

This problem has a boundary layer near $t = 1$ (see Fig. 1.10). In Fig. 1.11 we plot the estimate of the stiffness ratio obtained by considering two different perturbations of the boundary conditions of the form $(1, 0)^T$ and $(0, 1)^T$. The parameter μ is increased from 1 to 50 and, as expected, the stiffness ratio increases as well: for $\mu = 50$, it reaches the value 1.74×10^{12}.

Acknowledgements The authors wish to thank the reviewers, for their comments and suggestions.

References

1. Ascher, U.M., Mattheij, R.M.M., Russell, R.D.: Numerical Solution of Boundary Value Problems for Ordinary Differential Equations. SIAM, Philadelphia (1995)
2. Brugnano, L., Trigiante, D.: On the characterization of stiffness for ODEs. Dyn. Contin. Discrete Impuls. Syst. **2**, 317–335 (1996)
3. Brugnano, L., Trigiante, D.: A new mesh selection strategy for ODEs. Appl. Numer. Math. **24**, 1–21 (1997)
4. Brugnano, L., Trigiante, D.: Solving Differential Problems by Multistep Initial and Boundary Value Methods. Gordon & Breach, Amsterdam (1998)
5. Butcher, J.C.: The Numerical Analysis of Ordinary Differential Equations. Wiley, Chichester (1987)
6. Cash, J.R.: Efficient numerical methods for the solution of stiff initial-value problems and differential algebraic equations. Proc. R. Soc. Lond. A **459**, 797–815 (2003)
7. Cash, J.R., Mazzia, F.: A new mesh selection algorithm, based on conditioning, for two-point boundary value codes. J. Comput. Appl. Math. **184**, 362–381 (2005)
8. Cash, J.R., Sumarti, N., Mazzia, F., Trigiante, D.: The role of conditioning in mesh selection algorithms for first order systems of linear two-point boundary value problems. J. Comput. Appl. Math. **185**, 212–224 (2006)
9. Corduneanu, C.: Principles of Differential and Integral Equations. Chelsea, New York (1971)
10. Crank, J., Nicolson, P.: A pratical method for numerical evaluation of solutions od partial differential equations of the heat-conduction type. Proc. Camb. Philos. Soc. **43**, 50–67 (1947)
11. Curtiss, G.F., Hirshfelder, J.O.: Integration of Stiff equations. Proc. Natl. Acad. Sci. US **38**, 235–243 (1952)

12. Dahlquist, G.: Problems related to the numerical treatment of Stiff differential equations. In: Günther, E., et al. (eds.) International Computing Symposium, 1973, pp. 307–314. North Holland, Amsterdam (1974)
13. Dahlquist, G.: A special stability problem for linear multistep methods. BIT **3**, 27–43 (1964)
14. Dahlquist, G.: Error analysis for a class a methods for stiff nonlinear initial value problems. In: Num. Anal., Dundee. Lect. Notes in Math., vol. 506, pp. 60–74. Springer, Berlin (1975)
15. Dahlquist, G.: On stability and error analysis for stiff nonlinear problems. Part 1. Report Trita-NA-7508 (1975)
16. Dahlquist, G.: G-stability is equivalent to A-stability. BIT **18**, 384–401 (1978)
17. Dahlquist, G.: 33 Years of instability, Part I. BIT **25**, 188–204 (1985)
18. Galbraith, J.K.: A Short History of Financial Euphoria. Whittle Direct Book (1990)
19. Goodwin, R.H.: A growth cycle. In: Feinstein, C.H. (ed.) Socialism, Capitalism and Economic Growth. Cambridge University Press, Cambridge (1967)
20. Guglielmi, N., Hairer, E.: Stiff delay equations. Scholarpedia **2**(11), 2850 (2007)
21. Hairer, E., Wanner, G.: Solving Ordinary Differential Equations II. Springer, Berlin (1996). 2nd rev. edn
22. Hahn, W.: Stability of Motions. Springer, New York (1967)
23. Higham, D.J., Trefethen, L.N.: Stiffness of ODE. BIT **33**, 286–303 (1993)
24. Hindmarsh, A.C.: On Numerical Methods for Stiff Differential Equations–Getting the Power to the People. Lawrence Livermore Laboratory report, UCRL-83259 (1979)
25. Hundsdorfer, W.H.: The numerical solution of stiff initial value problems: an analysis of one step methods. CWI Tracts 12, Amsterdam (1980)
26. Iavernaro, F., Mazzia, F., Trigiante, D.: Stability and conditioning in numerical analysis. JNA-IAM **1**, 91–112 (2006)
27. Lakshmikantham, V., Leela, S.: Differential and Integral Inequalities. Academic Press, New York (1969)
28. Lakshikantham, V., Trigiante, D.: Theory of Difference Equations. Numerical Methods and Applications, 2nd edn. Marcel Dekker, New York (2002)
29. Lambert, J.D.: Numerical Methods for Ordinary Differential Equations. Wiley, New York (1991)
30. Le Veque, R.J.: Finite Difference Methods for Ordinary and Partial Differential Equations: Steady-State and Time-Dependent Problems. SIAM, Philadelphia (2007)
31. Liniger, W.: Solution Numériques des Équations Différentielle et au derivées partielle. Unpublished Lecture Notes of a course taught at Swiss Federal Institute of Technology, Lausanne, Switzerland (1972–1973)
32. Mazzia, F., Trigiante, D.: A hybrid mesh selection strategy based on conditioning for boundary value ODEs problems. Numer. Algorithms **36**(2), 169–187 (2004)
33. Mazzia, F., Trigiante, D.: Efficient strategies for solving nonlinear problems in BVPs codes. Nonlinear Studies (in press)
34. Miranker, W.L.: The Computational Theory of Stiff Differential Equations. Pubblicazioni IAC Roma Ser. III N. 102 (1975)
35. Rouche, N., Mawhin, J.: Équations Differentielle Ordinaire, vol. 2. Masson et Cie, Paris (1973)
36. Shampine, L.F., Thompson, S.: Stiff systems. Scholarpedia **2**(3), 2855 (2007)
37. Söderlind, G.: The logarithmic norm. History and modern theory. BIT **46**, 631–652 (2006)
38. Yoshizawa, T.: Stability Theory by Liapunov's Second Method. The Mathematical Soc. of Japan (1966)

Chapter 2
Efficient Global Methods for the Numerical Solution of Nonlinear Systems of Two Point Boundary Value Problems

Jeff R. Cash and Francesca Mazzia

Abstract In this paper we will be concerned with numerical methods for the solution of nonlinear systems of two point boundary value problems in ordinary differential equations. In particular we will consider the question "which codes are currently available for solving these problems and which of these codes might we consider as being state of the art". In answering these questions we impose the restrictions that the codes we consider should be widely available (preferably written in MATLAB and/or FORTRAN) they should have reached a fairly steady state in that they are seldom, if ever, updated, they try to achieve broadly the same aims and, of course, it is relatively inexpensive to purchase the site licence. In addition we will be concerned exclusively with so called boundary value (or global) methods so that, in particular, we will not include shooting codes or Shishkin mesh methods in our survey. Having identified such codes we go on to discuss the possibility of comparing the performance of these codes on a standard test set. Of course we recognise that the comparison of different codes can be a contentious and difficult task. However the aim of carrying out a comparison is to eliminate bad methods from consideration and to guide a potential user who has a boundary value problem to solve to the most effective way of achieving his aim. We feel that this is a very worthwhile objective to pursue. Finally we note that in this paper we include some new codes for BVP's which are written in MATLAB. These have not been available before and allow for the first time the possibility of comparing some powerful MATLAB codes for solving boundary value problems. The introduction of these new codes is an important feature of the present paper.

Work developed within the project "Numerical methods and software for differential equations".

J.R. Cash (✉)
Department of Mathematics, Imperial College, South Kensington, London SW7, UK
e-mail: j.cash@imperial.ac.uk

F. Mazzia
Dipartimento di Matematica, Università di Bari, Via Orabona 4, 70125 Bari, Italy

T.E. Simos (ed.), *Recent Advances in Computational and Applied Mathematics*,
DOI 10.1007/978-90-481-9981-5_2,

Keywords Two-point boundary value problems · Mesh selection · State of the art codes · Numerical comparisons

Mathematics Subject Classification (2000) 65L10 · 65L06

2.1 Introduction

An important task in many areas of Numerical Analysis is to carry out meaningful comparisons of numerical algorithms which attempt to achieve roughly the same well defined objectives. This is of particular interest to a user who wishes to find an appropriate code to solve his problem as well as to someone who has developed a new algorithm and wishes to compare it with existing state of the art codes. An obvious example of a particular case where such comparisons have been carried out leading to the development of codes which are very much more efficient is in the numerical solution of initial value problems of the form

$$\frac{dy}{dx} = f(x, y), \quad x \geq a, \quad y(a) = y_a. \tag{2.1.1}$$

The normal course of events which was followed by, for example, algorithms for initial value problems (and also by most linear algebra routines) is that first the mathematical theory behind the problems to be solved is well understood, then the mathematical theory behind the numerical algorithms is established, then the problems involved in the writing of high quality, robust codes are identified and overcome and finally some sort of numerical comparison is carried out to highlight the strengths and weaknesses of the codes.

It is particularly relevant to us in this paper to focus our attention for the present on what has been done for initial value problems of the general form (2.1.1) and this we now do. Traditionally when a new code was proposed for the numerical solution of (2.1.1) the code was illustrated by comparing it, on a small set of test problems, with other codes also designed to solve (2.1.1). This proved to be less than satisfactory since there is the danger that potential disadvantages of the new code are not discovered if such limited testing is done. The first widely used test set for general initial value problems was DETEST which was proposed by Enright, Hull and Lindberg [18]. After this test set first appeared, users proposing a new method were often expected to run their code on this test set (which contains 30 problems) and this proved very useful in eliminating poor codes. However as codes became even more powerful they tended to find the problems in DETEST to be too easy. A particular cause for concern was that many of the problems in DETEST were of very small dimension and so they were often solved extremely quickly even by relatively poor codes. Following an appraisal by Shampine [35] this test set was modified to make it much more challenging but the major disadvantage still was the small dimension of the problems. A big step forward in the quality of numerical testing came with the book of Hairer and Wanner [20]. They proposed a test set which was considerably more challenging than DETEST and, in particular, contained some problems of very large dimension (for example, the Brusselator problem on p. 151 of [20] is of dimension 32768). They also considerably advanced the methodology of testing and set new standards in attempting to make the tests as fair as possible. Based on the work of Hairer and Wanner an even more demanding test set was derived in Amsterdam by Lioen and de Swart [23]. They considerably broadened the aims and scope of the test set by adding differential algebraic equations with index ≤ 3. More recently this test set was taken over by Francesca Mazzia and her co-workers at the University of Bari [25] and it now plays a central role in providing realistic test problems for IVP solvers. It is important to realise, however, that this test set now has several additional facilities which can be extremely useful to users and which considerably strengthens the case for carrying out sophisticated numerical testing. For example, the user is easily able to run several state of the art codes either on his own test problems or on problems that appear in the test set. It also allows the user to test his own code against those appearing in the test set on really challenging problems and as output it produces highly relevant, and easy to understand, statistics. For these reasons the Bari test set now plays an important role in the development of powerful codes for the numerical solution of initial value problems of the form (1.1) and it also has several other important features available, which are much more important than was anticipated when this project was first begun. To see exactly what is available (and to note the very high level of use of this facility) the reader is urged to log into the Bari web page [25].

2.2 Boundary Value Problems

Having explained what has been done in developing test sets for initial value problems of the form (2.1.1), and witnessing the central role that is taken by test sets in deriving efficient codes for initial value problems, it clearly would be a worthwhile aim to extend some of these ideas to codes for the numerical solution of two-point boundary value problems. Ideally we would like to be able to answer the question "What are the best codes currently available for solving a large class of nonlinear two-point boundary value problems and what properties of the boundary value problem need to be taken into account when deriving efficient numerical methods for its solution". Here the situation is very different than it is for initial value problems simply because boundary value problems tend to be much harder to solve and much more diverse than initial value problems. However it could be argued that in these circumstances a user is in even more need of guidance. Due to the many forms that can be taken by two-point boundary value problems, it is clear that we will need to impose some important restrictions: firstly, on which classes of problems we will attempt to solve; secondly, on which classes of numerical methods we will consider. We need to decide for example whether or not we will consider non-separated boundary conditions, eigenvalues and other parameter dependent problems, singular problems, problems of the special form $y'' = f(x, y)$ where there is no first derivative present) and so on. What we will in fact consider in this paper is global methods, i.e. not shooting methods or Shishkin methods, for first order systems of nonlinear two-point boundary value problems with separated boundary conditions. This means that we will restrict the class of problems that we are interested in to

$$\frac{dy}{dx} = f(x, y), \quad a \le x \le b, \quad g(y(a), y(b)) = 0. \tag{2.2.1}$$

In this paper our aim will be to identify which codes are at present suitable to be used for the numerical solution of (2.1). In order to do this, we need to reconsider the codes that were presented in a previous article [5], where one of the present authors sought to identify those codes designed for the numerical solution of two point boundary value problems of the form (2.2.1), which were in some sense efficient for the solution of this class of problems. This has naturally involved some repetition of what was considered in [5] but we feel that this is justified since it is appropriate, for the sake of completeness, for us to collect together in a single article those codes which we feel are efficient and are suitable for inclusion in a numerical comparison. Note however that all of these codes in [5] are written in FORTRAN. In due course we will hope to provide a comparison of the performance of various codes on (2.2.1) but in the present paper we will be mainly interested in identifying 'state of the art' codes which will be suitable for inclusion in a comparison. Some interesting thoughts concerning this can be found in [2, p. 515]. In [5] three codes were identified as having the possibility of being considered state of the art codes and these were COLNEW.f/COLSYS.f, MIRKDC.f, and TWPBVP.f/TWPBVPL.f. Also identified in [5] as being powerful continuation codes were COLMOD.f and ACDC.f [13]. The first of these two codes is based on COLSYS.f and the second is an adaptation of TWPBVPL.f. These codes allow COLSYS.f and TWPBVPL.f to be run in a continuation framework. This makes these codes much more able to deal with really difficult singular perturbation problems than is the case when continuation is not used. The important questions we need to consider in this section are: whether these codes have stabilised in the time since [5] was written; whether these new codes can still be considered as state of the art; whether new codes which are competitive with these three codes have been developed in the interim; whether more recent codes written in MATLAB are now suitable. It is important to be aware of the fact that the development of a test set for initial value problems took a lot of time, and considerable effort, and many researchers contributed to this project. We would expect the same to be the case for BVPs and so the present paper, which aims to identify which codes should have a place in a numerical comparison, should be regarded as the first step in an on going project.

The first code we consider is COLSYS.f/COLNEW.f [1, 2]. This code is based on the approximation of the solution of the differential equation (2.2.1) by a piecewise polynomial and it uses collocation at Gauss points to define this polynomial uniquely. In fact, the COLSYS.f codes are applicable directly to mixed order systems (i.e., there is no need to reduce the problem to a first order system before attempting to solve it) and in what follows we describe how COLSYS.f can deal with such systems. This code was considered in [5] but for completeness we include it here. However, there is a major difference in that in [5] we considered a single high order equation whereas here we consider a mixed order system. We consider the mixed order system of ODEs with separated boundary conditions

$$u_i^{(m_i)} = f_i(x, u_1, \ldots, u_1^{(m_1-1)}, u_2, \ldots, u_d^{(m_d-1)}) \tag{2.2.2}$$

$$= f_i(x, z(u)), \quad 1 \le i \le d,\ a \le x \le b. \tag{2.2.3}$$

The boundary conditions are

$$g_j(z(u(\eta_j))) = 0, \quad 1 \le j \le m^*, \tag{2.2.4}$$

where

$$u(x) = [u_1(x), u_2(x), \ldots, u_d(x)]^T, \tag{2.2.5}$$

$$m^* = \sum_{i=1}^{d} m_i, \quad a = \eta_1 \le \eta_2 \le \cdots \le \eta_{m^*} = b, \tag{2.2.6}$$

and

$$z(u(x)) = (u_1(x), u_1'(x), \ldots, u_1^{(m_1-1)}(x), u_2(x), \ldots, u_2^{(m_2-1)}(x), \ldots, u_d^{(m_d-1)}(x))^T. \tag{2.2.7}$$

There are some important points to be noted about COLSYS.f. The first is that multipoint boundary conditions are allowed but all boundary conditions must be separated. The second point is that, as mentioned earlier, collocation codes such as COLSYS.f do not require the problem to be reduced to first order form before it can be solved. The codes COLSYS.f and COLNEW.f approximate the solution by using collocation at Gauss points. This requires

$$u(x) \in C^{m_i-1}[a, b], \quad \text{for } i = 1, 2, \ldots, d. \tag{2.2.8}$$

In this approach, an approximate solution of the form

$$u_\pi(x) = \sum_{j=1}^{M} \alpha_j \phi_j(x), \quad a \le x \le b, \tag{2.2.9}$$

is sought. Here the $\phi_j(x)$ are known linearly independent functions, defined on the range $[a, b]$, and the α_j are parameters that remain to be chosen. The M parameters are determined by the requirement that $u_\pi(x)$ should satisfy the following M conditions: It should satisfy the m boundary conditions and it must also satisfy the ODE at $M - m$ points in $[a, b]$. These $M - m$ points are called the collocation points. A popular choice for the $\phi_j(x)$ is to let them be piecewise polynomial functions and this is exactly what is done in COLSYS.f. It follows that the M conditions imposed on $u_\pi(x)$, to allow (2.2.9) to be uniquely defined are

(1) u_π satisfies the m boundary conditions (2.2.4)
(2) u_π satisfies the ODE at k points in each of the N mesh subintervals and this defines $Nk + m$ unknowns.

In order that the solution should satisfy the differential equation at Nk points we require that

$$0 = u_\pi^{(m)}(x_{ij}) - f(x_{ij}, u_\pi(x_{ij})), \quad 1 \le j \le k, 1 \le i \le N. \tag{2.2.10}$$

If we define the mesh with maximum mesh size h, and we derive our collocation method so that it is based on Gauss points with s collocation points per subinterval, then the global error is uniformly $O(h^s)$ while at the mesh points we have superconvergence and the error is $O(h^{2s})$. It is important to realise that COLSYS.f derives a continuous solution in the form of a piecewise polynomial and it also attempts to compute the error in this continuous solution. Of course, computing a continuous solution is a much more difficult task than just computing the solution at a discrete set of points as is done by several other codes. On the other hand, COLSYS.f performs the expensive task of computing a continuous solution even if the user only requires the solution to be computed at a few points. Thus, in any numerical comparison, we have to consider if the user will be provided with a continuous solution and if that is what he requires.

As mentioned previously, COLSYS.f attempts to provide an estimate of the error in the continuous solution and it does this using equi-distribution. The idea is that mesh points are distributed so that an equal error is made in each mesh interval. A description of this technique can be found in [2, p. 62].

The second code highlighted in [5] was MIRKDC.f. This code is based on Mono Implicit Runge-Kutta methods [4, 11, 16] which have been widely used for the numerical solution of two point boundary value problems. These methods have the special property that they can be implemented very efficiently for boundary value problems, due to a certain explicitness appearing in the MIRK equations. In what follows, we will describe exactly what MIRK formulae attempt to achieve and we show how this is done. Following Enright and Muir [17], we rewrite MIRK schemes in the slightly different form

$$y_{n+1} = y_n + h\sum_{i=1}^{s} b_i\, f(x_n + c_i\, h, Y_i),$$

$$Y_i = (1 - v_i)\, y_n + v_i\, y_{n+1} + h\sum_{j=1}^{s} z_{ij}\, f(x_n + c_j\, h,\, Y_j),\ i = 1, \ldots, s. \tag{2.2.11}$$

This is of course just a standard implicit Runge-Kutta method which can be expressed using the celebrated Butcher notation. A convenient way of expressing this formula as an array, is to write it in the form

$$\begin{array}{c|c|cccc} c_1 & v_1 & z_{11} & z_{12} & \cdots & z_{1s} \\ c_2 & v_2 & z_{21} & z_{22} & \cdots & z_{2s} \\ \vdots & \vdots & \vdots & \vdots & \vdots & \\ c_s & v_s & z_{s1} & z_{s2} & \cdots & z_{ss} \\ \hline & & b_1 & b_2 & \cdots & b_s \end{array} \tag{2.2.12}$$

This is written to emphasise the fact that both y_n and y_{n+1} appear in the Runge-Kutta equations written in the form (2.2.11). The link between (2.2.12) and the standard Butcher notation is

$$A = Z + vb^T. \tag{2.2.13}$$

Having defined the Runge-Kutta method in this particular way, Enright and Muir then compute a solution of the given two point boundary value problem on a discrete set of points. Once convergence has been obtained to a discrete solution, Enright and Muir consider how this can be transformed to a continuous solution by the computation of additional function evaluations. Muir and Owren consider schemes of the form

$$u(x) = u(x_i + \theta h_i) = y_i + h \sum_{r=1}^{s^*} b_r(\theta) k_r \tag{2.2.14}$$

where

$$\theta = (x - x_i)/h_i, \quad 0 \leq \theta \leq 1, \ x_i \leq x \leq x_{i+1}, \tag{2.2.15}$$

which they call CMIRK schemes [33]. Note that this defines a continuous solution on the ith subinterval. The number of stages required to compute this continuous solution is s^*, with $s < s^*$, and the first s stages of the interpolating polynomial (2.2.14) are the same as the first s stages of the discrete solution. The main difference between discrete MIRK schemes and continuous CMIRK schemes is that the weight coefficients of the MIRK scheme are replaced by weight polynomials in (2.2.14). Having derived a continuous solution $u(x)$ by using the CMIRK framework (2.2.14), (2.2.15), Enright and Muir use defect control both for mesh selection and accuracy control, in order to define a continuous defect. The (continuous) defect is defined as

$$\delta(x) = u'(x) - f(x, u(x)) \tag{2.2.16}$$

and Enright and Muir estimate the maximum value of this defect on each subinterval. For further discussion of this defect correction approach, the reader is referred to [5, p. 13]. In summary, we should point out that COLSYS.f controls the global error while MIRKDC.f controls the defect and these two tasks are not equivalent. The theory underlying the code MIRKDC.f [17] has remained largely unchanged since the MIRKDC.f code was written. What has changed, since [17] was written, is that there has been a major update to the code rather than to the theory. Concerned by the long calling sequence of many boundary value codes (in particular calling sequences for boundary value problems tend to be much longer than for initial value problems) and worried that many potential users may be put off using standard codes because of this, Shampine and Muir wrote a MIRKDC code which they called 'user friendly' [37]. This new code is written in FORTRAN 90/95, it cuts down drastically on the number of calling parameters required, it extends the capabilities of the MIRKDC code to the solution of problems with unknown parameters, it is able to deal with eigenvalue problems and some singular problems, and it allows the required Jacobian matrices to be computed using numerical differences.

There is also a discussion explaining why it is desirable for the code to be written in FORTRAN 90/95. Also the possibility of using the code in a continuation framework is discussed. For the purpose of this paper we will consider this code to be state of the art and, for more details, the reader is referred to [37].

2.3 Deferred Correction Codes

The most convenient way to describe our deferred correction approach is to write our methods in a Runge-Kutta framework. As mentioned earlier, if we are using MIRK formulae we can write our Runge-Kutta formulae in the form (2.2.11). In what follows we consider the deferred correction framework originally proposed by Fox [19]. In this approach we first need to define two Runge-Kutta methods which will be used to define our basic algorithm. The first Runge-Kutta method, which we will denote by ϕ, computes a cheap low order approximation to the solution of (2.2.1) while the second, denoted by ψ, computes an estimate of the local error in ϕ. This then allows us to define the deferred correction scheme in the basic form

$$\begin{aligned} \phi(\eta) &= 0, \\ \phi(\overline{\eta}) &= \psi(\eta). \end{aligned} \tag{2.3.1}$$

For a long time the question concerning the order of accuracy of (2.3.1) was unresolved. This problem was solved by Skeel [38] and in what follows we give his theorem. In the theorem below the first two conditions have been known for some time, it was the third condition that was elusive. However the theorem is quite intuitive if we regard the true solution as being the sum of the numerical solution and the global error. To state Skeel's theorem consider the deferred correction scheme (2.3.1) defined on the grid

$$\pi : a = x_0 < x_1 < \cdots < x_N = b. \tag{2.3.2}$$

Denote by Δy the restriction of the continuous solution $y(x)$ to the final grid π. Then Skeel's theorem says the following:

Theorem *Let ϕ be a stable numerical method and assume that the following conditions hold for the deferred correction scheme (2.3.1):*

$$\begin{aligned} \|\eta - \Delta y\| &= O(h^p), \\ \|\psi(\Delta y) - \phi(\Delta y)\| &= O(h^{r+p}), \\ \psi(\Delta w) &= O(h^r), \end{aligned} \tag{2.3.3}$$

for arbitrary functions w having at least r continuous derivatives. Then if $\phi(\overline{\eta}) = \psi(\eta)$ it follows that

$$\|\overline{\eta} - \Delta y\| = O(h^{r+p}). \tag{2.3.4}$$

The main problem was how to define the two operators ϕ and ψ. There are many ways in which the deferred correction schemes can be refined depending on the choices of ϕ and ψ and in what follows we use a particular form of deferred correction which was proposed by Fox [19] and later refined by Lindberg [22]. Their proposal was to consider two Runge-Kutta formulae of order i and j, respectively, where $i < j$. Having defined these formulae (and of course there is an extremely wide choice that we have), we consider the algorithm defined by

$$\begin{aligned}\phi_i(\eta) &= 0,\\ \phi_i(\overline{\eta}) &= -\phi_j(\eta).\end{aligned}$$

It is clear that the first two conditions of Skeel's theorem are trivially satisfied if we use this deferred correction approach, with $p = i$ and $r + p = j$. It is much more complicated to verify that the final condition is satisfied and this requires us to present our Runge-Kutta methods in a special way and to explain this the reader is referred to [10]. There are many ways in which these deferred correction methods can be refined. For example we could consider the deferred correction scheme

$$\begin{aligned}\phi_i(\eta) &= 0,\\ \phi_i(\overline{\eta}) &= -\phi_j(\eta),\\ \phi_i(\overline{\overline{\eta}}) &= -\phi_j(\eta) - \phi_k(\overline{\eta}),\end{aligned} \tag{2.3.5}$$

where an extra symmetric Runge-Kutta formula is introduced in an attempt to define a scheme of order k where $k > j$. Again the order of accuracy of this scheme can be determined by Skeel's theorem. A popular choice is to take $i = 4$, $j = 6$, $k = 8$. It is easy to show that if the ϕ are either MIRK formulae or Lobatto formulae then η, $\overline{\eta}$, and $\overline{\overline{\eta}}$ are of order 4, 6 and 8 respectively. If we use the deferred correction scheme based on MIRK methods then this defines the code TWPBVP.f and if we base our codes on Lobatto formulae then we have the code TWPBVPL.f. This leads to two widely used codes which use the basic framework (2.3.5). For an illustration of how these formulae perform on a wide range of test problems the reader is referred to the web page of one of the authors [8]. We emphasise that the schemes we have described are just some of many that can be defined in the Runge-Kutta framework. Another possibility which certainly seems worth investigating is to use methods of different orders in different parts of the mesh. Another possibility is to use fixed order but include methods of order 10 rather than stopping at order 8. An important point to note is that these schemes, which are based on Runge-Kutta methods, have the usual problem that there is no convenient error estimate available and, also, the obtained solution is a discrete one. In order to define a continuous solution, if it is needed, we have to derive a suitable interpolant and as an estimate of the error we compute

$$\phi_i(\overline{\overline{\eta}}) - \phi_i(\overline{\eta}). \tag{2.3.6}$$

This gives us an error estimate in the jth order solution $\overline{\eta}$ and this is used for grid refinement [39]. There is also an important difference in the way in which the mesh is formed when using iterated deferred correction methods. As mentioned earlier, COLSYS.f uses equi-distribution and this results in mesh points 'sliding about'. The deferred correction methods discussed in [5], namely TWPBVPL.f which is based on Lobatto formulae and TWPBVP.f which is based on MIRK formulae, either add in or take out mesh points depending on the error estimate and this results in a lot of the mesh points not moving during a grid refinement. These codes are on the authors web page and can be considered as being state of the art codes. In particular, the code TWPBVP.f tends to be very effective for non-stiff problems while TWPBVPL.f is particularly efficient for stiff problems. The main reason for this behaviour is that the Lobatto codes, which use implicit deferred corrections, are much more reliable than the codes based on MIRK formulae which use explicit deferred corrections [12].

An important way in which deferred correction codes have changed since TWPBVP and TWPBVPL were written is that they now have the option of taking into account the conditioning of the problem. The importance of having this facility was demonstrated in a paper of Shampine and Muir [36] who wanted to test out the diagnostic capabilities of the two-point boundary value code BVP4C. They did this by considering Bratu's problem:

$$\begin{aligned} y'' &= \lambda \exp(y), \\ y(0) &= y(1) = 0. \end{aligned} \tag{2.3.7}$$

Davis [14] has shown that if $0 \le \lambda < \lambda^* = 3.51383\ldots$ then there are 2 solutions to this problem and both are parabolic and are concave down in nature. If $\lambda = \lambda^*$ then there is just one solution and for $\lambda > \lambda^*$ there are no solutions. In their numerical experiments Shampine and Muir first described the performance of the code for $\lambda = 3.45$ and plotted one of the solutions which they obtained in a perfectly satisfactory way. However they did note that their estimate of the conditioning constant was quite high (the error tolerance imposed was 10^{-3}) and the estimate of the conditioning constant was 3.14×10^3. The code found this problem so easy that the solution was obtained on the initial mesh of 10 points. Having done this, Shampine and Muir solved Bratu's problem with $\lambda = 3.55$ for which the problem has no solution. The expectation was that BVP4C would fail to find a solution and would send this message back to the user. However the code did not do this, instead it returned a solution which had all the characteristics of being parabolic in nature and concave downwards. The solver provided no warning message. However the code did need 179 points to compute the solution and the estimate of the conditioning constant was 10^6 which can be considered as being very large given the precision that is required. The reason for this poor performance of BVP4C is not hard to understand. The problem arises from the fact that the solution is obtained, and the mesh is refined, using a local error estimate. We do this on the assumption that if the local error estimate is sufficiently small, then the global error will also be small. However a backward error analysis shows that this may not be the case if the problem is ill conditioned. All of this makes a very powerful case, when solving two point boundary value problems, for computing the conditioning constant of the problem as well as the solution. Without a conditioning estimate we are not able to have any confidence in a solution that we compute based on local errors and we feel that this is a really important problem which requires considerable further investigation.

Early work on the estimation of conditioning for boundary value problems was carried out by Trigiante, Brugnano and Mazzia [3, 28]. The approach of these researchers was rather different from that of Shampine and Muir. The Shampine Muir approach on the detection of poor conditioning was to warn the user of this and to warn that there may be a severe loss of correct digits in the solution computed. In contrast, in a recent series of papers [6, 7] Cash, Mazzia and their co-workers derived efficient algorithms for estimating the condition numbers of BVPs and they developed a mesh choosing algorithm which takes into account the conditioning of the problem. In this approach the aim is to choose the mesh so that a local error tolerance is satisfied and also so that the problem remains well conditioned. This led to the derivation of codes TWPBVPLC.f and TWPBVPC.f which are based on Lobatto and MIRK methods respectively and which give an estimate of the conditioning of the problem along with the numerical solution computed by the code. The existence of these new codes gives the user the option of taking into account the conditioning of the problem and, if this option is used, the original codes TWPBVP.f and TWPBVPL.f require the change of just one input parameter. Extensive numerical testing on these four deferred correction codes appear on the web page of one of the authors [8]. What is found, for most problems where conditioning is not an issue, is that there is little difference between the performance of TWPBVP.f/TWPBVPC.f and TWPBVPL.f/TWPBVPLC.f. However for problems where conditioning is important there is often a considerable gain in efficiency if the codes TWPBVPC.f and TWPBVPLC.f are used. Finally, another important feature, is that the estimate of the conditioning constant can be used in a backward error analysis to ensure that the final solution, which is based on local error estimation, gives a solution with a satisfactory global error. Although considerable advances have been made in the computation of conditioning constants we feel that there is still more work to be done. However the approach of Trigiante, Brugnano, Cash and Mazzia which chooses the mesh so that a local error tolerance is satisfied and so that the conditioning of the continuous and discrete problems remains roughly the same is a very powerful one.

2.4 Boundary Value Methods

In this section we describe a class of methods known as Boundary Value Methods (BVMs) [3]. These are linear multistep methods used in a special way that allows us to generate stable discrete boundary values schemes. In the case of symmetric schemes, given the grid π defined by (2.3.2) with $h_i = x_i - x_{i-1}$, $1 \le i \le N$, the numerical scheme generated by a k-step BVM is defined by the following equations:

$$\begin{cases} g(y_0, y_N) = 0, \\ \sum_{j=-i}^{k-i} \alpha_{j+i}^{(i)} y_{i+j} = h_i \sum_{j=-i}^{k-i} \beta_{j+i}^{(i)} f_{i+j}, \\ \quad i = 1, \dots, k_1 - 1 \\ \quad \text{(additional initial methods)}, \\ \sum_{j=-k_1}^{k_2} \alpha_{j+k_1}^{(i)} y_{i+j} = h_i \sum_{j=-k_1}^{k_2} \beta_{j+k_1}^{(i)} f_{i+j}, \\ \quad i = k_1, \dots, N - k_2 \\ \quad \text{(main methods)}, \\ \sum_{j=N-i-k}^{N-i} \alpha_{j-N+i+k}^{(i)} y_{i+j} = h_i \sum_{j=N-i-k}^{N-i} \beta_{j-N+i+k}^{(i)} f_{i+j}, \\ \quad i = N + 1 - k_2, \dots, N \\ \quad \text{(additional final methods)}, \end{cases} \tag{2.4.1}$$

where y_i is the approximation of $y(x_i)$, $f_i = f(x_i, y_i)$, k is odd, $k_1 = (k+1)/2$, $k_2 = k - k_1$ and $\boldsymbol{\alpha}^{(i)} = (\alpha_0^{(i)}, \dots, \alpha_k^{(i)})^T$ and $\boldsymbol{\beta}^{(i)} = (\beta_0^{(i)}, \dots, \beta_k^{(i)})^T$ are the coefficient vectors characterising the method. The so called "additional initial methods" and "additional final methods" i.e. $k_1 - 1$ initial and k_2 final equations, are derived by using appropriate discretisation schemes.

The main code currently available, which implements these methods, is known as TOM. This is a general purpose code for the solution of BVPs and is rather different from other codes that have been derived. The first release of the code was written in MATLAB in 2003 [24], and was based on a class of symmetric Boundary Value Methods (BVMs) [3]. The Top Order Methods (TOM) are k-step linear multistep methods with the highest possible order $2k$ and in the code TOM we use $k = 3$ to give a sixth order method. A complete description of these methods, along with their stability properties, can be found in [3, Sect. 7.4]. Recently the code has been updated by implementing another class of BVMs, namely the BS linear multistep methods [30–32]. The new version of the code is available in MATLAB.

The BS methods are derived by imposing the restriction that the numerical solution of the general k-step linear multistep formula is the same as is given by the collocation procedure using the B-spline basis. The coefficients are computed by solving special linear systems. The k-step BS scheme provides a C^k continuous solution that is kth-order accurate uniformly in $[a, b]$ and collocating the differential equation at the mesh points [32]. In order to introduce the continuous extension, some further notation is needed. First, we represent any spline function s of degree $d = (k+1)$ and knot $x_i, i = 0, \dots, N$ using the associated (dth-degree) B-spline basis $B_j(x), j = -d, \dots, N-1$. This can be defined after prescribing two additional sets of d knots, $\{x_i, i = -d, \dots, -1\}$ (*left auxiliary knots*), with $x_{-d} \le \dots \le x_0$, and $\{x_i, i = N+1, \dots, N+d\}$ (*right auxiliary knots*), with $x_N \le x_{N+1} \le \dots \le x_{N+d}$ [15]. Using this notation, the spline collocating the differential equation can be represented as follows,

$$Q_d^{(BS)}(y) = \sum_{j=-d}^{N-1} \mu_j^{(BS)}(y) B_j, \tag{2.4.2}$$

where $\mu_j^{(BS)}(y)$ are linear combinations of the values of y_i and f_i, $i = 0, \ldots, N$ in π. We note that, if the values y_i and f_i have been computed with a different scheme, the continuous approximation is the quasi-interpolation spline described in [26]. This means that this continuous extension could safely be used for any discretisation method.

The main feature of the code is that it implements a hybrid mesh selection strategy based on both the conditioning parameters of the problem and on a suitable approximation of the local error. This strategy was introduced in [28] and was first implemented in the code TOM. Subsequently it was modified in [6, 7] for use in TWPBVPC and TWPBVPLC. As is the case for the codes based on deferred correction, by changing just one input parameter, it is possible for the mesh selection to use a standard strategy based only on the local error for the mesh selection. Instead of using a damped Newton method for the solution of the nonlinear equations, the code TOM implements a quasi-linearisation technique [27, 29, 32]. This means that a sequence of continuous linear BVPs is solved to a suitable tolerance. This allows us to use very efficiently the mesh selection based on conditioning for non linear problems as well as linear ones.

We note that the conditioning parameters estimated by the code can be defined both for the continuous problem and for the discrete one giving the possibility to *measure* the reliability of the discrete problem with respect to the continuous one. In other words, we must be suspicious if the discrete problem provides parameters which are very different from the continuous ones, and this could be checked if the conditioning parameters do not converge. Since such parameters, for the discrete problem, depend on the chosen mesh, it is possible, for a fixed method, to vary the latter in order to be sure that the discrete parameters converge. This allows us, for example, to recognise if the solution has two or more time scales associated with it. This is the idea on which the mesh strategy is based.

Since the BS are collocation methods, the code provides as output a continuous solution, which could be used to estimate the error at any chosen point, or a set of discrete points, if, for example, the user requires only the solution at the mesh points. A continuous solution is also computed, if needed, when using the TOM code, by the quasi-interpolation technique based on the BS methods described in [26].

2.5 Interpolation

As mentioned earlier, the deferred correction codes TWPBVP.f and TWPBVPL.f are the only ones described in this paper which do not seek to provide a continuous solution. The obvious way of getting around this problem is to derive an a posteriori interpolant which can be computed using just a few extra function evaluations. This is rather similar to what the code MIRKDC does, although this code needs a continuous solution from the start so that it can estimate the defect. Here a discrete solution is computed first of all and this is accepted once the defect is sufficiently small. After the discrete solution has been accepted a continuous MIRK scheme is formed using an approach of Muir and Owren [33]. Note that, if a continuous solution is not required, for example the solution may be required only at a few mesh points, then this interpolant is not needed for error estimation. Numerical experience has shown that it is advisable to compute an interpolant using data from only one sub-interval, and this avoids problems at the end of the mesh. If the interpolant is a local one, that is it is defined over a single interval, then the interpolant is identical over each sub-interval and it also has the desirable property that it is symmetric. The problem of deriving high order interpolants for MIRK formulae has recently been considered in some detail. It has been shown in [9] that a sixth order MIRK interpolant can be computed using just one extra function evaluation. To compute an eighth order MIRK interpolant, we need four extra function evaluations and the way in which this is done is described in [9]. By deriving an a posteriori interpolant, we are effectively introducing a continuous solution and this allows a fair comparison with other codes discussed in this paper to be carried out.

2.6 Conclusion

The purpose of the present paper was to answer the question 'Which codes for the numerical solution of two point boundary value problems of the form (2.2.1) can be considered as being state of the art'. The codes identified as falling into this class were based on: collocation methods (COLSYS/COLNEW); defect control methods based on MIRK formulae (MIRKDC); deferred correction methods (TWPBVP/TWPBVPL) and boundary value methods (TOM). There is considerable numerical evidence to suggest that these codes are amongst the most efficient global methods currently available. In addition these codes have reached what we have called a "steady state" and this makes them ideal for use in a numerical comparison. However there are two important points that we need to bear in mind.

Firstly, these codes attempt to solve slightly different problems. In particular COLSYS/COLNEW and TOM attempt to define a continuous solution by computing the error in a polynomial approximation to the solution. MIRKDC also seeks to provide a continuous solution but controls the error in the defect. In contrast the deferred correction codes compute a discrete solution initialy and the continuous solution is obtained using an a posteriori interpolant.

The second point to note is that it is important to take into account the conditioning of the problem when solving a problem of the form (2.2.1). A standard backward error analysis which links the global error to the local one relies on the problem being well conditioned and, if it is not, a solution which has not got the required accuracy may be accepted. A considerable amount of effort has been applied to the estimation of the conditioning of a problem and the reader is referred to [3, 6, 7, 28]. The boundary vale methods and deferred correction codes allow conditioning estimation and an estimate of the conditioning can be obtained by changing just one input parameter for these codes. We expect the estimation of the conditioning of a problem to become a very important topic in the future. In addition we feel that the four FORTRAN codes we have discussed are a very firm basis for carrying out numerical comparisons of different methods and we hope to be able to investigate this in the near future. Finally, we list those codes in FORTRAN and MATLAB which would be candidates for use in a numerical comparison. We feel it appropriate to divide the codes into two catagories, namely those which are written in FORTRAN and those which are written in MATLAB. Generally speaking, if run time is an issue then the user should probably be using a FORTRAN code. If it is not, then the user may find a MATLAB code to be more convenient. The codes that we feel either are or will become state of the art codes are

- **FORTRAN codes**: TWPBVPC, TWPBVPLC, ACDC, COLNEW, COLMOD, MIRKDC, BVP_SOLVER.
- **MATLAB codes**: BVP4C, BVP5C ([21]), TOM, TWPBVPC, TWPBVPLC.

Note that the MATLAB codes TWPBVPC/TWPBVPLC have only just been released but are based on widely used codes. We finish this section with the observation that codes written in languages other than FORTRAN or MATLAB are starting to appear. As an example we mention the code TWPBVP which is now available in R (package `bvpSolve`) [34].

References

1. Ascher, U., Christiansen, J., Russell, R.D.: Collocation software for boundary-value odes. ACM Trans. Math. Softw. **7**(2), 209–222 (1981)
2. Ascher, U.M., Mattheij, R.M.M., Russell, R.D.: Numerical Solution of Boundary Value Problems for Ordinary Differential Equations. Classics in Applied Mathematics, vol. 13. SIAM, Philadelphia (1995). Corrected reprint of the 1988 original
3. Brugnano, L., Trigiante, D.: Solving Differential Problems by Multistep Initial and Boundary Value Methods. Stability and Control: Theory, Methods and Applications, vol. 6. Gordon and Breach, Amsterdam (1998)
4. Cash, J.R.: A class of implicit Runge-Kutta methods for the numerical integration of stiff ordinary differential equations. J. ACM **22**(4), 504–511 (1975)
5. Cash, J.R.: A survey of some global methods for solving two-point BVPs. Appl. Numer. Anal. Comput. Math. **1**(1–2), 7–17 (2004)
6. Cash, J.R., Mazzia, F.: A new mesh selection algorithm, based on conditioning, for two-point boundary value codes. J. Comput. Appl. Math. **184**(2), 362–381 (2005)
7. Cash, J.R., Mazzia, F.: Hybrid mesh selection algorithms based on conditioning for two-point boundary value problems. J. Numer. Anal. Ind. Appl. Math. **1**(1), 81–90 (2006)
8. Cash, J.R., Mazzia, F.: Algorithms for the solution of two-point boundary value problems. http://www.ma.ic.ac.uk/jcash/BVP_software/twpbvp.php
9. Cash, J.R., Moore, D.R.: High-order interpolants for solutions of two-point boundary value problems using MIRK methods. Comput. Math. Appl. **48**(10–11), 1749–1763 (2004)
10. Cash, J.R., Silva, H.H.M.: Iterated deferred correction for linear two-point boundary value problems. Comput. Appl. Math. **15**(1), 55–75 (1996)
11. Cash, J.R., Singhal, A.: High order methods for the numerical solution of two-point boundary value problems. BIT **22**(2), 184–199 (1982)
12. Cash, J.R., Wright, M.H.: A deferred correction method for nonlinear two-point boundary value problems: implementation and numerical evaluation. SIAM J. Sci. Stat. Comput. **12**(4), 971–989 (1991)

13. Cash, J.R., Moore, G., Wright, R.W.: An automatic continuation strategy for the solution of singularly perturbed linear two-point boundary value problems. J. Comput. Phys. **122**(2), 266–279 (1995)
14. Davis, H.T.: Introduction to Nonlinear Differential and Integral Equations. Dover, New York (1962)
15. de Boor, C.: A Practical Guide to Splines. Applied Mathematical Sciences, vol. 27. Springer, New York (2001). Revised edition
16. Enright, W.H., Muir, P.H.: Efficient classes of Runge-Kutta methods for two-point boundary value problems. Computing **37**(4), 315–334 (1986)
17. Enright, W.H., Muir, P.H.: Runge-Kutta software with defect control for boundary value ODEs. SIAM J. Sci. Comput. **17**(2), 479–497 (1996)
18. Enright, W.H., Hull, T.E., Lindberg, B.: Comparing numerical methods for stiff systems of O.D.Es. BIT **15**(2), 10–48 (1975)
19. Fox, L.: The Numerical Solution of Two-Point Boundary Problems in Ordinary Differential Equations. Oxford University Press, New York (1957)
20. Hairer, E., Wanner, G.: Solving Ordinary Differential Equations. II. Springer Series in Computational Mathematics, vol. 14. Springer, Berlin (1991). Stiff and differential-algebraic problems
21. Kierzenka, J., Shampine, L.F.: A BVP solver that controls residual and error. J. Numer. Anal. Ind. Appl. Math. **3**(1–2), 27–41 (2008)
22. Lindberg, B.: Error estimation and iterative improvement for discretization algorithms. BIT **20**(4), 486–500 (1980)
23. Lioen, W.M., de Swart, J.J.B.: Test set for IVP solvers. Technical Report MAS-R9832. CWI, Amsterdam (1998)
24. Mazzia, F.: Software for boundary value problems. Department of Mathematics, University of Bari and INdAM, Research Unit of Bari, February 2003. Available at http://www.dm.uniba.it/mazzia/bvp/index.html
25. Mazzia, F., Magherini, C.: Test set for initial value problem solvers, release 2.4. Department of Mathematics, University of Bari and INdAM, Research Unit of Bari, February 2008. Available at http://www.dm.uniba.it/testset
26. Mazzia, F., Sestini, A.: The BS class of hermite spline quasi-interpolants on nonuniform knot distributions. BIT **49**(3), 611–628 (2009)
27. Mazzia, F., Sgura, I.: Numerical approximation of nonlinear BVPs by means of BVMs. Appl. Numer. Math. **42**(1–3), 337–352 (2002). Ninth Seminar on Numerical Solution of Differential and Differential-Algebraic Equations (Halle, 2000)
28. Mazzia, F., Trigiante, D.: A hybrid mesh selection strategy based on conditioning for boundary value ODE problems. Numer. Algorithms **36**(2), 169–187 (2004)
29. Mazzia, F., Trigiante, D.: Efficient strategies for solving nonlinear problems in bvps codes. Nonlinear Stud. in press
30. Mazzia, F., Sestini, A., Trigiante, D.: B-spline linear multistep methods and their continuous extensions. SIAM J. Numer. Anal. **44**(5), 1954–1973 (2006) (electronic)
31. Mazzia, F., Sestini, A., Trigiante, D.: BS linear multistep methods on non-uniform meshes. J. Numer. Anal. Ind. Appl. Math. **1**(1), 131–144 (2006)
32. Mazzia, F., Sestini, A., Trigiante, D.: The continous extension of the B-spline linear multistep metods for BVPs on non-uniform meshes. Appl. Numer. Math. **59**(3–4), 723–738 (2009)
33. Muir, P., Owren, B.: Order barriers and characterizations for continuous mono-implicit Runge-Kutta schemes. Math. Comput. **61**(204), 675–699 (1993)
34. R Development Core Team: R: A Language and Environment for Statistical Computing. R Foundation for Statistical Computing, Vienna (2009). ISBN 3-900051-07-0
35. Shampine, L.F.: Evaluation of a test set for stiff ode solvers. ACM Trans. Math. Softw. **7**(4), 409–420 (1981)
36. Shampine, L.F., Muir, P.H.: Estimating conditioning of BVPs for ODEs. Math. Comput. Model. **40**(11–12), 1309–1321 (2004)
37. Shampine, L.F., Muir, P.H., Xu, H.: A user-friendly Fortran BVP solver. J. Numer. Anal. Ind. Appl. Math. **1**(2), 201–217 (2006)

38. Skeel, R.D.: A theoretical framework for proving accuracy results for deferred corrections. SIAM J. Numer. Anal. **19**(1), 171–196 (1982)
39. Wright, R., Cash, J., Moore, G.: Mesh selection for stiff two-point boundary value problems. Numer. Algorithms **7**(2–4), 205–224 (1994)

Chapter 3
Advances on Collocation Based Numerical Methods for Ordinary Differential Equations and Volterra Integral Equations

Dajana Conte, Raffaele D'Ambrosio, and Beatrice Paternoster

Abstract We present a survey on collocation based methods for the numerical integration of Ordinary Differential Equations (ODEs) and Volterra Integral Equations (VIEs), starting from the classical collocation methods, to arrive to the most important modifications appeared in the literature, also considering the multistep case and the usage of basis of functions other than polynomials.

Keywords Collocation · Two-step collocation · Runge–Kutta methods · Two-step Runge–Kutta methods · Mixed collocation

Mathematics Subject Classification (2000) 65L05 · 65L60 · 65R20

3.1 Introduction

Collocation is a widely applied and powerful technique in the construction of numerical methods for ODEs and VIEs. As it is well known, a collocation method is based on the idea of approximating the exact solution of a given functional equation with a suitable approximant belonging to a chosen finite dimensional space, usually a piecewise algebraic polynomial, which exactly satisfies the equation on a certain subset of the integration interval (i.e. the set of the so-called *collocation points*).

D. Conte · R. D'Ambrosio · B. Paternoster (✉)
Dipartimento di Matematica e Informatica, Università di Salerno, Fisciano, Italy
e-mail: beapat@unisa.it

D. Conte
e-mail: dajconte@unisa.it

R. D'Ambrosio
e-mail: rdambrosio@unisa.it

T.E. Simos (ed.), *Recent Advances in Computational and Applied Mathematics*,
DOI 10.1007/978-90-481-9981-5_3,

This technique, when applied to problems based on functional equations, allows the derivation of methods having many desirable properties. In fact, collocation methods provide an approximation over the entire integration interval to the solution of the equation. Moreover, the collocation function can be expressed as a linear combination of functions ad hoc for the problem we are integrating, in order to better reproduce the qualitative behavior of the solution.

The systematic study of collocation methods for initial value problems in ODEs, VIEs, and Volterra integro-differential equations (VIDEs) has its origin, respectively, in the late '60, the early '70 and the early '80s. The idea of multistep collocation was first introduced by Lie and Norsett in [58], and further extended and investigated by several authors [12, 24–26, 28, 31, 34–36, 38, 42, 57, 62].

Multistep collocation methods depend on more parameters than classical ones, without any significant increase in the computational cost, by regarding them as special case of multistep Runge–Kutta methods: therefore, there are much more degrees of freedom to be spent in order to obtain strong stability properties and an higher order and stage order of convergence. As a direct consequence the effective order of multistep collocation methods is generally higher with respect to one step collocation methods with the same number of stages. Moreover, as they generally have high stage order, they do not suffer from the order reduction phenomenon (see [11, 43]), which occurs in the integration of stiff systems.

The purpose of this paper is to present a review of recently introduced families of collocation and modified collocation methods for ODEs and VIEs. In particular we aim to present the main results obtained in the context of multistep collocation and almost collocation methods, i.e. methods obtained by relaxing some collocation and/or interpolation conditions in order to obtain desirable stability properties.

The paper is organized as follows: Sect. 3.2 reviews the main results concerning classical one-step and multistep collocation methods for ODEs and their recent extensions and modifications; Sect. 3.3 is dedicated to collocation methods for second order initial value problems and also collocation methods based on functional basis other than polynomials; in Sect. 3.4 we consider the evolution of the collocation technique for Volterra integral equations.

3.2 Collocation Based Methods for First Order ODEs

In this section we focus our attention on the hystorical background and more recent results concerning the collocation technique, its modifications and extensions for the derivation of highly stable continuous methods for the numerical solution of initial value problems based on first order ODEs

$$\begin{cases} y'(x) = f(x, y(x)), & x \in [x_0, X], \\ y(x_0) = y_0 \in \mathbb{R}^d, \end{cases} \tag{3.2.1}$$

with $f : [x_0, X] \times \mathbb{R}^d \to \mathbb{R}^d$. It is assumed that the function f is sufficiently smooth, in such a way that the problem (3.2.1) is well-posed.

3.2.1 Classical One-Step Collocation Methods

Let us suppose that the integration interval $[x_0, X]$ is discretized in an uniform grid $x_0 < x_1 < \cdots < x_N = X$. Classical collocation methods (see [6, 10, 11, 44, 45, 55, 78]) are determined by means of a continuous approximant, generally an algebraic polynomial $P(x)$, satisfying some opportune conditions: in order to advance from x_n to x_{n+1}, the polynomial $P(x)$ interpolates the numerical solution in x_n, and exactly satisfies the ODE (3.2.1)—i.e. *co-locates*—in the set of points $\{x_n + c_i h, i = 1, 2, \dots, m\}$, where $c_1, c_2, \dots, c_m$ are m real numbers (named *collocation nodes*), that is

$$\begin{cases} P(x_n) = y_n, \\ P'(x_n + c_i h) = f(x_n + c_i h, P(x_n + c_i h)), \quad i = 1, 2, \dots, m. \end{cases} \tag{3.2.2}$$

The solution in x_{n+1} can then be computed from the function evaluation

$$y_{n+1} = P(x_{n+1}). \tag{3.2.3}$$

The classical framework in which collocation methods must be placed is certainly constituted by implicit Runge–Kutta methods (IRK). In fact, Guillou and Soule in [42] and Wright in [78] independently proved that one step collocation methods form a subset of implicit Runge–Kutta methods

$$y_{n+1} = y_n + h \sum_{i=1}^{m} b_i f(x_n + c_i h, Y_i), \tag{3.2.4}$$

$$Y_i = y_n + h \sum_{j=1}^{m} a_{ij} f(x_n + c_j h, Y_j), \quad i = 1, 2, \dots, m, \tag{3.2.5}$$

where

$$a_{ij} = \int_0^{c_i} L_j(s)ds, \qquad b_j = \int_0^1 L_j(s)ds, \quad i, j = 1, 2, \dots, m \tag{3.2.6}$$

and $L_j(s)$, $j = 1, \dots, m$, are fundamental Lagrange polynomials. The maximum attainable order of such methods is at most $2m$, and it is obtained by using Gaussian collocation points [45, 55]. Anyway, unfortunately, the order $2m$ is gained only at the mesh points: the uniform order of convergence over the entire integration interval is only m. As a consequence, they suffer from order reduction showing effective order equal to m (see [10, 11, 43, 45]).

Butcher (see [10] and references therein) gave an interesting characterization of collocation methods in terms of easy algebraic conditions, and analogous results are also reported in [45, 55]. This characterization, together with many other several results regarding the main properties of collocation methods, comes out as natural consequence of an interesting interpretation of collocation methods in terms of quadrature formulae. In fact, if $f(x, y(x)) = f(x)$, (3.2.4)–(3.2.5) can be respectively interpreted as quadrature formulae for $\int_{x_n}^{x_n+h} f(x)dx$ and $\int_{x_n}^{x_n+c_i h} f(x)dx$, for $i = 1, 2, \ldots, m$. We next consider the following linear systems

$$A(q) : \sum_{j=1}^{m} a_{ij} c_j^{k-1} = \frac{c_i^k}{k}, \quad k = 1, 2, \ldots, q, \; i = 1, 2, \ldots, m, \tag{3.2.7}$$

$$B(p) : \sum_{i=1}^{m} b_i c_i^{k-1} = \frac{1}{k}, \quad k = 1, 2, \ldots, p. \tag{3.2.8}$$

Then, the following result holds (see [44, 55]):

Theorem 3.1 *If the condition $B(p)$ holds for some $p \geq m$, then the collocation method (3.2.2) has order p.*

As a consequence, a collocation method has the same order of the underlying quadrature formula (see [44], p. 28). Finally, the following result characterizing classical collocation methods arises (see [10, 44, 45, 55]).

Theorem 3.2 *An implicit m-stage Runge–Kutta method, satisfying $B(m)$ and having distinct collocation abscissae, is a collocation method if and only if conditions $A(m)$ holds.*

The most used collocation methods are those based on the zeros of some orthogonal polynomials, that is Gauss, Radau, Lobatto [10, 11, 43, 45, 55], having respectively order of convergence $2m$, $2m - 1$, $2m - 2$, where m is the number of collocation points (or the number of stages, regarding the collocation method as an implicit Runge–Kutta). Concerning their stability properties, it is known that Runge–Kutta methods based on Gaussian collocation points are A-stable, while the ones based on Radau IIA points are L-stable and, moreover, they are also both algebraically stable (see [11, 43, 49] and references therein contained); Runge–Kutta methods based on Lobatto IIIA collocation points, instead, are A-stable but they are not algebraically stable (see [10, 44, 45, 55]).

3.2.2 *Perturbed Collocation*

As remarked by Hairer and Wanner in [43], only some IRK methods are of collocation type, i.e. Gauss, Radau IIA, and Lobatto IIIA methods. An extension of the collocation idea, the so-called *perturbed collocation* is due to Norsett and Wanner (see [63, 64]), which applies to all IRK methods.

We denote by Π_m the linear space of polynomials of degree at most m and consider the polynomial $N_j \in \Pi_m$ defined by

$$N_j(x) = \frac{1}{j!}\sum_{i=0}^{m}(p_{ij} - \delta_{ij})x^i, \quad j = 1, 2, \dots, m,$$

where δ_{ij} is the usual Kronecker delta. We next define the *perturbation operator* $P_{x_0,h} : \Pi_m \to \Pi_m$ by

$$(P_{x_0,h}u)(x) = u(x) + \sum_{j=1}^{n} N_j\left(\frac{x - x_0}{h}\right)u^{(j)}(x_0)h^j.$$

Next, the following definition is given (see [63, 64]).

Definition 3.3 Let $c_1, \dots, c_m$ be given distinct collocation points. Then the corresponding *perturbed collocation method* is defined by

$$\begin{aligned} u(x_0) &= y_0, \quad u \in \Pi_m, \\ u'(x_0 + c_i h) &= f(x_0 + c_i h, (P_{x_0,h}u)(x_0 + c_i h)), \quad i = 1, 2, \dots, m, \\ y_1 &= u(x_0 + h). \end{aligned}$$

As the authors remark in [64], if all N_j's are identically zero, then $P_{x_0,h}$ is the identical map and the definition coincides with classical collocation. In the same paper the authors provide the equivalence result between the family of perturbed collocation methods and Runge–Kutta methods (see [64]). The interest of this results, as again is stated in [64], is that the properties of collocation methods, especially in terms of order, linear and nonlinear stability, can be derived in a reasonable short, natural and very elegant way, while it is known that, in general, these properties are very difficult to handle and investigate outside collocation.

3.2.3 Discontinuous Collocation

In the literature, perturbed collocation has been considered as a modification of the classical collocation technique, in such a way that much more Runge–Kutta methods could be regarded as perturbed collocation based methods, rather than classically collocation based. There are other possible extensions of the collocation idea, which apply to wider classes of Runge–Kutta methods, such as the so-called *discontinuous collocation* (see [44]).

Definition 3.4 Let $c_2, \dots, c_{m-1}$ be distinct real numbers (usually between 0 and 1), and let b_1, b_m be two arbitrary real numbers. The corresponding *discontinuous method* is then defined via a polynomial of degree $m - 2$ satisfying

$$u(x_0) = y_0 - hb_1(\dot{u}(x_0) - f(x_0, u(x_0))),$$

$$\dot{u}(x_0 + c_i h) = f(x_0 + c_i h, u(x_0 + c_i h)), \quad i = 2, 3, \ldots, m-1,$$

$$y_1 = u(x_1) - h b_s(\dot{u}(x_1) - f(x_1, u(x_1))).$$

Discontinuous collocation methods fall inside a large class of implicit Runge–Kutta methods, as stated by the following result (see [44]).

Theorem 3.5 *The discontinuous collocation method given in Definition 3.4 is equivalent to an m-stage Runge–Kutta method with coefficients determined by* $c_1 = 0$, $c_m = 1$ *and*

$$a_{i1} = b_1, \qquad a_{im} = 0, \quad i = 1, 2, \ldots, m,$$

while the other coefficients result as solutions of the linear systems $A(m-2)$ *and* $B(m-2)$ *defined in* (3.2.7) *and* (3.2.8).

As a consequence of this result, if $b_1 = 0$ and $b_m = 0$, then the discontinuous collocation method in Definition 3.4 is equivalent to the $(m-2)$-collocation method based on $c_2, \ldots, c_{m-1}$. An interesting example of implicit Runge–Kutta method which is not collocation based but is of discontinuous collocation type is the Lobatto IIIB method (see [10, 44, 45, 55]), which plays an important role in the context of geometric numerical integration, together with Lobatto IIIA method (see [44], p. 33). They are both nonsymplectic methods (see Theorem 4.3 in [44]) but, considered as a pair, the resulting method is symplectic. This is a nice example of methods which possess very strong properties, but are difficult to investigate as discrete scheme (they cannot be studied as collocation methods, because they are not both collocation based); however, re-casted as discontinuous collocation based methods, their analysis is reasonably simplified and very elegant [44].

3.2.4 Multistep Collocation

The successive results which appeared in literature (see [22, 42, 43, 57, 58]) have been devoted to the construction of multistep collocation methods. Guillou and Soulé introduced multistep collocation methods [42], by adding interpolation conditions in the previous k step points, so that the collocation polynomial is defined by

$$\begin{cases} P(x_{n-i}) = y_{n-i}, & i = 0, 1, \ldots, k-1, \\ P'(x_n + c_j h) = f(x_n + c_j h, P(x_n + c_j h)), & j = 1, 2, \ldots, m. \end{cases} \tag{3.2.9}$$

The numerical solution is given, as usual by

$$y_{n+1} = P(x_{n+1}). \tag{3.2.10}$$

Hairer–Wanner [43] and Lie–Norsett [58] derived different strategies to obtain multistep collocation methods. In [43] the Hermite problem with incomplete data (3.2.9) is solved by means of the introduction of a generalized Lagrange basis

$$\{\varphi_i(s),\ \psi_j(s),\ i=1,2,\dots,k,\ j=1,2,\dots,m\}$$

and, correspondingly, the collocation polynomial is expressed as linear combination of this set of functions, i.e.

$$P(x_n+sh)=\sum_{i=1}^{k}\varphi_i(s)y_{n-k+i}+h\sum_{i=1}^{s}\psi_i(s)P'(x_n+c_ih),$$

where $s=\frac{x-x_n}{h}$. Therefore, the problem (3.2.9) is transformed in the problem of deriving $\{\varphi_i,\psi_j, i=1,2,\dots,k,\ j=1,2,\dots,m\}$ in such a way that the corresponding polynomial $P(x_n+sh)$ satisfies the conditions (3.2.9).

Lie–Norsett in [58] completely characterized multistep collocation methods, giving the expressions of the coefficients of collocation based multistep Runge–Kutta methods in closed form, as stated by the following

Theorem 3.6 *The multistep collocation method* (3.2.9)–(3.2.10) *is equivalent to a multistep Runge–Kutta method*

$$\begin{aligned}Y_j&=\sum_{i=0}^{k-1}\varphi_i(c_j)y_{n+k-1-i}\\&\quad+h\sum_{i=1}^{m}\psi_i(c_j)f(x_{n+k-1}+c_ih,Y_i),\qquad j=1,2,\dots,m,\\y_{n+k}&=\sum_{i=0}^{k-1}\varphi_i(1)y_{n+k-1-i}+h\sum_{i=1}^{m}\psi_i(1)f(x_{n+k-1}+c_ih,Y_i),\end{aligned}$$

where the expression of the polynomials $\varphi_i(s)$, $\psi_i(s)$ are provided in Lemma 1 *of* [58].

Lie and Norsett in [58] also provided a complete study of the order of the resulting methods, stating order conditions by means of the study of variational matrices, and showing that the maximum attainable order of a k-step m-stage collocation method is $2m+k-1$. They also proved that there exist $\binom{m+k-1}{k-1}$ nodes that allow superconvergence and, in analogy with Runge–Kutta methods, they are named *multistep Gaussian* collocation points. As Hairer–Wanner stated in [43], these methods are not stiffly stable and, therefore, they are not suited for stiff problems: in order to obtain better stability properties, they derived methods of highest order $2m+k-2$, imposing $c_m=1$ and deriving the other collocation abscissa in a suited way to achieve this highest order and named the corresponding methods of "Radau"-type, studied their stability properties, deriving also many A-stable methods.

3.2.5 Two-Step Collocation and Almost Collocation Methods

In more recent times, our strenghts have been devoted to extend the multistep collocation technique to the class of two-step Runge–Kutta methods (TSRK)

$$\begin{cases} y_{n+1} = \theta y_{n-1} + \tilde{\theta} y_n + h \sum_{j=1}^{m} (v_j f(Y_j^{[n-1]}) + w_j f(Y_j^{[n]})), \\ Y_i^{[n]} = u_i y_{n-1} + \tilde{u}_i y_n + h \sum_{j=1}^{m} (a_{ij} f(Y_j^{[n-1]}) + b_{ij} f(Y_j^{[n]})), \end{cases} \tag{3.2.11}$$

introduced by Jackiewicz and Tracogna [50] and further investigated by several authors (see [49] and references therein contained). Two-step Runge–Kutta methods (3.2.11) differ from the multistep Runge–Kutta methods above described, because they also depend on the stage derivatives at two consecutive step points: as a consequence, "we gain extra degrees of freedom associated with a two-step scheme without the need for extra function evaluations" (see [50]), because the function evaluations $f(Y_j^{[n-1]})$ are completely inherited from the previous step. Therefore, the computational cost of these formulae only depends on the structure of the matrix B. Different approaches to the construction of continuous TSRK methods outside collocation are presented in [1, 2] and [51].

The continuous approximant

$$\begin{cases} P(x_n + sh) = \varphi_0(s) y_{n-1} + \varphi_1(s) y_n + h \sum_{j=1}^{m} (\chi_j(s) f(P(x_{n-1} + c_j h)) \\ \quad + \psi_j(s) f(P(x_n + c_j h))), \\ y_{n+1} = P(x_{n+1}), \end{cases} \tag{3.2.12}$$

expressed as linear combination of the basis functions

$$\{\varphi_0(s),\ \varphi_1(s),\ \chi_j(s),\ \psi_j(s),\ j = 1, 2, \dots, m\},$$

is an algebraic polynomial which is derived in order to satisfy some interpolation and collocation conditions, i.e.

$$\begin{aligned} &P(x_{n-1}) = y_{n-1}, \quad P(x_n) = y_n, \\ &P'(x_{n-1} + c_i h) = f(x_{n-1} + c_i h, P(x_{n-1} + c_i h)), \quad i = 1, 2, \dots, m, \\ &P'(x_n + c_i h) = f(x_n + c_i h, P(x_n + c_i h)), \quad i = 1, 2, \dots, m. \end{aligned} \tag{3.2.13}$$

As a first attempt, we have generalized in [34, 38] the techniques introduced by Guillou–Soulé [42], Hairer–Wanner [43] and Lie–Norsett [58], adapting and extending this technique to TSRK methods. Using the techniques introduced in [58], we have derived in [38] the coefficients of (3.2.12) in closed form: the corresponding results are reported in the following theorem (see [38]).

Theorem 3.7 *The method* (3.2.12) *is equivalent to a TSRK method in the form*

$$Y_j^{[n]} = \varphi_0(c_j) y_{n-1} + \varphi_1(c_j) y_n + h \sum_{i=1}^{m} [\chi_j(c_i) f(x_{n-1} + c_i h, Y_i^{[n-1]})$$

$$+ \psi_j(c_i) f(x_n + c_i h, Y_i^{[n]})], \quad j = 1, 2, \ldots, m,$$

$$y_{n+1} = \varphi_0(1) y_{n-1} + \varphi_1(1) y_n + h \sum_{j=1}^{m} [\chi_j(1) f(x_{n-1} + c_j h, Y_j^{[n-1]})$$

$$+ \psi_j(1) f(x_n + c_j h, Y_j^{[n]})],$$

where

$$\psi_j(s) = \int_0^s l_j(\tau)d\tau - \frac{\int_{-1}^0 l_j(\tau)d\tau}{\int_{-1}^0 M(\tau)d\tau} \int_0^s M(\tau)d\tau, \quad j = 1, 2, \ldots, m,$$

$$\chi_j(s) = \int_0^s \tilde{l}_j(\tau)d\tau - \frac{\int_{-1}^0 \tilde{l}_j(\tau)d\tau}{\int_{-1}^0 M(\tau)d\tau} \int_0^s M(\tau)d\tau, \quad j = 1, 2, \ldots, m,$$

$$\varphi_0(s) = -\frac{\int_0^s M(\tau)d\tau}{\int_{-1}^0 M(\tau)d\tau},$$

$$\varphi_1(s) = 1 + \frac{\int_0^s M(\tau)d\tau}{\int_{-1}^0 M(\tau)d\tau}.$$

with

$$l_i(s) = \prod_{j=1, j\neq i}^{2m} \frac{s - d_j}{d_i - d_j}, \qquad M(s) = \prod_{j=1}^{2m} (s - d_j), \qquad \begin{cases} d_i = c_i, \\ d_{m+i} = c_i - 1, \\ i = 1, 2, \ldots, m, \end{cases}$$

$$\tilde{l}_j(s) = \prod_{i=1, i\neq j}^{2m} \frac{s - e_i}{e_j - e_i}, \qquad \begin{cases} e_i = c_i - 1, \\ e_{m+i} = c_i, \end{cases} \quad i = 1, 2, \ldots, m.$$

We proved in [38] that the resulting methods have uniform order $2m + 1$ but such an high order enforces these methods to have bounded stability regions only. For this reason, in order to derive highly stable methods (i.e. A-stable and L-stable), we have introduced in [27, 28, 36] the class of *almost collocation methods*, which are obtained in such a way that only some of the above interpolation and collocation conditions are satisfied. Relaxing the above conditions, we obtain more degrees of freedom, which have been used in order to derive many A-stable and L-stable methods of order $m + r$, $r = 0, 1, \ldots, m$. Therefore the highest attainable order is $2m$ which, in principle, can seem the same of standard Runge–Kutta methods. As a matter of fact, this is not true: in fact, Runge–Kutta–Gauss methods have order $2m$ in the grid points, while the stage order is equal to m, therefore they suffer from order reduction in the integration of stiff problems (see [10, 11, 43]), i.e. the effective order of convergence in presence of stiffness is only m. Our methods, instead, do not suffer from order reduction, i.e. the effective order of convergence in the integration of stiff problems is $2m$, because they have high stage order. In [36] we have studied the existence of such methods, derived continuous order conditions, provided characterization results and studied their stability properties. A complete analysis of m-stage two-step continuous methods, with $m = 1, 2, 3, 4$, has been provided in [32], while the analysis of the implementation issues for two-step collocation methods has been provided in [33]. The construction of algebraically stable two-step collocation methods is object of a current research project.

3.3 Collocation Methods for Second Order ODEs of Special Type

We now concentrate our attention on the hystorical evolution of the collocation technique for the numerical solution of initial value problems based on second order ODEs with periodic and oscillating solution

$$\begin{cases} y'(x) = f(x, y(x)), & x \in [x_0, X], \\ y'(x_0) = y_0' \in \mathbb{R}^d, \\ y(x_0) = y_0, \end{cases} \tag{3.3.1}$$

where $f : [x_0, X] \times \mathbb{R}^d \rightarrow \mathbb{R}^d$ is assumed to be a is sufficiently smooth function, in order to ensure the existence and the uniqueness of the solution.

3.3.1 Direct and Indirect Collocation Methods

In the context of collocation methods for second order ODEs, two possibilities have been taken into account in the literature, i.e. methods based on *indirect* or *direct collocation* [77]. Indirect collocation methods are generated by applying a collocation based Runge–Kutta method to the first order representation of (3.3.1), which has doubled dimension. If

$$\begin{array}{c|c} c & A \\ \hline & b^T \end{array}$$

is the Butcher array of a collocation Runge–Kutta method, the tableau of the corresponding indirect collocation method is

$$\begin{array}{c|c} c & A^2 \\ \hline & A^T b \\ & b^T \end{array}$$

which results in a Runge–Kutta–Nyström method [45]. The theory of indirect collocation methods completely parallels the well-known theory of collocation methods for first order equations (see [77]) and, therefore, the properties of a collocation method are totally inherited by the corresponding indirect collocation method. Thus, the maximum attainable order is $2m$, where m is the number of stages, and it is achieved by Gauss-type methods, which are also A-stable, while L-stability is achieved by Radau IIA-type methods, of order $2m-1$.

In the case of direct collocation methods, the collocation polynomial is derived directly for the second order problem. Van der Houwen et al. in [77] studied the order, stage order of direct collocation methods and also provided their stability analysis, extending the results of Kramarz [54]. Concerning order and stage order, the following result holds (see [77]):

Theorem 3.8 *Direct and indirect collocation methods with the same collocation nodes have the same order. The stage order of direct collocation methods is one higher whenever*

$$\int_0^1 \prod_{i=1}^m (s-c_i)ds = 0.$$

Therefore, while indirect and direct collocation methods have the same order, their stage order is different and, in particular, direct methods have higher stage order. However, they are not competitive in terms of stability. Van der Houwen et al. in [77] clearly state that "From a practical point of view, direct collocation methods based on Gauss, Radau and Lobatto collocation points are of limited value, because the rather small stability or periodicity boundaries make them unsuitable for stiff problems. The A-stable indirect analogues are clearly more suitable for integrating stiff problems".

Moreover, Coleman [17] proved that no P-stable one step symmetric collocation methods exist. P-stability (see Lambert-Watson paper [56]) is a very relevant property for the numerical treatment of a second order system whose theoretical solution is periodic with a moderate frequency and a high frequency oscillation of small amplitude superimposed. This phenomenon is known in literature as *periodic stiffness* [73], which can be reasonably faced using P-stable methods, exactly as A-stable methods are suitable for stiff problems. In other terms, P-stability ensures that the choice of the stepsize is independent from the values of the frequencies, but it only depends on the desired accuracy [21, 68].

In [56], the authors proved that P-stable linear multistep methods

$$\sum_{j=0}^{p} \alpha_j y_{n+j} = h^2 \sum_{j=0}^{p} \beta_j f_{n+j}$$

can achieve maximum order 2. In the context of Runge–Kutta–Nyström methods

$$y_{n+1} = y_n + hy'_n + h^2 \sum_{i=1}^{m} \bar{b}_i f(x_n + c_i h, Y_i),$$

$$y'_{n+1} = y'_n + h \sum_{i=1}^{m} b_i f(x_n + c_i h, Y_i),$$

$$Y_i = y_n + c_i h y_n' + h^2 \sum_{j=1}^{m} a_{ij} f(x_n + c_j h, Y_j), \quad i = 1, 2, \dots, m,$$

many A-stable and P-stable methods exist, but the ones falling in the subclass of collocation methods, whose coefficients (see [45]) are of the form

$$a_{ij} = \int_0^{c_i} L_j(s) ds,$$

$$b_i = \int_0^1 L_i(s) ds,$$

$$\bar{b}_i = \int_0^1 (1 - s) L_i(s) ds,$$

have only bounded stability intervals and are not P-stable [68].

3.3.2 Two-Step Runge–Kutta–Nyström Methods

We have observed in the previous paragraph that P-stability is a desirable property that only few methods in the context of linear multistep methods and Runge–Kutta–Nyström methods possess. In order to create a good balance between high order and strong stability properties, further steps in the literature have been devoted to the development of multistep Runge–Kutta–Nystrom methods for second order problems. Much of this work has been done by Paternoster (see [61, 68–72]). In particular, the author proved that no P-stable methods can be found in the class of indirect collocation TSRK methods, while it is possible to find P-stable methods in the context of *two-step Runge–Kutta–Nyström methods*

$$Y_j^{[n-1]} = y_{n-1} + c_j h y_{n-1}' + h^2 \sum_{k=1}^{m} a_{jk} f(x_{n-1} + c_k h, Y_k^{[n-1]}), \quad j = 1, 2, \dots, m,$$

$$Y_j^{[n]} = y_n + c_j h y_n' + h^2 \sum_{k=1}^{m} a_{jk} f(x_n + c_k h, Y_k^{[n]}), \quad j = 1, 2, \dots, m,$$

$$y_{n+1} = (1 - \theta) y_n + \theta y_{n-1} + h \sum_{j=1}^{m} (v_j y_{n-1}' + w_j y_n')$$

$$+ h^2 \sum_{j=1}^{m} \bar{v}_j f(x_{n-1} + c_j h, Y_j^{[n-1]}) + \bar{w}_j f(x_n + c_j h, Y_j^{[n]}),$$

$$y_{n+1}' = (1 - \theta) y_n' + \theta y_{n-1}' + h \sum_{j=1}^{m} (v_j f(x_{n-1} + c_j h, Y_j^{[n-1]})$$

$$+ w_j f(x_n + c_j h, Y_j^{[n]})),$$

which represent the extension to second order problems of the two-step Runge–Kutta methods introduced in [52] for first order ODEs.

3.3.3 Collocation Based Two-Step Hybrid Methods

In the numerical integration of second order ODEs through collocation, many possibilities can be taken into account: for example, Runge–Kutta–Nyström methods provide an approximation to the solution and its first derivative at each step point. However, as Henrici observed in [46], "If one is not particularly interested in the values of the first derivatives, it seems unnatural to introduce them artificially". For this reason, other types of methods have been taken into account in the literature, i.e. methods which provide an approximation to the solution without computing any approximation to the first derivative. Coleman introduced in [19] the following class of two-step hybrid methods for second order equations:

$$Y_i^{[n]} = u_i y_{n-1} + (1-u_i) y_n + h^2 \sum_{j=1}^{m} a_{ij} f(x_n + c_j h, Y_j^{[n]}),$$

$$i = 1, 2, \ldots, m, \tag{3.3.2}$$

$$y_{n+1} = \theta y_{n-1} + (1-\theta) y_n + h^2 \sum_{j=1}^{m} w_j f(x_n + c_j h, Y_j^{[n]}). \tag{3.3.3}$$

This class of methods has been further investigated in [16, 37, 40, 75, 76]. In more recent times, we derived in [35] collocation based methods belonging to the class of Coleman hybrid methods (3.3.2)–(3.3.3), extending the technique introduced by Hairer and Wanner in [43] for first order problems. The collocation polynomial takes the form

$$P(x_n + sh) = \varphi_1(s) y_{n-1} + \varphi_2(s) y_n + h^2 \sum_{j=1}^{m} \chi_j(s) P''(x_n + c_j h), \tag{3.3.4}$$

where $s = \frac{x - x_n}{h} \in [0, 1]$, and the unknown basis functions

$$\{\varphi_1(s),\ \varphi_2(s),\ \chi_j(s),\ j = 1,\ 2, \ldots, m\},$$

are derived imposing the following $m+2$ conditions

$$\begin{aligned} P(x_{n-1}) &= y_{n-1}, \\ P(x_n) &= y_n, \\ P''(x_n + c_j h) &= f(x_n + c_j h, P(x_n + c_j h)), \quad j = 1, 2, \ldots, m. \end{aligned}$$

After computing the basis functions as solutions of $m+2$ linear systems (see [62]), the resulting class of methods takes the following form

$$Y_i^{[n]} = \varphi_1(c_i)y_{n-1} + \varphi_2(c_i)y_n + h^2 \sum_{j=1}^{m} \chi_j(c_i)P''(x_n + c_j h), \tag{3.3.5}$$

$$y_{n+1} = \varphi_1(1)y_{n-1} + \varphi_2(1)y_n + h^2 \sum_{j=1}^{m} \chi_j(1)P''(x_n + c_j h). \tag{3.3.6}$$

In [35] we have provided the study of stability and periodicity properties and derived continuous order conditions for (3.3.6)–(3.3.5), which are object of the following result.

Theorem 3.9 *Assume that the function f is sufficiently smooth. The collocation method associated to* (3.3.4) *has uniform order p if the following conditions are satisfied*:

$$1 - \varphi_1(s) - \varphi_2(s) = 0, \qquad s + \varphi_1(s) = 0,$$

$$s^k + (-1)^{k+1}\varphi_1(s) - k(k-1)\sum_{j=1}^{m} \chi_j(s)c_j^{k-2} = 0, \quad k = 2, 3, \ldots, p, \ s \in [0,1].$$

Theorem 3.9 allows us to prove that every two-step collocation method associated to (3.3.4), has order $p = m$ on the whole integration interval, and this result is in keeping with [19].

3.3.4 Mixed Collocation Methods

The development of classical collocation methods (i.e. methods based on algebraic polynomials), even if it is not the most suitable choice for second order problems that do not possess solutions with polynomial behavior, it is the first necessary step in order to construct collocation methods whose collocation function is expressed as linear combination of different functions, e.g. trigonometric polynomials, mixed or exponential basis (see, for instance, [20, 48]), which can better follow the qualitative behavior of the solution. It is indeed more realistic to choose basis functions which are not polynomials.

Many authors have considered in literature different functional basis, instead of the polynomial one, e.g. [8, 18, 21, 37, 39, 41, 48, 53, 65, 67, 69, 71, 74]. In particular we mention here the work by Coleman and Duxbury [20], where the authors introduced mixed collocation methods applied to the Runge–Kutta–Nyström scheme, where the collocation function is expressed as linear combination of trigonometric functions and powers, in order to provide better approximations for oscillatory solutions. The methods are derived in order to exactly integrate the armonic oscillator

$$y'' = -k^2 y,$$

where k is a constant, a feature which is not achievable by algebraic polynomial collocation. The term *mixed interpolation* appeared for the first time in [39] to describe interpolation by a linear combination of a sine and cosine of a given frequency, and powers of the relevant variable, and later used by Brunner et al. in [8] in the context of Volterra integral equations. The solution on the generic integration interval $[x_n, x_{n+1}]$ is approximated by the collocating function

$$u(x_n + sh) = a\cos\theta s + b\sin\theta s + \sum_{i=0}^{m-1} \Gamma_i s^i, \tag{3.3.7}$$

which satisfies the following collocation and interpolation conditions

$$\begin{aligned} u(x_n) = y_n, \qquad u'(x_n) = y'_n, \\ u''(x_n + c_j h) = f(x_n + c_j h, u(x_n + c_j h)), \quad j = 1, 2, \ldots, m. \end{aligned}$$

Integrating (3.3.7) twice, we obtain the Runge–Kutta–Nyström formulation of the methods, i.e.

$$u'(x_n + sh) = y'_n + h \sum_{i=1}^{m} \alpha_i(s) f(x_n + c_i h),$$

$$u(x_n + sh) = y_n + shy'_n + h^2 \sum_{i=1}^{m} \beta_i(s) f(x_n + c_i h),$$

where

$$\alpha_i(s) = \int_0^s L_i(\tau)d\tau, \qquad \beta_i(s) = \int_0^s (s - \tau)L_i(\tau)d\tau.$$

Outside collocation, many authors derived methods having frequency dependent parameters (see, for instance, [48, 53, 66, 74] and references therein contained). The linear stability analysis of these methods is carried out in [21]. In [37] also a method with parameters depending on two frequencies is presented, and the modification in the stability analysis is performed, leading to a three dimensional region.

3.4 Collocation Methods for VIEs

Piecewise polynomial collocation methods for Volterra Integral Equations introduce a number of aspects not present when solving ODEs. In this section we will present the main results in the context of collocation and almost collocation methods for

VIEs of the form

$$y(x) = g(x) + \int_0^x k(x, \tau, y(\tau))d\tau, \quad x \in I := [0, X], \tag{3.4.1}$$

where $k \in C(D \times \mathbb{R})$, with $D := \{(x, \tau) : 0 \le \tau \le x \le X\}$, and $g \in C(I)$, also underlying connections and differences with the case of ODEs.

3.4.1 Classical One-Step Collocation Methods

Let us discretize the interval I by introducing a uniform mesh

$$I_h = \{x_n := nh, n = 0, \dots, N, h \ge 0, Nh = X\}.$$

Equation (3.4.1) can be rewritten, by relating it to this mesh, as

$$y(x) = F_n(x) + \Phi_n(x), \quad x \in [x_n, x_{n+1}],$$

where

$$F_n(x) := g(x) + \int_0^{x_n} k(x, \tau, y(\tau))d\tau$$

and

$$\Phi_n(x) := \int_{x_n}^{x} k(x, \tau, y(\tau))d\tau$$

represent respectively the *lag term* and the *increment function*. Let us fix m collocation parameters $0 \le c_1 < \cdots < c_m \le 1$ and denote by $x_{nj} = x_n + c_j h$ the collocation points. The collocation polynomial, restricted to the interval $[x_n, x_{n+1}]$, is of the form:

$$u_n(x_n + sh) = \sum_{j=1}^{m} L_j(s)U_{nj}, \quad s \in [0, 1], \ n = 0, 1, \dots, N-1, \tag{3.4.2}$$

where $L_j(s)$ is the j-th Lagrange fundamental polynomial with respect to the collocation parameters and $U_{nj} := u_n(x_{nj})$. *Exact* collocation methods are obtained by imposing that the collocation polynomial (3.4.2) exactly satisfies the VIE (3.4.1) in the collocation points x_{ni} and by computing $y_{n+1} = u_n(x_{n+1})$:

$$\begin{cases} U_{ni} = F_{ni} + \Phi_{ni}, \\ y_{n+1} = \sum_{j=1}^{m} L_j(1)U_{nj}, \end{cases} \tag{3.4.3}$$

where

$$F_{ni} = g(x_{ni}) + h \sum_{\nu=0}^{n-1} \int_0^1 k(x_{ni}, x_\nu + sh, u_\nu(x_\nu + sh))ds, \quad i = 1, 2, \dots, m, \tag{3.4.4}$$

$$\Phi_{ni} = h \int_0^{c_i} k(x_{ni}, x_n + sh, u_n(x_n + sh))ds, \quad i = 1, 2, \ldots, m. \tag{3.4.5}$$

Note that the first equation in (3.4.3) represents a system of m nonlinear equations in the m unknowns U_{ni}. We obtain an approximation $u(x)$ of the solution $y(x)$ of the integral equation (3.4.1) in $[0, X]$, by considering

$$u(x)|_{(x_n, x_{n+1}]} = u_n(x), \tag{3.4.6}$$

where $u_n(x)$ given by (3.4.2).

We recall that, in contrast with what happens in the case of ODEs, generally $u(x)$ is not continuous in the mesh points, as

$$u(x) \in S_{m-1}^{(-1)}(I_h), \tag{3.4.7}$$

where

$$S_\mu^{(d)}(I_h) = \{v \in C^d(I) : v|_{(x_n, x_{n+1}]} \in \Pi_\mu \ (0 \le n \le N-1)\}.$$

Here, Π_μ denotes the space of (real) polynomials of degree not exceeding μ. A complete analysis of collocation methods for linear and nonlinear Volterra integral and integro–differential equations, with smooth and weakly singular kernels is given in [6]. In particular, as shown in [6, 7], the classical one–step collocation methods for a second-kind VIE do no longer exhibit $O(h^{2m})$ superconvergence at the mesh points if collocation is at the Gauss points, in fact they have uniform order m for any choice of the collocation parameters and local superconvergence order in the mesh points of $2m-2$ (m Lobatto points or $m-1$ Gauss points with $c_m = 1$) or $2m-1$ (m Radau II points). The optimal order is recovered only in the iterated collocation solution.

We observe that, differently from the case of ODEs, the collocation equations are in general not yet in a form amenable to numerical computation, due to the presence of the memory term given by the Volterra integral operator. Thus, another discretization step, based on quadrature formulas $\bar{F}_{ni} \simeq F_{ni}$ and $\bar{\Phi}_{ni} \simeq \Phi_{ni}$ for approximating the lag term (3.4.4) and the increment function (3.4.5), is necessary to obtain the fully discretized collocation scheme, thus leading to *discretized* collocation methods. Such methods preserve, under suitable hypothesis on the quadrature formulas, the same order of the exact collocation methods [7].

The connection between collocation and implicit Runge–Kutta methods for VIEs (the so called VRK methods) is not immediate: a collocation method for VIEs is equivalent to a VRK method if and only if $c_m = 1$ (see Theorem 5.2.2 in [7]). Some other continuous extensions of Runge–Kutta methods for VIEs, which do not necessarily lead to collocation methods, have been introduced in [3].

Many efforts have been made in the literature with the aim of obtaining fast collocation and more general Runge–Kutta methods for the numerical solution of VIEs. It is known that the numerical treatment of VIEs is very expensive from computational point of view because of presence of the lag-term, which contains the entire history of the phenomenon. To this cost, it has also to be added the one due to the increment term which leads, for implicit methods (generally possessing the best stability properties), to the resolution of a system of nonlinear equations at each step of integration. In order to reduce the computational effort in the lag–term computation, fast collocation and Runge–Kutta methods have been constructed for convolution VIEs of Hammerstein type, see [13, 23, 59, 60].

The stability analysis of collocation and Runge–Kutta methods for VIEs can be found in [4, 7, 14, 30] and the related bibliography. In particular a collocation method for VIEs is A-stable if the corresponding method for ODEs is A-stable.

3.4.2 Multistep Collocation

Multistep collocation and Runge–Kutta methods for VIEs, have been introduced in order to bring down the computational cost related to the resolution of non-linear systems for the computation of the increment term. As a matter of fact such methods, showing a dependence on stages and steps in more consecutive grid points, permit to raise the order of convergence of the classical methods, without inflating the computational cost or, equivalently, having the same order at a lower computational cost.

A first analysis of multistep collocation methods for VIEs appeared in [24, 25], where the methods are obtained by introducing in the collocation polynomial the dependence from r previous time steps; namely we seek for a collocation polynomial, whose restriction to the interval $[x_n, x_{n+1}]$ takes the form

$$P_n(x_n + sh) = \sum_{k=0}^{r-1} \varphi_k(s) y_{n-k} + \sum_{j=1}^{m} \psi_j(s) Y_{nj},$$
$$s \in [0, 1], \; n = 0, 1, \ldots, N-1, \tag{3.4.8}$$

where

$$Y_{nj} := P_n(x_{nj}) \tag{3.4.9}$$

and $\varphi_k(s)$, $\psi_j(s)$ are polynomials of degree $m + r - 1$ to be determined by imposing the interpolation conditions at the points x_{n-k}, that is $u_n(x_{n-k}) = y_{n-k}$, and by satisfying (3.4.9). It is proved in [26, 29] that, assuming $c_i \neq c_j$ and $c_1 \neq 0$, the polynomials $\varphi_k(s)$, $\psi_j(s)$ have the form:

$$\varphi_k(s) = \prod_{i=1}^{m} \frac{s - c_i}{-k - c_i} \cdot \prod_{\substack{i=0 \\ i \neq k}}^{r-1} \frac{s + i}{-k + i},$$
$$\psi_j(s) = \prod_{i=0}^{r-1} \frac{s + i}{c_j + i} \cdot \prod_{\substack{i=1 \\ i \neq j}}^{m} \frac{s - c_i}{c_j - c_i}. \tag{3.4.10}$$

Then the discretized multistep collocation method assumes the form:

$$\begin{cases} Y_{ni} = \bar{F}_{ni} + \bar{\Phi}_{ni}, \\ y_{n+1} = \sum_{k=0}^{r-1} \varphi_k(1) y_{n-k} + \sum_{j=1}^{m} \psi_j(1) Y_{nj}. \end{cases} \tag{3.4.11}$$

The lag-term and increment–term approximations

$$\bar{F}_{ni} = g(x_{ni}) + h \sum_{\nu=0}^{n-1} \sum_{l=0}^{\mu_1} b_l k(x_{ni}, x_\nu + \xi_l h, P_\nu(x_\nu + \xi_l h)),$$

$$i = 1, 2, \ldots, m \tag{3.4.12}$$

$$\bar{\Phi}_{ni} = h \sum_{l=0}^{\mu_0} w_{il} k(x_{ni}, x_n + d_{il} h, P_n(x_n + d_{il} h)), \quad i = 1, 2, \ldots, m \tag{3.4.13}$$

are obtained by using quadrature formulas of the form

$$(\xi_l, b_l)_{l=1}^{\mu_1}, \quad (d_{il}, w_{il})_{l=1}^{\mu_0}, \quad i = 1, 2, \ldots, m, \tag{3.4.14}$$

where the quadrature nodes ξ_l and d_{il} satisfy $0 \le \xi_1 < \cdots < \xi_{\mu_1} \le 1$ and $0 \le d_{i1} < \cdots < d_{i\mu_0} \le 1$, μ_0 and μ_1 are positive integers and w_{il}, b_l are suitable weights.

The discretized multistep collocation method (3.4.8)–(3.4.11) provides a continuous approximation $P(x)$ of the solution $y(x)$ of the integral equation (3.4.1) in $[0, X]$, by considering

$$P(x)|_{(x_n, x_{n+1}]} = P_n(x), \tag{3.4.15}$$

where $P_n(x)$ is given by (3.4.8). We note that usually the polynomial constructed in the collocation methods for VIEs doesn't interpolate the numerical solution in the previous step points, resulting a discontinuous approximation of the solution (3.4.7). In this multistep extension, the collocation polynomial is instead a continuous approximation to the solution, i.e. $u(x) \in S_{m+r-1}^{(0)}(I_h)$.

The discretized multistep collocation method (3.4.8)–(3.4.11) can be regarded as a multistep Runge–Kutta method for VIEs:

$$\begin{cases} Y_{ni} = \bar{F}_n(x_{ni}) \\ \qquad + h \sum_{l=1}^{\mu_0} w_{il} k\big(x_n + e_{il} h,\ x_n + d_{il} h, \sum_{k=0}^{r-1} \gamma_{ilk} y_{n-k} + \sum_{j=1}^{m} \beta_{ilj} Y_{nj}\big), \\ y_{n+1} = \sum_{k=0}^{r-1} \theta_k y_{n-k} + \sum_{j=1}^{m} \lambda_j Y_{nj}, \end{cases} \tag{3.4.16}$$

where

$$\bar{F}_n(x) = g(x) + h \sum_{\nu=0}^{n-1} \sum_{l=1}^{\mu_1} b_l k\left(x, x_\nu + \xi_l h, \sum_{k=0}^{r-1} \delta_{lk} y_{\nu-k} + \sum_{j=1}^{m} \eta_{lj} Y_{\nu,j}\right) \tag{3.4.17}$$

and

$$e_{il} = c_i, \quad \gamma_{ilk} = \varphi_k(d_{il}), \quad \beta_{ilj} = \psi_j(d_{il}),$$
$$\theta_k = \varphi_k(1), \quad \lambda_j = \psi_j(1),$$
$$\delta_{lk} = \varphi_k(\xi_l), \quad \eta_{lj} = \psi_j(\xi_l).$$

The reason of interest of the multistep collocation methods lies in the fact that they increase the order of convergence of collocation methods without increasing the computational cost, except for the cost due to the starting procedure. As a matter of fact, in advancing from x_n to x_{n+1}, we make use of the approximations y_{n-k}, $k = 0, 1, \ldots, r-1$, which have already been evaluated at the previous steps. This permits to increase the order, by maintaining in (3.4.11) the same dimension m of the nonlinear system (3.4.3).

The r-steps m-points collocation methods have uniform order $m+r$, and order of local superconvergence $2m+r-1$. The knowledge of the collocation polynomial, which provides a continuous approximation of uniform order to the solution, will allow a cheap variable stepsize implementation. Indeed, when the stepsize changes, the new approximation values can be computed by simply evaluating the collocation polynomial, without running into problems of order reduction, as a consequence of the uniform order.

3.4.3 Two-Step Collocation and Almost Collocation Methods

Unfortunately multistep methods of the form (3.4.8)–(3.4.11) do not lead to a good balance between high order and strong stability properties, infact, altough methods with unbounded stability regions exist, no A-stable methods have been found. Therefore in [25] a modification in the technique has been introduced, thus obtaining two-step *almost* collocation methods, also for systems of VIEs, by relaxing some of the collocation conditions and by introducing some previous stage values, in order to further increase the order and to have free parameters in the method, to be used to get A-stability.

The methods are defined by

$$\begin{cases} P(x_n + sh) = \varphi_0(s) y_{n-1} + \varphi_1(s) y_n + \sum_{j=1}^m \chi_j(s) P(x_{n-1,j}) \\ \qquad\qquad\quad + \sum_{j=1}^m \psi_j(s)(\bar{F}_{nj} + \bar{\Phi}_{nj}), \\ y_{n+1} = P(x_{n+1}), \end{cases} \tag{3.4.18}$$

$s \in (0, 1]$, $n = 1, 2, \ldots, N-1$.

If the polynomials $\varphi_0(s)$, $\varphi_1(s)$, $\chi_j(s)$ and $\psi_j(s)$, $j = 1, 2, \ldots, m$ satisfy the interpolation conditions

$$\varphi_0(0) = 0, \quad \varphi_1(0) = 1, \quad \chi_j(0) = 0, \quad \psi_j(0) = 0,$$
$$\varphi_0(-1) = 1, \quad \varphi_1(-1) = 0, \quad \chi_j(-1) = 0, \quad \psi_j(-1) = 0,$$

and the collocation conditions

$$\begin{array}{llll} \varphi_0(c_i) = 0, & \varphi_1(c_i) = 0, & \chi_j(c_i) = 0, & \psi_j(c_i) = \delta_{ij}, \\ \varphi_0(c_i - 1) = 0, & \varphi_1(c_i - 1) = 0, & \chi_j(c_i - 1) = \delta_{ij}, & \psi_j(c_i - 1) = 0, \end{array}$$

$i = 1, 2, \ldots, m$, then we obtain order $p = 2m + 1$.

In our search for A-stable methods we will have been mainly concerned with methods of order $p = 2m + 1 - r$, where $r = 1$ or $r = 2$ is the number of relaxed conditions. Namely we have chosen $\varphi_0(s)$ as a polynomial of degree $\leq 2m + 1 - r$, which satisfies the collocation conditions

$$\varphi_0(c_i) = 0, \quad i = 1, 2, \ldots, m. \tag{3.4.19}$$

This leads to the polynomial $\varphi_0(s)$ of the form

$$\varphi_0(s) = (q_0 + q_1 s + \cdots + q_{m+1-r} s^{m+1-r}) \prod_{i=1}^{m} (s - c_i), \tag{3.4.20}$$

where $q_0, q_1, \ldots, q_{m+1-r}$ are free parameters. Moreover, for $p = 2m - 1$ we have chosen $\varphi_1(s)$ as a polynomial of degree $\leq 2m - 1$ which satisfies the collocation conditions

$$\varphi_1(c_i) = 0, \quad i = 1, 2, \ldots, m. \tag{3.4.21}$$

This leads to the polynomial $\varphi_1(s)$ of the form

$$\varphi_1(s) = (p_0 + p_1 s + \cdots + p_{m-1} s^{m-1}) \prod_{i=1}^{m} (s - c_i), \tag{3.4.22}$$

where $p_0, p_1, \ldots, p_{m-1}$ are free parameters.

The methods have uniform order of convergence $p = 2m + 1 - r$, and are therefore suitable for an efficient variable stepsize implementation. Moreover methods which are A-stable with respect to the basic test equation and have unbounded stability regions with respect to the convolution test equation have been provided.

3.4.4 Mixed Collocation

In the case of VIEs with periodic highly oscillatory solutions, traditional methods may be inefficient, as they may require the use of a small stepsize in order to follow accurately the oscillations of high frequency. As in the case of ODEs "ad hoc" numerical methods have been constructed, incorporating the a priori knowledge of the behavior of the solution, in order to use wider stepsizes with respect to classical methods and simultaneously to simulate with high accuracy the oscillations of the solution.

A first work on the numerical treatment of VIEs with periodic solution is [5], where numerical methods were constructed by means of mixed interpolation. Recently, mixed collocation methods have been introduced in [8, 9] for VIEs and VIDEs. In particular in [8], mixed collocation methods have been introduced for linear convolution VIEs of the form

$$y(x) = g(x) + \int_{-\infty}^{x} k(x-\tau)y(\tau)d\tau, \quad x \in [0, X], \tag{3.4.23}$$

with

$$y(x) = \psi(x), \quad x \in [-\infty, 0],$$

where $k \in L^1(0, \infty)$, g is a continuous periodic function and ψ is a given bounded and continuous function. The collocation polynomial is taken in the form

$$P_n(x_n + sh) = \sum_{k=0}^{m} B_k(s)Y_{n,k},$$

where the $B_k(s)$ are combinations of trigonometric functions and algebraic polynomials given in [8]. The numerical method is of the form

$$\begin{cases} Y_{ni} = \bar{F}_{ni} + \bar{\Phi}_{ni}, \\ y_{n+1} = \sum_{k=0}^{m} B_k(1)Y_{n,k}, \end{cases} \tag{3.4.24}$$

where the lag-term and increment term approximations are given by

$$\bar{F}_{ni} = g(x_{ni}) + \int_{-\infty}^{0} k(x_{ni}-\tau)\psi(\tau)d\tau + h\sum_{\nu=0}^{n-1}\sum_{l=0}^{m} w_l(1)k(x_{nj} - x_{\nu,l}P_\nu(x_{\nu,l}),$$

$$\bar{\Phi}_{ni} = hc_i \sum_{l=0}^{m} w_l(1)k(x_{ni} - x_n - hc_i c_l)\left(\sum_{k=0}^{m} B_k(c_i c_l)Y_{n,k}\right)$$

with

$$w_l(s) = \int_0^s B_l(\tau)d\tau.$$

With some suitable choices for collocation parameters such methods accurately integrate systems for which the period of oscillation of the solution is known. In the paper [15] the authors introduce a family of linear methods, namely Direct Quadrature (DQ) methods, specially tuned on the specific feature of the problem, based on the exponential fitting [47, 48], which is extremely flexible when periodic functions are treated. Such methods are based on a three-term quadrature formula, that is of the same form as the usual Simpson rule, but specially tuned on integrands of the form $k(s)y(s)$ where k and y are of type

$$k(x) = e^{\alpha x}, \qquad y(x) = a + b\cos(\omega x) + c\sin(\omega x), \tag{3.4.25}$$

where $\alpha, \omega, a, b, c \in \mathbb{R}$. The coefficients of the new quadrature rule depend on the parameters of the integrand, i.e. α and ω. It has been shown as the use of exponentially fitted based three-point quadrature rules produces a definite improvement in the accuracy when compared with the results from the classical Simpson rule, and that the magnitude of the gain depends on how good is the knowledge of the true frequencies. The results also indicate that, as a rule, if the input accuracy is up to 10 percent, then the accuracy gain in the output is substantial.

3.5 Conclusions and Future Perspectives

In this paper we have described, at the best of our knowledge, some of the collocation methods appeared in the literature for ODEs and VIEs. Some interesting properties of collocation-based methods are, in our opinion, still to be exploited. For instance, the knowledge of the collocation function on the whole interval of integration might allow cheap and realiable error estimators, to be used in a variable stepsize-variable order environment, also for problems with delay. Therefore, although collocation technique is an old idea in Numerical Analysis, we strongly believe that it will constitute building blocks for the development of modern software for an efficient and accurate integration of evolutionary problems.

References

1. Bartoszewski, Z., Jackiewicz, Z.: Derivation of continuous explicit two-step Runge–Kutta methods of order three. J. Comput. Appl. Math. **205**, 764–776 (2007)
2. Bartoszewski, Z., Podhaisky, H., Weiner, R.: Construction of stiffly accurate two-step Runge–Kutta methods of order three and their continuous extensions using Nordsieck representation. Report No. 01, Martin-Luther-Universität Halle-Wittenberg, Fachbereich Mathematik und Informatik (2007)
3. Bellen, A., Jackiewicz, Z., Vermiglio, R., Zennaro, M.: Natural continuous extensions of Runge–Kutta methods for Volterra integral equations of the second kind and their applications. Math. Comput. **52**(185), 49–63 (1989)
4. Bellen, A., Jackiewicz, Z., Vermiglio, R., Zennaro, M.: Stability analysis of Runge–Kutta methods for Volterra integral equations of the second kind. IMA J. Numer. Anal. **10**(1), 103–118 (1990)
5. Bother, P., De Meyer, H., Vanden Berghe, G.: Numerical solution of Volterra equations based on mixed interpolation. Comput. Math. Appl. **27**, 1–11 (1994)
6. Brunner, H.: Collocation Methods for Volterra Integral and Related Functional Equations. Cambridge University Press, Cambridge (2004)
7. Brunner, H., van der Houwen, P.J.: The Numerical Solution of Volterra Equations. CWI Monographs, vol. 3. North-Holland, Amsterdam (1986)
8. Brunner, H., Makroglou, A., Miller, R.K.: On mixed collocation methods for Volterra integral equations with periodic solution. Appl. Numer. Math. **24**, 115–130 (1997)
9. Brunner, H., Makroglou, A., Miller, R.K.: Mixed interpolation collocation methods for first and second order Volterra integro-differential equations with periodic solution. Appl. Numer. Math. **23**, 381–402 (1997)
10. Butcher, J.C.: The Numerical Analysis of Ordinary Differential Equations. Runge–Kutta and General Linear Methods. Wiley, Chichester/New York (1987)
11. Butcher, J.C.: Numerical Methods for Ordinary Differential Equations, 2nd edn. Wiley, Chichester (2008)

12. Butcher, J.C., Tracogna, S.: Order conditions for two-step Runge–Kutta methods. Appl. Numer. Math. **24**, 351–364 (1997)
13. Capobianco, G., Conte, D., Del Prete, I., Russo, E.: Fast Runge–Kutta methods for nonlinear convolution systems of Volterra integral equations. BIT **47**(2), 259–275 (2007)
14. Capobianco, G., Conte, D., Del Prete, I., Russo, E.: Stability analysis of fast numerical methods for Volterra integral equations. Electron. Trans. Numer. Anal. **30**, 305–322 (2008)
15. Cardone, A., Ixaru, L.Gr., Paternoster, B.: Exponential fitting Direct Quadrature methods for Volterra integral equations. Numer. Algorithms (2010). doi:10.1007/s11075-010-9365-1
16. Chan, R.P.K., Leone, P., Tsai, A.: Order conditions and symmetry for two-step hybrid methods. Int. J. Comput. Math. **81**(12), 1519–1536 (2004)
17. Coleman, J.P.: Rational approximations for the cosine function; P-acceptability and order. Numer. Algorithms **3**(1–4), 143–158 (1992)
18. Coleman, J.P.: Mixed interpolation methods with arbitrary nodes. J. Comput. Appl. Math. **92**, 69–83 (1998)
19. Coleman, J.P.: Order conditions for a class of two-step methods for $y'' = f(x, y)$. IMA J. Numer. Anal. **23**, 197–220 (2003)
20. Coleman, J.P., Duxbury, S.C.: Mixed collocation methods for $y'' = f(x, y)$. J. Comput. Appl. Math. **126**(1–2), 47–75 (2000)
21. Coleman, J.P., Ixaru, L.Gr.: P-stability and exponential-fitting methods for $y'' = f(x, y)$. IMA J. Numer. Anal. **16**(2), 179–199 (1996)
22. Cong, N.H., Mitsui, T.: Collocation-based two-step Runge–Kutta methods. Jpn. J. Ind. Appl. Math. **13**(1), 171–183 (1996)
23. Conte, D., Del Prete, I.: Fast collocation methods for Volterra integral equations of convolution type. J. Comput. Appl. Math. **196**(2), 652–663 (2006)
24. Conte, D., Paternoster, B.: A family of multistep collocation methods for Volterra integral equations. In: Simos, T.E., Psihoyios, G., Tsitouras, Ch. (eds.) Numerical Analysis and Applied Mathematics. AIP Conference Proceedings, vol. 936, pp. 128–131. Springer, Berlin (2007)
25. Conte, D., Paternoster, B.: Multistep collocation methods for Volterra integral equations. Appl. Numer. Math. **59**, 1721–1736 (2009)
26. Conte, D., Jackiewicz, Z., Paternoster, B.: Two-step almost collocation methods for Volterra integral equations. Appl. Math. Comput. **204**, 839–853 (2008)
27. Conte, D., D'Ambrosio, R., Ferro, M., Paternoster, B.: Piecewise-polynomial approximants for solutions of functional equations, in press on Collana Scientifica di Ateneo, Univ. degli Studi di Salerno
28. Conte, D., D'Ambrosio, R., Ferro, M., Paternoster, B.: Practical construction of two-step collocation Runge–Kutta methods for ordinary differential equations. In: Applied and Industrial Mathematics in Italy III. Series on Advances in Mathematics for Applied Sciences, pp. 278–288. World Scientific, Singapore (2009)
29. Conte, D., D'Ambrosio, R., Ferro, M., Paternoster, B.: Modified collocation techniques for Volterra integral equations. In: Applied and Industrial Mathematics in Italy III. Series on Advances in Mathematics for Applied Sciences, pp. 268–277. World Scientific, Singapore (2009)
30. Crisci, M.R., Russo, E., Vecchio, A.: Stability results for one-step discretized collocation methods in the numerical treatment of Volterra integral equations. Math. Comput. **58**(197), 119–134 (1992)
31. D'Ambrosio, R.: Highly stable multistage numerical methods for functional equations: theory and implementation issues. Bi-Nationally supervised Ph.D. Thesis in Mathematics, University of Salerno, Arizona State University (2010)
32. D'Ambrosio, R., Jackiewicz, Z.: Continuous two-step Runge–Kutta methods for ordinary differential equations. Numer. Algorithms **54**(2), 169–193 (2010)
33. D'Ambrosio, R., Jackiewicz, Z.: Construction and implementation of highly stable two-step collocation methods, submitted
34. D'Ambrosio, R., Ferro, M., Paternoster, B.: A general family of two step collocation methods for ordinary differential equations. In: Simos, T.E., Psihoyios, G., Tsitouras, Ch. (eds.)

Numerical Analysis and Applied Mathematics. AIP Conference Proceedings, vol. 936, pp. 45–48. Springer, Berlin (2007)
35. D'Ambrosio, R., Ferro, M., Paternoster, B.: Two-step hybrid collocation methods for $y'' = f(x, y)$. Appl. Math. Lett. **22**, 1076–1080 (2009)
36. D'Ambrosio, R., Ferro, M., Jackiewicz, Z., Paternoster, B.: Two-step almost collocation methods for ordinary differential equations. Numer. Algorithms **53**(2–3), 195–217 (2010)
37. D'Ambrosio, R., Ferro, M., Paternoster, B.: Trigonometrically fitted two-step hybrid methods for second order ordinary differential equations with one or two frequencies, to appear on Math. Comput. Simul.
38. D'Ambrosio, R., Ferro, M., Paternoster, B.: Collocation based two step Runge–Kutta methods for ordinary differential equations. In: Gervasi, O., et al. (eds.) ICCSA 2008. Lecture Notes in Comput. Sci., Part II, vol. 5073, pp. 736–751. Springer, New York (2008)
39. De Meyer, H., Vanthournout, J., Vanden Berghe, G.: On a new type of mixed interpolation. J. Comput. Appl. Math. **30**, 55–69 (1990)
40. Franco, J.M.: A class of explicit two-step hybrid methods for second-order IVPs. J. Comput. Appl. Math. **187**(1), 41–57 (2006)
41. Gautschi, W.: Numerical integration of ordinary differential equations based on trigonometric polynomials. Numer. Math. **3**, 381–397 (1961)
42. Guillou, A., Soulé, F.L.: La résolution numérique des problèmes differentiels aux conditions par des méthodes de collocation. RAIRO Anal. Numér. Ser. Rouge **R-3**, 17–44 (1969)
43. Hairer, E., Wanner, G.: Solving Ordinary Differential Equations II—Stiff and Differential–Algebraic Problems. Springer Series in Computational Mathematics, vol. 14. Springer, Berlin (2002)
44. Hairer, E., Lubich, C., Wanner, G.: Geometric Numerical Integration—Structure-Preserving Algorithms for Ordinary Differential Equations. Springer Series in Computational Mathematics. Springer, Berlin (2000)
45. Hairer, E., Norsett, S.P., Wanner, G.: Solving Ordinary Differential Equations I—Nonstiff Problems. Springer Series in Computational Mathematics, vol. 8. Springer, Berlin (2000)
46. Henrici, P.: Discrete Variable Methods in Ordinary Differential Equations. Wiley, New York (1962)
47. Ixaru, L.Gr.: Operations on oscillatory functions. Comput. Phys. Commun. **105**, 1–19 (1997)
48. Ixaru, L.Gr., Vanden Berghe, G.: Exponential Fitting. Kluwer Academic, Dordrecht (2004)
49. Jackiewicz, Z.: General Linear Methods for Ordinary Differential Equations. Wiley, Hoboken (2009)
50. Jackiewicz, Z., Tracogna, S.: A general class of two step Runge–Kutta methods for ordinary differential equations. SIAM J. Numer. Anal. **32**, 1390–1427 (1995)
51. Jackiewicz, Z., Tracogna, S.: Variable stepsize continuous two-step Runge–Kutta methods for ordinary differential equations. Numer. Algorithms **12**, 347–368 (1996)
52. Jackiewicz, Z., Renaut, R., Feldstein, A.: Two-step Runge–Kutta methods. SIAM J. Numer. Anal. **28**(4), 1165–1182 (1991)
53. Konguetsof, A., Simos, T.E.: An exponentially-fitted and trigonometrically-fitted method for the numerical solution of periodic initial-value problems. Numerical methods in physics, chemistry, and engineering. Comput. Math. Appl. **45**(1–3), 547–554 (2003)
54. Kramarz, L.: Stability of collocation methods for the numerical solution of $y'' = f(t, y)$. BIT **20**(2), 215–222 (1980)
55. Lambert, J.D.: Numerical Methods for Ordinary Differential Systems: The Initial Value Problem. Wiley, Chichester (1991)
56. Lambert, J.D., Watson, I.A.: Symmetric multistep methods for periodic initial value problems. J. Inst. Math. Appl. **18**, 189–202 (1976)
57. Lie, I.: The stability function for multistep collocation methods. Numer. Math. **57**(8), 779–787 (1990)
58. Lie, I., Norsett, S.P.: Superconvergence for multistep collocation. Math. Comput. **52**(185), 65–79 (1989)
59. Lopez-Fernandez, M., Lubich, C., Schädle, A.: Fast and oblivious convolution quadrature. SIAM J. Sci. Comput. **28**, 421–438 (2006)

60. Lopez-Fernandez, M., Lubich, C., Schädle, A.: Adaptive, fast, and oblivious convolution in evolution equations with memory. SIAM J. Sci. Comput. **30**, 1015–1037 (2008)
61. Martucci, S., Paternoster, B.: General two step collocation methods for special second order ordinary differential equations. Paper Proceedings of the 17th IMACS World Congress Scientific Computation, Applied Mathematics and Simulation, Paris, July 11–15 (2005)
62. Martucci, S., Paternoster, B.: Vandermonde-type matrices in two step collocation methods for special second order ordinary differential equations. In: Bubak, M., et al. (eds.) Computational Science, ICCS 2004. Lecture Notes in Comput. Sci., Part IV, vol. 3039, pp. 418–425. Springer, Berlin (2004)
63. Norsett, S.P.: Collocation and perturbed collocation methods. In: Numerical Analysis, Proc. 8th Biennial Conf., Univ. Dundee, Dundee, 1979. Lecture Notes in Math., vol. 773, pp. 119–132. Springer, Berlin (1980)
64. Norsett, S.P., Wanner, G.: Perturbed collocation and Runge Kutta methods. Numer. Math. **38**(2), 193–208 (1981)
65. Paternoster, B.: Runge–Kutta(–Nyström) methods for ODEs with periodic solutions based on trigonometric polynomials. Appl. Numer. Math. **28**, 401–412 (1998)
66. Paternoster, B.: A phase-fitted collocation-based Runge–Kutta–Nyström methods. Appl. Numer. Math. **35**(4), 339–355 (2000)
67. Paternoster, B.: General two-step Runge–Kutta methods based on algebraic and trigonometric polynomials. Int. J. Appl. Math. **6**(4), 347–362 (2001)
68. Paternoster, B.: Two step Runge–Kutta–Nystrom methods for $y'' = f(x, y)$ and P-stability. In: Sloot, P.M.A., Tan, C.J.K., Dongarra, J.J., Hoekstra, A.G. (eds.) Computational Science, ICCS 2002. Lecture Notes in Computer Science, Part III, vol. 2331, pp. 459–466. Springer, Amsterdam (2002)
69. Paternoster, B.: Two step Runge–Kutta–Nystrom methods for oscillatory problems based on mixed polynomials. In: Sloot, P.M.A., Abramson, D., Bogdanov, A.V., Dongarra, J.J., Zomaya, A.Y., Gorbachev, Y.E. (eds.) Computational Science, ICCS 2003. Lecture Notes in Computer Science, Part II, vol. 2658, pp. 131–138. Springer, Berlin/Heidelberg (2003)
70. Paternoster, B.: Two step Runge–Kutta–Nystrom methods based on algebraic polynomials. Rend. Mat. Appl., Ser. VII **23**, 277–288 (2003)
71. Paternoster, B.: Two step Runge–Kutta–Nystrom methods for oscillatory problems based on mixed polynomials. In: Sloot, P.M.A., Abramson, D., Bogdanov, A.V., Dongarra, J.J., Zomaya, A.Y., Gorbachev, Y.E. (eds.) Computational Science, ICCS 2003. Lecture Notes in Computer Science, Part II, vol. 2658, pp. 131–138. Springer, Berlin/Heidelberg (2003)
72. Paternoster, B.: A general family of two step Runge–Kutta–Nyström methods for $y'' = f(x, y)$ based on algebraic polynomials. In: Alexandrov, V.N., van Albada, G.D., Sloot, P.M.A., Dongarra, J.J. (eds.) Computational Science, ICCS 2003. Lecture Notes in Computer Science, Part IV, vol. 3994, pp. 700–707. Springer, Berlin (2006)
73. Petzold, L.R., Jay, L.O., Yen, J.: Numerical solution of highly oscillatory ordinary differential equations. Acta Numer. **6**, 437–483 (1997)
74. Simos, T.E.: Dissipative trigonometrically-fitted methods for linear second-order IVPs with oscillating solution. Appl. Math. Lett. **17**(5), 601–607 (2004)
75. Van de Vyver, H.: A phase-fitted and amplification-fitted explicit two-step hybrid method for second-order periodic initial value problems. Int. J. Mod. Phys. C **17**(5), 663–675 (2006)
76. Van de Vyver, H.: Phase-fitted and amplification-fitted two-step hybrid methods for $y'' = f(x, y)$. J. Comput. Appl. Math. **209**(1), 33–53 (2007)
77. Van den Houwen, P.J., Sommeijer, B.P., Nguyen, H.C.: Stability of collocation-based Runge–Kutta–Nyström methods. BIT **31**(3), 469–481 (1991)
78. Wright, K.: Some relationships between implicit Runge–Kutta, collocation and Lanczos τ-methods, and their stability properties. BIT **10**, 217–227 (1970)

Chapter 4
Basic Methods for Computing Special Functions

Amparo Gil, Javier Segura, and Nico M. Temme

Abstract This paper gives an overview of methods for the numerical evaluation of special functions, that is, the functions that arise in many problems from mathematical physics, engineering, probability theory, and other applied sciences. We consider in detail a selection of basic methods which are frequently used in the numerical evaluation of special functions: converging and asymptotic series, including Chebyshev expansions, linear recurrence relations, and numerical quadrature. Several other methods are available and some of these will be discussed in less detail. We give examples of recent software for special functions where these methods are used. We mention a list of new publications on computational aspects of special functions available on our website.

Keywords Numerical evaluation of special functions · Chebyshev expansions · Quadrature methods · Transformation of series · Continued fractions · Asymptotic analysis

Mathematics Subject Classification (2000) 65D20 · 41A60 · 33C05 · 33C10 · 33C15

A. Gil
Departamento de Matemática Aplicada y CC. de la Computación, ETSI Caminos, Universidad de Cantabria, 39005 Santander, Spain
e-mail: amparo.gil@unican.es

J. Segura (✉)
Departamento de Matemáticas, Estadística y Computación, Universidad de Cantabria, 39005 Santander, Spain
e-mail: javier.segura@unican.es

N.M. Temme
CWI, Science Park 123, 1098 XG Amsterdam, The Netherlands
e-mail: Nico.Temme@cwi.nl

T.E. Simos (ed.), *Recent Advances in Computational and Applied Mathematics*,
DOI 10.1007/978-90-481-9981-5_4,

4.1 Introduction

For workers in the applied sciences the *Handbook of Mathematical Functions with Formulas, Graphs, and Mathematical Tables* [1], edited by Milton Abramowitz and Irene Stegun, and published in 1964 is usually the first source of information about the properties of special functions. It may be the most cited book in mathematics. These days the Handbook is being updated as a Digital Library of Mathematical Functions (DLMF), and will be freely accessible in a Web version. Other sources for collections of formulas for special functions on the web are Wolfram MathWorld[1] and Wikipedia[2].

These sources give many properties of special functions of which a number can be used for their numerical evaluation, sometimes with references to suitable algorithms. However, it is not always clear how to write efficient and reliable algorithms.

The present paper gives an overview on numerical methods for special functions. It is based on our recent book [54] in which we consider four Basic Methods, namely

1. Convergent and divergent series.
2. Chebyshev expansions.
3. Linear recurrence relations and associated continued fractions.
4. Quadrature methods.

In addition we give a selection of published algorithms for special functions.

There are many other methods, which are also discussed in our book, and some of them will be discussed in this overview as well. For example, the use of differential equations will be discussed in connection with the Taylor expansion method for initial boundary problems.

Our general point of view in connection with special functions is to remain modest in the number of parameters. It is possible to design straightforward algorithms for the generalized hypergeometric function

$$ {}_pF_q\left(\begin{matrix} a_1,\ldots,a_p \\ b_1,\ldots,b_q \end{matrix}; z\right) = \sum_{n=0}^{\infty} \frac{(a_1)_n \cdots (a_p)_n}{(b_1)_n \cdots (b_q)_n} \frac{z^n}{n!}, \tag{4.1.1} $$

where $p \le q+1$ and $(a)_n$ is the Pochhammer symbol, also called the shifted factorial, defined by

$$ (a)_0 = 1, \qquad (a)_n = a(a+1)\cdots(a+n-1) \quad (n \ge 1), \qquad (a)_n = \frac{\Gamma(a+n)}{\Gamma(a)}. \tag{4.1.2} $$

Many special functions can be written in terms of this function, with the main cases given by $p = 1, q = 1$ (Kummer functions, with convergence of the series for all complex z) and $p = 2, q = 1$ (Gauss hypergeometric functions, with convergence if

[1] http://mathworld.wolfram.com/.

[2] http://en.wikipedia.org/.

$|z| < 1$). For efficient and reliable algorithms the series in (4.1.1) is only of limited use.

Also, differential equations, especially those that arise in the physical sciences, are the reservoir that generates many special functions. One may define a general second order equation, make an algorithm for this equation, and expect that solutions of simpler equations follow from the general solution. However, very difficult problems may arise then. Consider as an example the equation

$$\frac{d^2}{dz^2}w(z) = (pz^2 + qz + r)w(z), \tag{4.1.3}$$

the solutions of which can be expressed in terms of the parabolic cylinder functions U and V, see [1, Chap. 19], which are solutions of the equation

$$\frac{d^2}{dz^2}w(z) = \left(\frac{1}{4}z^2 + a\right)w(z). \tag{4.1.4}$$

When $p = r = 0$ in (4.1.3) that equation reduces to the Airy equation, whereas the Airy functions are not special cases of the parabolic cylinder functions U and V (in the sense that Airy functions are Bessel functions for certain values of the order of these functions). A nontrivial limiting process with $p \to 0$ and $r \to 0$ is needed to get Airy functions from a linear combination of the solutions of (4.1.3).

In its turn, the parabolic cylinder function $U(a, z)$ is a special case of the confluent hypergeometric function (also called Kummer functions) $U(a, c, z)$. We have two forms [1, p. 691]:

$$\begin{aligned} U(a, z) &= 2^{-\frac{1}{4}-\frac{1}{2}a} e^{-\frac{1}{4}z^2} U\left(\frac{1}{2}a + \frac{1}{4}, \frac{1}{2}, \frac{1}{2}z^2\right) \\ &= 2^{-\frac{3}{4}-\frac{1}{2}a} z e^{-\frac{1}{4}z^2} U\left(\frac{1}{2}a + \frac{3}{4}, \frac{3}{2}, \frac{1}{2}z^2\right). \end{aligned} \tag{4.1.5}$$

The first form suggests that the function $U(a, z)$ is an even function of z, the second one that it is odd. The point is that this Kummer function is multi-valued, and the representation

$$\begin{aligned} U(a, \pm z) = &\frac{\sqrt{\pi}\, 2^{-\frac{1}{4}-\frac{1}{2}a} e^{-\frac{1}{4}z^2}}{\Gamma(\frac{3}{4} + \frac{1}{2}a)}\, {}_1F_1\left(\begin{matrix}\frac{1}{2}a + \frac{1}{4} \\ \frac{1}{2}\end{matrix}; \frac{1}{2}z^2\right) \\ &\mp \frac{\sqrt{\pi}\, 2^{\frac{1}{4}-\frac{1}{2}a} z e^{-\frac{1}{4}z^2}}{\Gamma(\frac{1}{4} + \frac{1}{2}a)}\, {}_1F_1\left(\begin{matrix}\frac{1}{2}a + \frac{3}{4} \\ \frac{3}{2}\end{matrix}; \frac{1}{2}z^2\right) \end{aligned} \tag{4.1.6}$$

gives a better insight. However, this form is extremely unstable for intermediate or large values of z.

In our opinion it is important to have codes that can be used for a limited class of functions. In this sense we have written algorithms for conical functions $P^{\mu}_{-1/2+i\tau}(x)$ [55] for real x, τ and μ, and not for Legendre functions of general complex degree. Also, we have written codes [51] for modified Bessel functions of purely imaginary order, that is for $K_{ia}(x)$ and a related function, and not a general code for Bessel functions with general complex order.

4.2 Convergent and Divergent Series

Convergent series for special functions usually arise in the form of hypergeometric series, with as general form the one shown in (4.1.1). The series is easy to evaluate because of the recursion $(a)_{n+1} = (a+n)(a)_n$, $n \geq 0$, of the Pochhammer symbols in (4.1.2). For certain special function, for example for the modified Bessel function

$$I_\nu(z) = \left(\frac{1}{2}z\right)^\nu \sum_{n=0}^{\infty} \frac{(\frac{1}{4}z^2)^n}{\Gamma(\nu+n+1)\,n!} = \left(\frac{1}{2}z\right)^\nu {}_0F_1\left({- \atop \nu+1}; \tfrac{1}{4}z^2\right) \tag{4.2.1}$$

it gives a stable representation when $z > 0$ and $\nu \geq 0$ and it is an efficient representation when z is not large compared with ν. However, when we use this expansion in the representation of the other modified Bessel function

$$K_\nu(z) = \frac{1}{2}\pi \frac{I_{-\nu}(z) - I_\nu(z)}{\sin \pi\nu}, \tag{4.2.2}$$

it can be used only for small values of z. This is because of the cancellation of numerical digits, which can be seen from the asymptotic estimates

$$I_\nu(z) \sim \frac{e^z}{\sqrt{2\pi z}}, \qquad K_\nu(z) \sim \sqrt{\frac{\pi}{2z}}e^{-z}, \quad z \to \infty, \tag{4.2.3}$$

which is valid for fixed values of ν.

There is another phenomenon when using combinations of hypergeometric functions. When ν is an integer, the form in (4.2.2) is well defined by a limiting process, but for numerical computations a special algorithm is needed. See [109], where it is shown that it is sufficient to treat the case $\nu \sim 0$ in detail and that the remaining integer values follow from recursion relations.

The case for confluent hypergeometric functions is more complicated. We have for the function $U(a,c,z)$ the representation

$$U(a,c,z) = \frac{\pi}{\sin \pi c}\left(\frac{{}_1F_1({a \atop c}; z)}{\Gamma(1+a-c)\Gamma(c)} - z^{1-c}\frac{{}_1F_1({1+a-c \atop 2-c}; z)}{\Gamma(a)\Gamma(2-c)}\right), \tag{4.2.4}$$

and it is useful for small values of z. Consider $c \sim 0$. We have

$$\lim_{c\to 0} \frac{{}_1F_1({a \atop c}; z)}{\Gamma(1+a-c)\Gamma(c)} = \frac{z\,{}_1F_1({a+1 \atop 2}; z)}{\Gamma(a)}. \tag{4.2.5}$$

So, apart from this limit, another (simultaneous) limiting process for $c \to 0$ needs to be controlled, and also the extra parameter a makes it more difficult to write a stable algorithm. A published algorithm seems not to be available for this case.

For Gauss hypergeometric functions similar problems arise in the connection formulas, say the one writing a function with argument z as a linear combination of two functions with argument $1-z$. See [36] for numerical algorithms.

Other instabilities occur when the parameters of the hypergeometric function become large and/or complex.

For Gauss and Kummer hypergeometric functions many other convergent expansions are available, for example in terms of Chebyshev polynomials and of Bessel functions; see [75, Sects. 9.3.4, 9.4.1, 9.4.3]. For a different type of expansion in terms of Bessel functions, with an application to the parabolic cylinder functions, see [78].

4.2.1 Divergent Expansions

With this we mean asymptotic expansions of the form

$$F(z) \sim \sum_{n=0}^{\infty} \frac{c_n}{z^n}, \qquad z \to \infty. \tag{4.2.6}$$

The series usually diverges, but it has the property

$$F(z) = \sum_{n=0}^{N-1} \frac{c_n}{z^n} + R_N(z), \qquad R_N(z) = \mathcal{O}(z^{-N}),\ z \to \infty, \tag{4.2.7}$$

for $N = 0, 1, 2, \ldots$, and the order estimate holds for fixed N. This is the Poincaré-type expansion and for special functions like the gamma and Bessel functions they are crucial for evaluating these functions. Other variants of the expansion are also important, in particular expansions that hold for a certain range of additional parameters (this leads to the uniform asymptotic expansions in terms of other special functions like Airy functions, which are useful in turning point problems).

Usually the optimal choice of N with a given (large) z occurs for the N that makes c_N/z^N the smallest term. And usually the error estimate in (4.2.7) may be exponentially small for this N. Say, with $z > 0$, the smallest term is achieved when $N \sim z$, then it may happen that $R_N(N) = \mathcal{O}(\exp(-N))$. Many asymptotic expansions for special functions share this property, and it makes asymptotic expansions very useful for numerical computations.

Convergent power series may be very unstable in certain parts of their convergence region, as the expansion of e^{-z} for $\Re z > 0$. In a similar way, asymptotic expansions may be very useful in a certain sector of the complex z-plane, but may become useless in other sectors. Other expansions may be available in these sectors.

For example, consider the compound expansion of the Kummer function

$$\frac{1}{\Gamma(c)}\,{}_1F_1\left({a \atop c};z\right)=\frac{e^z z^{a-c}}{\Gamma(a)}\left[\sum_{n=0}^{R-1}\frac{(c-a)_n(1-a)_n}{n!\,z^n}+\mathcal{O}\left(|z|^{-R}\right)\right]$$
$$+\frac{z^{-a}e^{\pm i\pi a}}{\Gamma(c-a)}\left[\sum_{n=0}^{S-1}\frac{(a)_n(1+a-c)_n}{n!\,(-z)^n}+\mathcal{O}\left(|z|^{-S}\right)\right],\qquad(4.2.8)$$

the upper sign being taken if $-\frac{1}{2}\pi<\operatorname{ph} z<\frac{3}{2}\pi$, the lower sign if $-\frac{3}{2}\pi<\operatorname{ph} z<\frac{1}{2}\pi$. When $\Re z>0$ the second term can be neglected because of e^z in front of the first term. We see that within a sector properly inside the sector $-\frac{1}{2}\pi<\operatorname{ph} z<\frac{1}{2}\pi$ we can work with one expansion, and in a sector containing the negative z-axis with another one. In sectors containing the imaginary axis we need both expansions.

The fact that an entire function, as is the Kummer function, does not have a unique asymptotic expansion valid for all phases of z will be explained in Sect. 4.6, where we discuss elements of the Stokes phenomenon.

A remarkable point is in this example that we have, say for $-\frac{1}{2}\pi+\delta<\operatorname{ph} z<\frac{1}{2}\pi-\delta$ (δ a small positive number), not only one expansion, but also an expansion that gives an exponentially small correction to the main expansion. For computations (and also for applications in physics) this may give interesting information. The role of exponentially small terms in asymptotics has been discussed in great detail the last twenty years. For many aspects from a physicists' point of view, we refer to *The Devil's Invention: Asymptotic, Superasymptotic and Hyperasymptotic Series* [13] for a lively introduction to this topic.[3] In Sect. 4.6 we also discuss aspects of the role of exponentially small terms.

4.3 Linear Recurrence Relations

Many special functions of mathematical physics satisfy three-term recurrence relations. We first give a simple relation and discuss stability and direction of recursion, which elements are important in the general theory.

4.3.1 A Simple Recurrence Relation

The recurrence relations

$$f_n=f_{n-1}-\frac{x^n}{n!},\qquad g_n=g_{n-1}+\frac{x^n}{n!},\qquad n=1,2,\ldots,\qquad(4.3.1)$$

[3]The "Devil's invention" refers to a quote from Niels Hendrik Abel (1828), who claimed "Divergent series are the invention of the devil, and it is shameful to base on them any demonstration whatsoever."

with initial values $f_0 = e^x - 1,\ g_0 = 1$ have solutions

$$f_n = \sum_{m=n+1}^{\infty} \frac{x^m}{m!}, \qquad g_n = e^x - f_n = \sum_{m=0}^{n} \frac{x^m}{m!}, \tag{4.3.2}$$

which are in fact special cases of the incomplete gamma functions:

$$f_n = e^x \frac{\gamma(n+1,x)}{n!}, \qquad g_n = e^x \frac{\Gamma(n+1,x)}{n!}, \qquad n = 0, 1, 2, \ldots. \tag{4.3.3}$$

Assume that $x > 0$. Then, following our intuition, the recursion for g_n will not cause any problem, since two positive numbers are always added during the recursion. For the recurrence relation of f_n it is not clear, but there is a potential danger owing to the subtraction of two positive quantities. Note that the computation of the initial value f_0, for small values of x, may produce a large relative error, when the quantities e^x and 1 are simply subtracted. This problem repeats itself for each subsequent f_n that is computed by using the recurrence relation: in each step the next term of the Taylor series is subtracted from the exponential function.

Apparently, this is a hopeless procedure for computing successive f_n (even when x is not small). On the other hand, the computation of successive g_n does not show any problem.

In the study of recurrence relations it may make sense to change the direction of the recursion. Writing the recursion for f_n and g_n in the backward direction:

$$f_{n-1} = f_n + \frac{x^n}{n!}, \qquad g_{n-1} = g_n - \frac{x^n}{n!} \tag{4.3.4}$$

then we note that for both solutions the roles are reversed: g_n is obtained by subtraction, whereas f_n is obtained by addition of positive numbers. In addition, $\lim_{n\to\infty} f_n = 0$.

It can be easily verified that both f_n and g_n satisfy the recurrence relation

$$(n+1)y_{n+1} - (x+n+1)y_n + xy_{n-1} = 0. \tag{4.3.5}$$

Again, this relation is stable for the computation of g_n in the forward direction; it is stable for f_n in the backward direction. Note that the solutions of this recursion satisfy $f_n \to 0,\ g_n \to e^x$ as $n \to \infty$. Apparently, the solution which becomes ultimately small in the forward direction (small compared to the other solution), is the victim. A similar phenomenon occurs in the backward direction. This phenomenon will be explained and put in a general framework in the following section.

4.3.2 Some Elements of the General Theory

For details on the theory of this topic we refer to [54, Chap. 4].

Consider the recurrence relation

$$y_{n+1} + b_n y_n + a_n y_{n-1} = 0, \quad n = 1, 2, 3, \ldots, \tag{4.3.6}$$

where a_n and b_n are given, with $a_n \neq 0$. Equation (4.3.6) is also called a linear homogeneous difference equation of the second order. In analogy with the theory of differential equations, two linearly independent solutions f_n, g_n exist in general, with the property that any solution y_n of (4.3.6) can be written in the form

$$y_n = A f_n + B g_n, \tag{4.3.7}$$

where A and B do not depend on n. We are interested in the special case that the pair $\{f_n, g_n\}$ satisfies

$$\lim_{n\to\infty} \frac{f_n}{g_n} = 0. \tag{4.3.8}$$

Then, for any solution (4.3.7) with $B \neq 0$, we have $f_n/y_n \to 0$ as $n \to \infty$. When $B = 0$ in (4.3.7), we call y_n a minimal solution; when $B \neq 0$, we call y_n a dominant solution. When we have two initial values y_0, y_1, assuming that f_0, f_1, g_0, g_1 are known as well, then we can compute A and B. That is,

$$A = \frac{g_1 y_0 - g_0 y_1}{f_0 g_1 - f_1 g_0}, \qquad B = \frac{y_0 f_1 - y_1 f_0}{g_0 f_1 - g_1 f_0}. \tag{4.3.9}$$

The denominators are different from 0 when the solutions f_n, g_n are linearly independent.

When we assume that the initial values y_0, y_1 are to be used for generating a dominant solution, then A may, or may not, vanish; B should not vanish: $y_0 f_1 \neq y_1 f_0$. When however the initial values are to be used for the computation of a minimal solution, then the much stronger condition $y_0 f_1 = y_1 f_0$ should hold. It follows that, in this case, one and only one initial value can be prescribed and the other one follows from the relation $y_0 f_1 = y_1 f_0$; in other words, the minimal solutions, if it exists, is unique up to a constant multiplicative factor. In the numerical approach this leads to the well-known instability phenomena for the computation of minimal solutions. The fact is that, when our initial values y_0, y_1 are not specified to an infinite precision,—and consequently B does not vanish exactly—the computed solution (4.3.7) always contains a fraction of a dominant solution g_n. Hence, in the long run, our solution y_n does not behave as a minimal solution, although we assumed that we were computing a minimal solution. This happens even if all further computations are done exactly.

In applications it is important to know whether a given equation (4.3.6) has dominant and minimal solutions. Often this can be easily concluded from the asymptotic behavior of the coefficients a_n and b_n.

Assume that for large values of n the coefficients a_n, b_n behave as follows:

$$a_n \sim a n^{\alpha}, \qquad b_n \sim b n^{\beta}, \qquad ab \neq 0 \tag{4.3.10}$$

with α and β real; assume that t_1, t_2 are the zeros of the characteristic polynomial $\Phi(t) = t^2 + bt + a$ with $|t_1| \geq |t_2|$. Then it follows from Perron's theorem [54, p. 93] that we have the following results.

1. If $\beta > \frac{1}{2}\alpha$ then the difference equation (4.3.6) has two linearly independent solutions $y_{n,1}$ and $y_{n,2}$, with the property

$$\frac{y_{n+1,1}}{y_{n,1}} \sim -bn^{\beta}, \qquad \frac{y_{n+1,2}}{y_{n,2}} \sim -\frac{a}{b}n^{\alpha-\beta}, \qquad n \to \infty. \tag{4.3.11}$$

2. If $\beta = \frac{1}{2}\alpha$ and $|t_1| > |t_2|$, then the difference equation (4.3.6) has two linear independent solutions $y_{n,1}$ and $y_{n,2}$, with the property

$$\frac{y_{n+1,1}}{y_{n,1}} \sim t_1 n^{\beta}, \qquad \frac{y_{n+1,2}}{y_{n,2}} \sim t_2 n^{\beta}, \qquad n \to \infty. \tag{4.3.12}$$

3. If $\beta = \frac{1}{2}\alpha$ and $|t_1| = |t_2|$, or if $\beta < \frac{1}{2}\alpha$, then some information is still available, but the theorem is inconclusive with respect to the existence of minimal and dominant solutions.

4.3.3 Miller's Algorithm

This algorithm can be used for calculating a sequence

$$f_0, \quad f_1, \quad \ldots, \quad f_N \tag{4.3.13}$$

of values of a minimal solution that satisfies (4.3.6); N is a non-negative integer. Such sequences frequently occur in expansions of special functions; see for example the expansions in terms of Chebyshev polynomials in (4.4.16).

When we use (4.3.6) in the backward direction we may start with two initial values f_N and f_{N-1}. But these are perhaps difficult to obtain. Miller's algorithm does not need these values, and uses a smart idea for the computation of the required sequence (4.3.13). The algorithm works for many interesting cases and gives an efficient method for computing such sequences.

Assume we have a relation of the form

$$S = \sum_{n=0}^{\infty} \lambda_n f_n, \qquad S \neq 0. \tag{4.3.14}$$

The series should be convergent and λ_n and S should be known. The series in (4.3.14) plays a role in normalizing the required minimal solution. The series may be finite; we only require that at least one coefficient λ_n with $n \leq N$ is different from zero. When just one coefficient, say λ_n, is different from zero, we assume that the value f_n is available.

Table 4.1 Computing the modified Bessel functions $I_n(x)$ for $x = 1$ by using (4.3.17) in the backward direction. The underlined digits in the *third column* are correct

n	y_n before normalization	$y_n \doteq I_n(1)$ after normalization
0	$2.2879\,49300\;_{10}{+8}$	$\underline{1.26606\,5878}01\;_{10}{-0}$
1	$1.0213\,17610\;_{10}{+8}$	$\underline{5.65159\,104}106\;_{10}{-1}$
2	$2.4531\,40800\;_{10}{+7}$	$\underline{1.35747\,6697}94\;_{10}{-1}$
3	$4.0061\,29000\;_{10}{+6}$	$\underline{2.21684\,2492}88\;_{10}{-2}$
4	$4.9434\,00000\;_{10}{+5}$	$\underline{2.73712\,0221}60\;_{10}{-3}$
5	$4.9057\,00000\;_{10}{+4}$	$\underline{2.71463\,1560}12\;_{10}{-4}$
6	$4.0640\,00000\;_{10}{+3}$	$\underline{2.24886\,6147}61\;_{10}{-5}$
7	$2.8900\,00000\;_{10}{+2}$	$\underline{1.59921\,82}9887\;_{10}{-6}$
8	$1.8000\,00000\;_{10}{+1}$	$\underline{9.960}52\,919710\;_{10}{-8}$
9	$1.0000\,00000\;_{10}{+0}$	$\underline{5.}53362\,733172\;_{10}{-9}$
10	$0.0000\,00000\;_{10}{+0}$	$0.00000\,000000\;_{10}{-0}$

In Miller's algorithm a starting value ν is chosen, $\nu > N$, and a solution $\{y_n^{(\nu)}\}$ of (4.3.6) is computed with the false initial values

$$y_{\nu+1}^{(\nu)} = 0, \qquad y_\nu^{(\nu)} = 1. \tag{4.3.15}$$

The right-hand sides may be replaced by other values; at least one value should be different from zero. In some cases a judicious choice of these values may improve the convergence of the algorithm.

The computed solution y_n, with (4.3.15) as initial values, is a linear combination of the solutions f_n and g_n, g_n being a dominant solution. When we choose ν large enough it follows that the wanted solution f_n satisfies $f_n \doteq \rho y_n$, $n = 0, 1, \ldots, N$, because the dominant solution g_n can be neglected in the backward direction. For details and proofs we refer to [37] and [54, Sect. 4.6]. The number ρ then follows from the normalizing sum (4.3.14). That is,

$$\rho \doteq \frac{1}{S}\sum_{n=0}^{\nu} \lambda_n y_n. \tag{4.3.16}$$

In [11] the above method was introduced for computing the modified Bessel functions $I_n(x)$. The recurrence relation for these functions reads

$$I_{n+1}(x) + \frac{2n}{x} I_n(x) - I_{n-1}(x) = 0. \tag{4.3.17}$$

A normalizing condition (4.3.14) is

$$e^x = I_0(x) + 2I_1(x) + 2I_2(x) + 2I_3(x) + \cdots. \tag{4.3.18}$$

That is, $S = e^x$, $\lambda_0 = 1$, $\lambda_n = 2\,(n \geq 1)$. We take $x = 1$ and initial values (4.3.15) with $\nu = 9$ and obtain the results given in Table 4.1.

The rightmost column in Table 4.1 is obtained by dividing the results of the middle column by

$$\rho \doteq \frac{1}{e}\sum_{n=0}^{9} \lambda_n y_n^{(9)} = 1.8071328986_{10}{}^{+8}. \tag{4.3.19}$$

When we take $N = 5$, which means we want to compute the sequence $I_0(1), I_1(1), \ldots, I_5(1)$, we see that these quantities are computed with at least 10 correct decimal digits.

4.3.4 *Examples of Hypergeometric Functions and Recursions*

We mention classes of functions of hypergeometric type that are of interest for applications and give a few details about their recursions.

4.3.4.1 Bessel Functions

In the case of ordinary Bessel functions, we have the recurrence relation

$$y_{n+1} - \frac{2n}{z} y_n + y_{n-1} = 0, \quad z \neq 0, \tag{4.3.20}$$

with solutions

$$f_n = J_n(z), \qquad g_n = Y_n(z). \tag{4.3.21}$$

This is covered by (4.3.11), with

$$a = 1, \quad \alpha = 0, \qquad b = -\frac{2}{z}, \quad \beta = 1. \tag{4.3.22}$$

In this case

$$\frac{f_{n+1}}{f_n} \sim \frac{z}{2n}, \qquad \frac{g_{n+1}}{g_n} \sim \frac{2n}{z}. \tag{4.3.23}$$

The known asymptotic behavior of the Bessel functions reads

$$f_n \sim \frac{1}{n!}\left(\frac{z}{2}\right)^n, \qquad g_n \sim -\frac{(n-1)!}{\pi}\left(\frac{2}{z}\right)^n, \qquad n \to \infty. \tag{4.3.24}$$

Similar results hold for the modified Bessel functions, with recurrence relation

$$y_{n+1} + \frac{2n}{z} y_n - y_{n-1} = 0, \quad z \neq 0, \tag{4.3.25}$$

with solutions $I_n(z)$ (minimal) and $K_n(z)$ (dominant).

There is an extensive literature on the use of recursion for evaluating Bessel functions, with [37] as pioneering paper; see also [5, 70, 101].

4.3.4.2 Kummer Functions

The Kummer functions (or confluent hypergeometric functions) ${}_1F_1$ and U do not satisfy the same recurrence relations, but by multiplying them with certain gamma functions they do. We assume $z > 0$. An overview of the relevant recurrence relations can be found in [1, Chap. 13].

Recursion with respect to *a*. The functions

$$\frac{\Gamma(a+n)}{\Gamma(a+n+1-c)}\,{}_1F_1\left({a+n \atop c};z\right) \quad \text{and} \quad \frac{\Gamma(a+n)}{\Gamma(a)}U(a+n,c,z) \qquad (4.3.26)$$

are respectively dominant and minimal.

Recursion with respect to *c*. The functions

$$\frac{\Gamma(c-a+n)}{\Gamma(c+n)}\,{}_1F_1\left({a \atop c+n};z\right) \quad \text{and} \quad U(a,c+n,z) \qquad (4.3.27)$$

are respectively minimal and dominant.

There are other interesting cases: recursion with respect to both a and c, and recursion with respect to negative n. All the possible cases are analyzed in [103], where it is shown that the Kummer recurrences always have a minimal solution except for the case of recursion over a when z is real and positive (for $a \to -\infty$) or negative real (for $a \to +\infty$). See also [30] and [54, Sect. 4.5.1].

4.3.4.3 Gauss Hypergeometric Functions

The recursions for the functions

$${}_2F_1\left({a+\epsilon_1 n,\ b+\epsilon_2 n \atop c+\epsilon_3 n};z\right), \qquad (4.3.28)$$

where $\epsilon_j = 0, \pm 1$, not all equal to zero, and z is complex are analyzed in [53, 59]. Of the 27 nontrivial cases, only a limited set of these recursions need to be considered. This is because of several relations between contiguous Gauss functions. Among other results, in [59] it is shown that the function (4.3.28) is minimal around $z = 0$ when $\epsilon_3 > 0$. An overview of the relevant recurrence relations can be found in [1, Chap. 15].

4.3.4.4 Legendre Functions

For definitions and properties, see [1, Chap. 8] and [111, Chap. 8]. Legendre functions are special cases of Gauss hypergeometric functions, but the recursions need special attention. When $\Re z > 0$, $z \notin (0,1]$, $P_\nu^\mu(z)$ is the minimal solution of the recursion with respect to positive order μ; $Q_\nu^\mu(z)$ is dominant. Particular cases are toroidal functions and conical functions. The latter have the form $P^{\mu}_{-1/2+i\tau}(z)$, $Q^{\mu}_{-1/2+i\tau}(z)$, which are real for $z > -1$ and real τ and μ.

For recursion with respect to the degree ν, $Q^{\mu}_{\nu}(z)$ is a minimal solution and $P^{\mu}_{\nu}(z)$ is dominant.

For further details on numerical aspects and algorithms we refer to [41–44, 46, 55] and [54, Sect. 12.3].

4.3.4.5 Coulomb Wave Functions

Information on these functions can be found in [1, Chap. 14]. Coulomb wave functions are special cases of the Kummer functions, and they can also be viewed as generalizations of Bessel functions. The regular function $F_{\lambda}(\eta, \rho)$ is the minimal solution with respect to increasing λ, while the irregular $G_{\lambda}(\eta, \rho)$ function is a dominant one. Algorithms based on recursions are discussed in [85]; in [96, 97] several types of series expansions are considered, with references to earlier algorithms.

4.3.4.6 Parabolic Cylinder Functions

For definitions and properties, see [1, Chap. 19]. The standard forms are $U(a, z)$ and $V(a, z)$, and, again, special cases of the Kummer functions. The function $U(a, x)$ is minimal in the forward a-recursion. For negative values of a the situation is quite different, and for $|a|$ large enough ($a \ll -z^2/4$), the solutions are neither minimal nor dominant. See [54, p. 102]. Algorithms using recursion can be found in [57, 58, 95, 102].

4.4 Chebyshev Expansions

Chebyshev expansions are examples of convergent expansions, considered earlier, but because of their special properties they deserve a separate discussion.

Chebyshev polynomials of the first kind $T_n(x)$ have the nice property $T_n(\cos\theta) = \cos(n\theta)$, giving an equal ripple in the θ-interval $[0, \pi]$ and in the x-interval $[-1, 1]$. Because of their excellent convergence properties, Chebyshev expansions may replace convergent power series and divergent asymptotic expansions, or they may be used for filling the gap between the domains where convergent and asymptotic expansions can be used.

The standard Chebyshev expansion is of the form

$$f(x) = \sum_{n=0}^{\infty}{}' c_n T_n(x), \quad -1 \le x \le 1, \tag{4.4.1}$$

where the prime means that the first term is to be halved. Provided that the coefficients c_k decrease in magnitude sufficiently rapidly, the error made by truncating the Chebyshev expansion after the terms $k = n$, that is,

$$E_n(x) = \sum_{k=n+1}^{\infty} c_k T_k(x), \tag{4.4.2}$$

will be given approximately by

$$E_n(x) \doteq c_{n+1} T_{n+1}(x), \tag{4.4.3}$$

that is, the error approximately satisfies the equioscillation property, which is happening in best-approximation (mini-max) methods.

4.4.1 Clenshaw's Summation Method

There is a very simple algorithm due to Clenshaw [24] for evaluating the finite sum

$$S_n(x) = \frac{1}{2}c_0 + \sum_{k=1}^{n} c_k T_k(x), \tag{4.4.4}$$

which is based on the recurrence relation

$$x T_k(x) = \frac{1}{2}\big(T_{k+1}(x) + T_{|k-1|}(x)\big). \tag{4.4.5}$$

The algorithm computes a sequence $b_1, b_2, \ldots, b_{n+1}$ and starts with putting $b_{n+1} = 0$ and $b_n = c_n$. Next,

$$b_k = 2x b_{k+1} - b_{k+2} + c_k, \quad k = n-1, n-2, \ldots, 1. \tag{4.4.6}$$

Then, $S_n(x) = x b_1 - b_2 + \frac{1}{2}c_0$.

4.4.2 Methods for Obtaining the Coefficients

The coefficients c_n in the expansion in (4.4.1) can be obtained in several ways, and we mention a few elements of the main methods. For details we refer to [54, Chap. 3].

4.4.2.1 Tabled Coefficients

In the case that the function f is an elementary or a one-variable special function, such as the error function $\operatorname{erf} x$, the Bessel function $J_0(x)$, and so on, the coefficients can be obtained from tables, see [75]. Usually 20D accuracy of the coefficients is given. In [93] 30D coefficients are given for the error function and the complementary error function. Nowadays, computer algebra systems can be used to obtain tables of high precision coefficients.

4.4.2.2 Discretizing the Integral

A numerical method uses discretization of the integral representation. That is,

$$c_k = \frac{2}{\pi}\int_{-1}^{1} \frac{f(x)T_k(x)}{\sqrt{1-x^2}}\,dx = \frac{2}{\pi}\int_0^{\pi} f(\cos\theta)\cos(k\theta)\,d\theta, \tag{4.4.7}$$

and discretization gives

$$c_k \doteq \frac{2}{n}\sum_{j=0}^{n}{}'' f\left(\cos\frac{\pi j}{n}\right)\cos\frac{\pi k j}{n}, \tag{4.4.8}$$

where the primes mean that the first and last terms is are to be halved. This is a discrete cosine transform, which can be computed by methods based on the fast Fourier transform [116].

4.4.2.3 Clenshaw's Method

This method can be used for functions satisfying linear ordinary differential equations with polynomial coefficients of the form

$$\sum_{k=0}^{m} p_k(x) f^{(k)}(x) = h(x), \tag{4.4.9}$$

with p_k polynomials and where the coefficients of the Chebyshev expansion of the function h are known. The idea is as follows. Next to the expansion in (4.4.1), we introduce expansions for the relevant derivatives:

$$f^{(s)}(x) = \sum_{n=0}^{\infty}{}' c_n^{(s)} T_n(x), \quad s = 0, 1, 2, \ldots, \tag{4.4.10}$$

and from known properties of the Chebyshev polynomials we have

$$2r c_r^{(s)} = c_{r-1}^{(s+1)} - c_{r+1}^{(s+1)}, \quad r \geq 1. \tag{4.4.11}$$

A next element in Clenshaw's method is to handle the powers of x occurring in the differential equation satisfied by f. For this we need the relation (4.4.5) and formulas for higher powers of x in the left-hand side. These can easily be obtained from the above one.

By substituting the expansion in (4.4.1) in the equation in (4.4.9) and using the formulas for the powers of x, a set of recursions for the coefficients $c_r^{(s)}$ can be obtained. Together with boundary values (or other known relations) this set of recursions can be solved numerically, and Clenshaw [25, 26] explained how this can be done by using a backward recursion method.

For example, the exponential function $y(x) = e^{ax}$ satisfies the differential equation $y' = ay$. Substituting expansions following from (4.4.1) we obtain $c_r^{(1)} = ac_r$. Using this in the form $c_{r+2}^{(1)} = ac_{r+2}$ and using (4.4.11) we have the recursion for the coefficients:

$$2(r+1)rc_{r+1} = a(c_r - c_{r+2}), \quad r \geq 0. \tag{4.4.12}$$

This is the recursion relation for the modified Bessel function $I_r(a)$, and so c_r is multiple of this function (the other modified Bessel function $K_r(a)$ being excluded because of its behavior for large r). The value $y(0) = 1$ gives $c_r = I_r(a)$. This result is known because of the expansion

$$e^{a\cos\theta} = {\sum_{r=0}^{\infty}}{}' I_r(a)\cos(r\theta). \tag{4.4.13}$$

It is not at all needed that we know the solution in terms of a known function; for numerical purposes it is enough to have (4.4.12), and to use a backward recursion scheme, the Miller algorithm, as explained in Sect. 4.3.3.

However, several questions arise in this successful method. The recursion given in (4.4.12) is very simple, and we can find its exact solution. In more complicated recursion schemes obtained for the coefficients $c_r^{(s)}$ this information is not available. The scheme may be of large order and may have several solutions of which the asymptotic behavior is unknown. So, in general, we don't know if Clenshaw's method for differential equations computes the solution that we want, and if for the wanted solution the scheme is stable in the backward direction.

Clenshaw's method goes wrong in another simple example. Consider $y(x) = e^{ax+bx^2}$ with differential equation $y' = (a+2bx)y$. It is again easy to give a recursion scheme for the coefficients $c_r^{(s)}$. It reads (we use also (4.4.5))

$$c_r^{(1)} = ac_r + b\big(c_{r+1} + c_{|r-1|}\big), \quad r \geq 0. \tag{4.4.14}$$

The coefficient $c_r^{(1)}$ can be eliminated simply by using this relation with r replaced with $r+2$, and invoking (4.4.11). This gives

$$2(r+1)c_{r+1} = a(c_r - c_{r+2}) + b(c_{r+1} - c_{r+3}), \quad r = 0, 1, 2, \ldots. \tag{4.4.15}$$

When applying a backward recursion algorithm for computing c_r it appears that the solution does not give the requested function $y(x)$. The problem is that the recurrence is a third order difference equation, with three independent solutions that can be chosen in such a way that two of them are minimal solutions while the third one is dominant. Straightforward application of Miller's backward algorithm explained in Sect. 4.3.3 gives a linear combination of such two minimal solutions. There are modifications of the Miller algorithm, which can be used for obtaining the requested minimal solution, eliminating the contamination introduced by the unwanted minimal solution.

In [56] a similar phenomenon has been discussed for the computation of the Weber function $W(a,x)$, solution of the equation $y'' + (x^2/4 - a)y = 0$. In that paper we describe a modification of Miller's algorithm in detail. See also [77] for an instability problem in Chebyshev expansions for special functions.

4.4.2.4 Known Coefficients in Terms of Special Functions

As we have seen in (4.4.13), the coefficients in a Chebyshev expansion for the exponential function are known in terms of special functions. There are many other examples, also for higher transcendental functions. For example, we have (see [75, p. 37])

$$\begin{aligned} J_0(ax) &= \sum_{n=0}^{\infty} \epsilon_n (-1)^n J_n^2(a/2)\, T_{2n}(x), \\ J_1(ax) &= 2\sum_{n=0}^{\infty} (-1)^n J_n(a/2) J_{n+1}(a/2)\, T_{2n+1}(x), \end{aligned} \tag{4.4.16}$$

where $-1 \le x \le 1$ and $\epsilon_0 = 1$, $\epsilon_n = 2$ if $n > 0$. The parameter a can be any complex number. Similar expansions are available for J-Bessel functions of any complex order, in which the coefficients are ${}_1F_2$-hypergeometric functions, and explicit recursion relations are available for computing the coefficients. For general integer order, the coefficients are products of two J-Bessel functions, as in (4.4.16). See again [75].

Another example is the expansion for the error function,

$$e^{a^2x^2}\,\mathrm{erf}(ax) = \sqrt{\pi}\, e^{\frac{1}{2}a^2} \sum_{n=0}^{\infty} I_{n+\frac{1}{2}}\left(\frac{1}{2}a^2\right) T_{2n+1}(x), \quad -1 \le x \le 1, \tag{4.4.17}$$

in which the modified Bessel function is used. Again, a can be any complex number.

The complexity of computing the coefficients of the expansions in (4.4.16) seems to be greater than the computation of the function that has been expanded. In some sense this is true, but the coefficients in (4.4.16), and those of many other examples for special functions, satisfy linear recurrence relations, and the coefficients satisfying such relations can usually be computed very efficiently by the backward recursion algorithm; see Sect. 4.3.3.

The expansions in (4.4.16) and (4.4.17) can be viewed as expansions near the origin. Other expansions are available that can be viewed as expansions at infinity, and these may be considered as alternatives for asymptotic expansions of special functions. For example, for the confluent hypergeometric U-functions we have the convergent expansion in terms of shifted Chebyshev polynomials $T_n^*(x) = T_n(2x-1)$:

$$(\omega z)^a U(a,c,\omega z) = \sum_{n=0}^{\infty} C_n(z) T_n^*(1/\omega), \tag{4.4.18}$$

where

$$z \neq 0, \quad |\mathrm{ph}\, z| < \frac{3}{2}\pi, \quad 1 \leq \omega \leq \infty. \tag{4.4.19}$$

Furthermore, $a, 1+a-c \neq 0, -1, -2, \ldots$. When equalities hold for these values of a and c, the Kummer U-function reduces to a Laguerre polynomial. This follows from

$$U(a,c,z) = z^{1-c} U(1+a-c, 2-c, z) \tag{4.4.20}$$

and

$$U(-n, \alpha+1, z) = (-1)^n n! L_n^{\alpha}(z), \quad n = 0, 1, 2, \ldots. \tag{4.4.21}$$

The expansion (4.4.18) is given in [75, p. 25]. The coefficients can be represented in terms of generalized hypergeometric functions, in fact, Meijer G-functions, and they can be computed from the recurrence relation

$$\frac{2C_n(z)}{\epsilon_n} = 2(n+1)A_1 C_{n+1}(z) + A_2 C_{n+2}(z) + A_3 C_{n+3}(z), \tag{4.4.22}$$

where $b = a+1-c$, $\epsilon_0 = \frac{1}{2}$, $\epsilon_n = 1$ $(n \geq 1)$, and

$$\begin{aligned} A_1 &= 1 - \frac{(2n+3)(n+a+1)(n+b+1)}{2(n+2)(n+a)(n+b)} - \frac{2z}{(n+a)(n+b)}, \\ A_2 &= 1 - \frac{2(n+1)(2n+3-z)}{(n+a)(n+b)}, \\ A_3 &= -\frac{(n+1)(n+3-a)(n+3-b)}{(n+2)(n+a)(n+b)}. \end{aligned} \tag{4.4.23}$$

For applying the backward recursion algorithm it is important to know that

$$\sum_{n=0}^{\infty} (-1)^n C_n(z) = 1, \quad |\mathrm{ph}\, z| < \frac{3}{2}\pi. \tag{4.4.24}$$

This follows from

$$\lim_{\omega \to \infty} (\omega z)^a U(a,c,\omega z) = 1 \quad \text{and} \quad T_n^*(0) = (-1)^n. \tag{4.4.25}$$

The standard backward recursion scheme (see Sect. 4.3) for computing the coefficients $C_n(z)$ works only for $|\mathrm{ph}\, z| < \pi$, and for $\mathrm{ph}\, z = \pm\pi$ a modification seems to be possible; see [75, p. 26].

Although the expansion in (4.4.18) converges for all $z \neq 0$ in the indicated sector, it is better to avoid small values of the argument of the U-function. Luke gives an estimate of the coefficients $C_n(z)$ of which the dominant factor that determines the speed of convergence is given by

$$C_n(z) = \mathcal{O}\left(n^{2(2a-c-1)/3} e^{-3n^{\frac{2}{3}} z^{\frac{1}{3}}}\right), \quad n \to \infty, \tag{4.4.26}$$

and we see that large values of $\Re z^{1/3}$ improve the convergence.

The expansion in (4.4.18) can be used for all special cases of the Kummer U-function, that is, for Bessel functions (Hankel functions and K-modified Bessel function), for the incomplete gamma function $\Gamma(a,z)$, with special cases the complementary error function and exponential integrals. In [54, Sect. 3.10] numerical coefficients are derived for expansions of the Airy function $\mathrm{Ai}(x)$ for $x \geq 7$ and for its derivative by using the expansion in (4.4.18).

4.5 Quadrature Methods

We start with a simple example in which an oscillatory integral can be transformed into a stable representation. Consider the integral

$$F(\lambda) = \int_{-\infty}^{\infty} e^{-t^2+2i\lambda t}\, dt = \sqrt{\pi}\, e^{-\lambda^2}. \tag{4.5.1}$$

Taking $\lambda = 10$ we get

$$F(\lambda) \doteq 0.6593662990\,{}_{10}^{-43}. \tag{4.5.2}$$

When we ask a well-known computer algebra system to do a numerical evaluation of the integral, without using the exact answer in (4.5.1) and with standard 10 digits accuracy, we obtain

$$F(\lambda) \doteq 0.24\,{}_{10}^{-12}. \tag{4.5.3}$$

We see that in this way this simple integral, with strong oscillations, cannot be evaluated correctly. Increasing the accuracy from 10 to 50 digits we obtain the answer

$$F(\lambda) \doteq 0.65936629906\,{}_{10}^{-43}, \tag{4.5.4}$$

the first 8 digits being correct. Observe that we can shift the path of integration upwards until we reach the point $t = i\lambda$, the saddle point, and we write

$$F(\lambda) = \int_{-\infty}^{\infty} e^{-(t-i\lambda)^2-\lambda^2}\, dt = e^{-\lambda^2} \int_{-\infty}^{\infty} e^{-s^2}\, ds. \tag{4.5.5}$$

Now the saddle point is at $s = 0$, we integrate through this point along a path where no oscillations occur, a steepest descent path. Moreover the small factor $e^{-\lambda^2}$ that causes the main problems in the standard quadrature method, is now in front of the s-integral.

Similar methods can be applied to more complicated functions, in particular to a wide class of special functions from mathematical physics. Much software has been developed for many of these functions, but for large parameters the software is not at all complete and reliable, in particular when the parameters are large and complex.

We have come to the conclusion that methods based on asymptotic analysis are important for evaluating integrals by quadrature. Choosing suitable paths of integration and scaling the functions by separating dominant factors are important steps in these methods.

In this section we discuss the application of a simple quadrature rule, namely, the trapezoidal rule. For integral representations of special functions it may perform very well. Not always the standard integral representations should be taken, but modifications obtained by transformations or by choosing contours in the complex plane.

4.5.1 The Trapezoidal Rule

Gauss quadrature is a well-known quadrature method for evaluating integrals. It has a very good performance for various types of integrals over real intervals, given that the quadrature has maximal degree of exactness. However, one of the drawbacks is that it is not very flexible in algorithms when we want adjustable precision or when additional parameters are present. Also, we need zeros and weights of a certain class of orthogonal polynomials. For high precision algorithms computing these numbers in advance may be time consuming and/or not reliable.

The n-point extended trapezoidal rule

$$\int_a^b f(t)\,dt = \frac{1}{2}h[f(a) + f(b)] + h\sum_{j=1}^{n-1} f(hj) + R_n, \quad h = \frac{b-a}{n}, \tag{4.5.6}$$

is more flexible, because we don't need precomputed zeros and weights; for this rule these numbers are trivial.

The error term has the form

$$R_n = -\frac{1}{12}(b-a)h^2 f''(\xi), \tag{4.5.7}$$

for some point $\xi \in (a, b)$, and for functions with continuous second derivative.

More insight in the error term follows from the Euler-Maclaurin summation rule [54, p. 131]. This rule gives the representation (for functions f having $2m + 2$

Table 4.2 The remainder R_n of the rule in (4.5.9) for several choices of n

n	R_n
4	$-0.12_{10}-0$
8	$-0.48_{10}-6$
16	$-0.11_{10}-21$
32	$-0.13_{10}-62$
64	$-0.13_{10}-163$
128	$-0.53_{10}-404$

continuous derivatives in $[a,b]$):

$$\begin{aligned} R_n = &\sum_{j=0}^{m} \frac{B_{2j}}{(2j)!} h^{2j} \left(f^{(2j-1)}(a) - f^{(2j-1)}(b) \right) \\ &- (b-a)h^{2m+2} \frac{B_{2m+2}}{(2m+2)!} f^{(2m+2)}(\xi), \end{aligned} \tag{4.5.8}$$

for some point $\xi \in (a,b)$. B_m are the Bernoulli numbers. The first numbers with even index are $B_0 = 1$, $B_2 = \frac{1}{6}$, $B_4 = -\frac{1}{30}$.

We take as an example the Bessel function

$$\pi\, J_0(x) = \int_0^{\pi} \cos(x \sin t)\, dt = h + h \sum_{j=1}^{n-1} \cos\bigl[x \sin(h\, j)\bigr] + R_n, \tag{4.5.9}$$

where $h = \pi/n$, and use this rule for $x = 5$. The results of computations are shown in Table 4.2.

We observe that the error R_n is much smaller than the upper bound that can be obtained from (4.5.7). The explanation comes from the periodicity in the integral for the Bessel function. Hence, all terms of the sum in (4.5.8) vanish, and we infer that now the error is $\mathcal{O}(h^{2m+2})$. And because for this integral this is true for any positive m, we conclude that the error is exponentially small as a function of h.

Another example is the Bessel function integral for general order

$$J_\nu(x) = \frac{1}{2\pi i} \int_{\mathcal{C}} e^{x \sinh t - \nu t}\, dt, \tag{4.5.10}$$

where $\mathcal{C}$ starts at $\infty - i\pi$ and terminates at $\infty + i\pi$; see [111, p. 222] and [118, p. 176]. Without further specifications, on such contour oscillations will occur, but we will select a special contour that is free of oscillations for the case $0 < x \le \nu$. This contour will run through a saddle point of the integrand. In particular when the parameters are large, strong oscillations occur on a general path, and numerical quadrature will be very unstable. When we select the contour that is free of oscillations we are also able to pick up the most relevant part in the asymptotic representation of this Bessel function.

We write $\nu = x\cosh\mu$, $\mu \ge 0$. The real saddle points of $x\sinh t - \nu t = x(\sinh t - t\cosh\mu)$ occur at $t = \pm\mu$, and at these saddle points the imaginary part of $x\sinh t - \nu t$ equals zero. It is possible to select a path free of oscillations (a steepest descent path) through the saddle point at $t = \mu$ (this is not possible when we would have selected the saddle point at $t = -\mu$). This path can be described by the equation $\Im(x\sinh t - \nu t) = 0$. Writing $t = \sigma + i\tau$ we obtain for the path the equation

$$\cosh\sigma = \cosh\mu\frac{\tau}{\sin\tau}, \qquad \mu \le \sigma,\ -\pi < \tau < \pi. \tag{4.5.11}$$

On this path we have

$$\Re(x\sinh t - \nu t) = x(\sinh\sigma\ \cos\tau - \sigma\cosh\mu). \tag{4.5.12}$$

Integrating with respect to τ, using $dt/d\tau = (d\sigma/d\tau + i)$ (where $d\sigma/d\tau$ is an odd function of τ, and does not contribute), we obtain

$$J_\nu(x) = \frac{1}{2\pi}\int_{-\pi}^{\pi} e^{x(\sinh\sigma\ \cos\tau - \sigma\cosh\mu)}\,d\tau, \qquad 0 < x \le \nu. \tag{4.5.13}$$

The integrand is analytic and vanishes with all its derivatives at the points $\tau = \pm\pi$. We can interpret the integrand as being a periodic C^∞ function with period 2π, and consider representation (4.5.8) of the remainder. Again, the error in the trapezoidal rule is exponentially small.

When $\nu \gg x$ the Bessel function becomes very small and we can take the dominant part $e^{x(\sinh\mu - \mu\cosh\mu)}$ in front of the integral. When $x \ge \nu$ (the oscillatory case), the Bessel function can be represented in a similar way, now by using two integrals (coming from the Hankel functions).

Integral representations in terms of non-oscillating integrals (starting from complex contours) can be obtained for many of other special functions with large parameters. In other cases it may be difficult to obtain a suitable parametrization of the path, in which case it still may be possible to choose a path through a relevant saddle point and running into the valleys of the saddle point. In that case the oscillations will be less harmful compared with a path not going through a saddle point. For more details we refer to [54, Sect. 5.5].

Our main conclusion of this section is that the trapezoidal rule may be very efficient and accurate when dealing with a certain class of integrals. Smoothness and periodicity of the integrand are the key properties here.

4.5.1.1 The Trapezoidal Rule on $\mathbb{R}$

In the previous section we considered integrals over finite intervals. For integrals over $\mathbb{R}$ the trapezoidal rule may again be very efficient and accurate.

We consider

$$\int_{-\infty}^{\infty} f(t)\,dt = h\sum_{j=-\infty}^{\infty} f(hj + d) + R_d(h), \tag{4.5.14}$$

Table 4.3 The remainder R_0 of the rule in (4.5.19) for several choices of h

h	j_0	$R_0(h)$
1	2	$-0.18_{10}{-1}$
1/2	5	$-0.24_{10}{-6}$
1/4	12	$-0.65_{10}{-15}$
1/8	29	$-0.44_{10}{-32}$
1/16	67	$-0.19_{10}{-66}$
1/32	156	$-0.55_{10}{-136}$
1/64	355	$-0.17_{10}{-272}$

where $h > 0$ and $0 \le d < h$. We apply this rule with even functions f analytic in a strip G_a of width $2a > 0$ around $\mathbb{R}$:

$$G_a = \{z = x + iy \mid x \in \mathbb{R},\ -a < y < a\}, \tag{4.5.15}$$

which are bounded in G_a and for which $\lim_{x\to\pm\infty} f(x + iy) = 0$ (uniformly in $|y| \le a$) and

$$M_a(f) = \int_{-\infty}^{\infty} |f(x + ia)|\, dx < \infty. \tag{4.5.16}$$

Then, for real functions f, the remainder $R_d(h)$ of (4.5.14) satisfies

$$|R_d(h)| \le \frac{e^{-\pi a/h}}{\sinh(\pi a/h)} M_a(f). \tag{4.5.17}$$

The proof is based on residue calculus; see [54, p. 151].

As an example we consider the modified Bessel function

$$K_0(x) = \frac{1}{2}\int_{-\infty}^{\infty} e^{-x\cosh t}\, dt. \tag{4.5.18}$$

We have, with $d = 0$,

$$e^x K_0(x) = \frac{1}{2}h + h\sum_{j=1}^{\infty} e^{-x(\cosh(hj)-1)} + R_0(h). \tag{4.5.19}$$

For $x = 5$ and several values of h we obtain the results given in Table 4.3 (j_0 denotes the number of terms used in the series in (4.5.19)).

We see in this example that halving the value of h gives a doubling of the number of correct significant digits (and, roughly speaking, a doubling of the number of terms needed in the series). When programming this method, observe that when halving h, previous function values can be used.

4.5.2 Complex Contours

In Sect. 4.5.1 we have transformed the complex contour integral (4.5.10) for the Bessel function $J_\nu(x)$ into a more suitable integral (4.5.13) by using saddle point methods. Here we explain how this works for the Airy function for applying the trapezoidal rule on the real line of Sect. 4.5.1.1. This gives a very flexible and efficient algorithm with adjustable precision.

We consider

$$\mathrm{Ai}(z) = \frac{1}{2\pi i}\int_{\mathcal{C}} e^{\frac{1}{3}w^3 - zw}\,dw, \tag{4.5.20}$$

where $\mathrm{ph}\,z \in [0, \frac{2}{3}\pi]$ and $\mathcal{C}$ is a contour starting at $\infty e^{-i\pi/3}$ and terminating at $\infty e^{+i\pi/3}$ (in the valleys of the integrand).

Let

$$\phi(w) = \frac{1}{3}w^3 - zw. \tag{4.5.21}$$

The saddle points are $w_0 = \sqrt{z}$ and $-w_0$ and follow from solving $\phi'(w) = w^2 - z = 0$.

The saddle point contour (the path of steepest descent) that runs through the saddle point w_0 is defined by

$$\Im[\phi(w)] = \Im[\phi(w_0)]. \tag{4.5.22}$$

We write

$$z = x + iy = re^{i\theta}, \qquad w = u + iv, \qquad w_0 = u_0 + iv_0. \tag{4.5.23}$$

Then

$$u_0 = \sqrt{r}\cos\frac{1}{2}\theta, \qquad v_0 = \sqrt{r}\sin\frac{1}{2}\theta, \qquad x = u_0^2 - v_0^2, \qquad y = 2u_0v_0. \tag{4.5.24}$$

The path of steepest descent through w_0 is given by the equation

$$u = u_0 + \frac{(v - v_0)(v + 2v_0)}{3\left[u_0 + \sqrt{\frac{1}{3}(v^2 + 2v_0v + 3u_0^2)}\right]}, \qquad -\infty < v < \infty. \tag{4.5.25}$$

Examples for $r = 5$ and a few θ-values are shown in Fig. 4.1. The relevant saddle points are located on the circle with radius $\sqrt{r}$ and are indicated by small dots.

The saddle point on the positive real axis corresponds with the case $\theta = 0$ and the two saddles on the imaginary axis with the case $\theta = \pi$. This is out of the range of present interest, but it is instructive to see that the contour may split up and run through both saddle points $\pm w_0$. When $\theta = \frac{2}{3}\pi$ both saddle points are on one path, and the half-line in the z-plane corresponding with this θ is called a Stokes line (see Sect. 4.6).

Fig. 4.1 Saddle point contours for $\theta = 0, \frac{1}{3}\pi, \frac{2}{3}\pi, \pi$ and $r = 5$

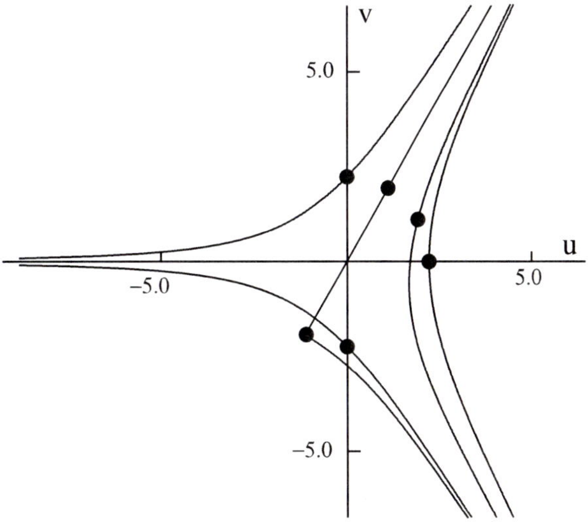

Integrating with respect to $\tau = u - u_0$ (and writing $\sigma = v - v_0$) we obtain

$$\mathrm{Ai}(z) = \frac{e^{-\zeta}}{2\pi i}\int_{-\infty}^{\infty} e^{\psi_r(\sigma,\tau)}\left(\frac{d\sigma}{d\tau} + i\right)d\tau, \tag{4.5.26}$$

where $\zeta = \frac{2}{3}z^{\frac{3}{2}}$, and

$$\sigma = \frac{\tau(\tau + 3v_0)}{3[u_0 + \sqrt{\frac{1}{3}(\tau^2 + 4v_0\tau + 3r)}]}, \qquad -\infty < \tau < \infty, \tag{4.5.27}$$

$$\psi_r(\sigma,\tau) = \Re[\phi(w) - \phi(w_0)] = u_0(\sigma^2 - \tau^2) - 2v_0\sigma\tau + \frac{1}{3}\sigma^3 - \sigma\tau^2. \tag{4.5.28}$$

Quadrature methods for evaluating complex Airy functions can be found in [39, 47, 48, 50, 94].

4.6 The Stokes Phenomenon

The Stokes phenomenon concerns the abrupt change across certain rays in the complex plane, known as *Stokes lines*, exhibited by the coefficients multiplying exponentially subdominant terms in compound asymptotic expansions. There is much recent interest in the Stokes phenomenon, and it fits in the present paper because it has to do with sudden changes in approximations when a certain parameter (in this case the phase of the large parameter) passes critical values.

4.6.1 The Airy Function

First we explain this phenomenon by using a simple example from differential equations. Consider Airy's equation

$$\frac{d^2 w}{dz^2} = z\, w, \tag{4.6.1}$$

the solutions of which are entire functions. When $|z|$ is large the solutions of (4.6.1) are approximated by linear combinations of

$$w_\pm = z^{-\frac{1}{4}} e^{\pm\xi}, \quad \xi = \frac{2}{3} z^{3/2}. \tag{4.6.2}$$

Obviously, $w_\pm$ are multivalued functions of the complex variable z with a branch point at $z = 0$. Therefore, as we go once around the origin, the solutions of (4.6.1) will return to their original values, but $w_\pm$ will not. It follows that the constants $c_\pm$ in the linear combination

$$w(z) \sim c_- w_-(z) + c_+ w_+(z), \quad z \to \infty, \tag{4.6.3}$$

are domain-dependent. The constants change when we cross certain lines, the boundaries of certain sectors in the z-plane.

In the above example one of the terms e^{ξ}, $e^{-\xi}$ maximally dominates the other one at the rays $\operatorname{ph} z = 0$, $\operatorname{ph} z = \pm 2\pi/3$. In this example these 3 rays are the Stokes lines. At the rays $\operatorname{ph} z = \pm\frac{1}{3}\pi$ and the negative z-axis the quantity ξ is purely imaginary, and, hence, the terms e^{ξ}, $e^{-\xi}$ are equal in magnitude. These three rays are called the *anti-Stokes lines*[4].

For the Airy function $\operatorname{Ai}(z)$ we have the full asymptotic expansion (see [1, Chap. 10])

$$\operatorname{Ai}(z) \sim c_- z^{-\frac{1}{4}} e^{-\xi} \sum_{n=0}^{\infty} (-1)^n c_n \xi^{-n}, \quad c_- = \frac{1}{2} \pi^{-\frac{1}{2}}, \ |\operatorname{ph} z| < \pi, \tag{4.6.4}$$

with coefficients

$$c_n = \frac{\Gamma(3n + \frac{1}{2})}{54^n\, n!\, \Gamma(n + \frac{1}{2})}, \quad n = 0, 1, 2, \ldots. \tag{4.6.5}$$

On the other hand, in another sector of the z-plane, we have

$$\operatorname{Ai}(-z) \sim c_-\, z^{-\frac{1}{4}} \left[e^{-\xi} \sum_{n=0}^{\infty} (-1)^n c_n \xi^{-n} + i e^{\xi} \sum_{n=0}^{\infty} c_n \xi^{-n} \right], \tag{4.6.6}$$

[4]This terminology is not the same in all branches of applied mathematics and mathematical physics: sometimes one sees a complete interchange of the names 'Stokes line' and 'anti-Stokes line'.

in which exactly the same term (with the same constant c_-) is involved as in (4.6.4), and there is another term corresponding with w_+. We can rewrite this in a more familiar expansion

$$\mathrm{Ai}(-z) \sim \pi^{-\frac{1}{2}} z^{-\frac{1}{4}} \left(\sin\left(\xi + \frac{1}{4}\pi\right) \sum_{n=0}^{\infty} (-1)^n \frac{c_{2n}}{\xi^{2n}} \right.$$
$$\left. - \cos\left(\xi + \frac{1}{4}\pi\right) \sum_{n=0}^{\infty} (-1)^n \frac{c_{2n+1}}{\xi^{2n+1}} \right), \tag{4.6.7}$$

valid in the sector $|\mathrm{ph}\, z| < \frac{2}{3}\pi$. In the overlapping domain of expansions (4.6.4) and (4.6.7), that is, when $\frac{1}{3}\pi < |\mathrm{ph}\, z| < \pi$, the term with w_+ is asymptotically small compared with w_-, and it suddenly appears in the asymptotic approximation when we cross with increasing values of $|\mathrm{ph}\, z|$ the Stokes lines at $\mathrm{ph}\, z = \pm\frac{2}{3}\pi$. It seems that, when going from (4.6.4) to (4.6.6), the constant multiplying w_+ changes discontinuously from zero values (when $|\mathrm{ph}\, z| < \frac{2}{3}\pi$) to a non-zero value when we cross the Stokes line. This sudden appearance of the term w_+ does not have much influence on the asymptotic behavior near the Stokes lines at $|\mathrm{ph}\, z| = \frac{2}{3}\pi$, because w_+ is dominated maximally by w_- at these rays. However, see Sect. 4.6.3 below. Observe also the special value $\theta = \frac{2}{3}\pi$ in Sect. 4.5.2.

4.6.2 The Recent Interest in the Stokes Phenomenon

This phenomenon of the *discontinuity of the constants* was discovered by Stokes and was discussed by him in a series of papers (on Airy functions in 1857, on Bessel functions in 1868). It is a phenomenon which is not confined to Airy or Bessel functions. The discovery by Stokes was, as Watson says, apparently one of those which are made at three o'clock in the morning. Stokes wrote in a 1902 retrospective paper: "The inferior term enters as it were into a mist, is hidden for a little from view, and comes out with its coefficients changed."

In 1989 the mathematical physicist Michael Berry provided a deeper explanation. He suggested that the coefficients of the subdominant expansion should be regarded not as a discontinuous constant but, for fixed $|z|$, as a continuous function of $\mathrm{ph}\, z$. Berry's innovative and insightful approach was followed by a series of papers by himself and other writers. In particular, Olver put the formal approach by Berry on a rigorous footing in papers with applications to confluent hypergeometric functions (including Airy functions, Bessel functions, and Weber parabolic functions).

At the same time interest arose in earlier work by Stieltjes [106], Airey [2], Dingle [32], and others, to expand remainders of asymptotic expansions at optimal values of the summation variable. This resulted in *exponentially-improved* asymptotic expansions, a method of improving asymptotic approximations by including small terms in the expansion that are in fact negligible compared with other terms in the expansion.

4.6.3 Exponentially Small Terms in the Airy Expansions

We conclude this discussion by pointing out the relation between the Stokes phenomenon and the exponentially small terms in the asymptotic expansion of the Airy function. Consider the terms in the expansions in (4.6.4)–(4.6.7). They have the asymptotic form

$$c_n\xi^{-n} = \mathcal{O}\big[\Gamma(n)\,(2\xi)^{-n}\big], \quad n\to\infty. \tag{4.6.8}$$

When z is large the terms decrease at first and then increase. The least term of the first series of (4.6.6) is near $n = n^* = [|2\xi|]$ and its size is of order $e^{-2|\xi|}$. At the Stokes lines at $|\mathrm{ph}\,z| = \frac{2}{3}\pi$ the quantity ξ is negative and the exponential term in front of the first series in (4.6.6) equals $e^{|\xi|}$. Hence the order of magnitude of $e^{-\xi}c_{n^*}\xi^{-n^*}$ is roughly of the same size as the second part in (4.6.7), that is, of the size of e^{ξ} that is present in front of the second series. It follows that near the Stokes lines (and of course when z turns to the negative axis) the second series in (4.6.7) is not at all negligible when we truncate the first series at the least term with index n^*.

At present we know, after Berry's observations, that near the Stokes lines one of the constants $c_\pm$ in the asymptotic representation in (4.6.2) in fact is a rapidly changing function of z. In the case of (4.6.6) we can write

$$\mathrm{Ai}(z) \sim c_-\,z^{-\frac14}\left[e^{-\xi}\sum_{n=0}^{\infty}(-1)^n c_n\xi^{-n} + i\,S(z)\,e^{\xi}\sum_{n=0}^{\infty}c_n\xi^{-n}\right], \tag{4.6.9}$$

where $S(z)$ switches rapidly but smoothly from 0 to 1 across the Stokes line at $\mathrm{ph}\,z = \frac{2}{3}\pi$. A good approximation to $S(z)$ involves the error function, which function can describe the fast transition in this asymptotic problem.

Many writers have contributed recently in this field, both for the Stokes phenomenon of integrals and that of differential equations. For more details see the survey paper [87].

4.7 A Selection of Other Methods

Many other methods are available for computing special functions. In this section we mention a selection. For all these topics more details can be found in [54].

4.7.1 Continued Fractions

For many elementary and special functions representations as continued fractions exist. We give examples for incomplete gamma functions and incomplete beta functions that are useful for numerical computations.

We introduce the following notation. Let $\{a_n\}_{n=1}^{\infty}$ and $\{b_n\}_{n=0}^{\infty}$ be two sequences of real or complex numbers. With these numbers we construct a continued fraction of the form

$$b_0 + \cfrac{a_1}{b_1 + \cfrac{a_2}{b_2 + \cfrac{a_3}{b_3 + \cfrac{a_4}{b_4 + \ddots}}}} \tag{4.7.1}$$

A more convenient notation is

$$b_0 + \frac{a_1}{b_1+}\,\frac{a_2}{b_2+}\,\frac{a_3}{b_3+}\,\frac{a_4}{b_4+\cdots}. \tag{4.7.2}$$

For convergence aspects, contractions, even and odd parts, equivalence transformations, and so on, we refer to the literature; see [54, Chap. 6] and a recent handbook with many details for special functions [29].

To evaluate the finite part (also called the convergent)

$$C_n = b_0 + \frac{a_1}{b_1+}\,\frac{a_2}{b_2+}\,\frac{a_3}{b_3+}\cdots\frac{a_n}{b_n}, \tag{4.7.3}$$

we can use recursion. Let

$$A_{-1} = 1, \qquad A_0 = b_0, \qquad B_{-1} = 0, \qquad B_0 = 1. \tag{4.7.4}$$

We compute A_n and B_n by using the following recursion

$$A_n = b_n A_{n-1} + a_n A_{n-2}, \qquad B_n = b_n B_{n-1} + a_n B_{n-2}, \quad n \ge 1. \tag{4.7.5}$$

Then $C_n = A_n/B_n$. Several other algorithms are available; see [54, Sect. 6.6]. The recursion for A_n and B_n may produce large numbers of these quantities, causing overflow. However, because only the ratio A_n/B_n is needed to compute the convergent C_n, scaling can be used to keep control.

4.7.1.1 Incomplete Gamma Functions

The incomplete gamma functions are defined by

$$\gamma(a,z) = \int_0^z t^{a-1}e^{-t}\,dt, \qquad \Gamma(a,z) = \int_z^{\infty} t^{a-1}e^{-t}\,dt, \tag{4.7.6}$$

where for the first form we require $\Re a > 0$ and for the second one $|\mathrm{ph}\, z| < \pi$.

We have

$$z^{-a}e^{z}\gamma(a,z) = b_0 + \frac{a_1}{b_1+}\,\frac{a_2}{b_2+}\,\frac{a_3}{b_3+}\,\frac{a_4}{b_4+\cdots}, \tag{4.7.7}$$

where z and a are complex, $a \neq 0, -1, -2, \ldots$, and

$$b_0 = \frac{1}{a-z}, \qquad a_m = mz, \qquad b_m = a + m - z, \quad m \geq 1. \tag{4.7.8}$$

This fraction corresponds to the power series

$$az^{-a}e^z\gamma(a,z) = \sum_{k=0}^{\infty} \frac{z^k}{(1+a)_k}. \tag{4.7.9}$$

For $\Gamma(a,z)$ we have

$$(z+1-a)z^{-a}e^z\Gamma(a,z) = \frac{1}{1+}\,\frac{\alpha_1}{1+}\,\frac{\alpha_2}{1+}\,\frac{\alpha_3}{1+\cdots}, \tag{4.7.10}$$

where

$$\alpha_n = \frac{n(a-n)}{(z+2n-1-a)(z+2n+1-a)}, \qquad n = 1, 2, 3, \ldots. \tag{4.7.11}$$

This form is used in [38] for computing the function $\Gamma(a,x)$, for $x > 1.5$ and $-\infty < a < \alpha^*(z)$, where $\alpha^*(x) \sim x$ for large x. The fraction in (4.7.10) is convergent for all $z \neq 0$ in the sector $|\mathrm{ph}\, z| < \pi$, and for computations it is an excellent alternative for the corresponding asymptotic expansion

$$z^{1-a}e^z\Gamma(a,z) \sim \sum_{k=0}^{\infty} (-1)^k \frac{(1-a)_k}{z^k}, \tag{4.7.12}$$

valid for $a \in \mathbb{C}$, $z \to \infty$, $|\mathrm{ph}\, z| < \frac{3}{2}\pi$.

4.7.1.2 Incomplete Beta Function

This function is defined by

$$B_x(p,q) = \int_0^x t^{p-1}(1-t)^{q-1}\,dt, \quad \Re p > 0,\ \Re q > 0, \tag{4.7.13}$$

and usually $0 \leq x \leq 1$; when $x < 1$ the condition on q can be omitted. The beta integral is obtained when we take $x = 1$, that is,

$$B(p,q) = \int_0^1 t^{p-1}(1-t)^{q-1}\,dt = \frac{\Gamma(p)\Gamma(q)}{\Gamma(p+q)}, \quad \Re p > 0,\ \Re q > 0. \tag{4.7.14}$$

We have the continued fraction

$$px^{-p}(1-x)^{-q}B_x(p,q) = \frac{1}{1+}\,\frac{d_1}{1+}\,\frac{d_2}{1+}\,\frac{d_3}{1+\cdots}, \tag{4.7.15}$$

where, for $n = 0, 1, 2, \ldots,$

$$
\begin{aligned}
d_{2n+1} &= -\frac{(p+n)(p+q+n)}{(p+2n)(p+2n+1)}x, \\
d_{2n+2} &= \frac{(n+1)(q-n-1)}{(p+2n+1)(p+2n+2)}x.
\end{aligned}
\tag{4.7.16}
$$

When $p > 1$ and $q > 1$, the maximum of the integrand in (4.7.13) occurs at $x_0 = (p-1)/(p+q-2)$, and the best numerical results are obtained when $x \le x_0$. When $x_0 < x \le 1$, we use the reflection relation with the beta integral (see (4.7.14))

$$B_x(p,q) = B(p,q) - B_{1-x}(q,p). \tag{4.7.17}$$

From a numerical point of view the continued fraction (4.7.15) has an interesting property of the convergents: C_{4n} and C_{4n+1} are less than this value of the continued fraction and C_{4n+2}, C_{4n+3} are greater than this value. This gives excellent control of the convergence of an algorithm that uses (4.7.15).

4.7.2 Sequence Transformations

When applying numerical techniques to physical problems, results are usually produced in the form of sequences. Examples are iterative methods, discretization methods, perturbation methods, and—most important in the context of special functions—series expansions. Often, the sequences that are produced in this way converge too slowly to be numerically useful. When dealing with asymptotic series, summation of the sequences may also be difficult.

Sequence transformations are tools to overcome convergence problems of that kind. A slowly convergent (or even divergent in the asymptotic sense) sequence $\{s_n\}_{n=0}^{\infty}$, whose elements may be the partial sums

$$s_n = \sum_{k=0}^{n} a_k \tag{4.7.18}$$

of a convergent or formal infinite series, is converted into a new sequence $\{s'_n\}_{n=0}^{\infty}$ with hopefully better numerical properties.

We discuss sequence transformations that are useful in the context of special functions. For many special functions convergent and divergent (asymptotic) power series are available. Consequently, the emphasis in this section will be on sequence transformations that are able either to accelerate the convergence of slowly convergent power series effectively or to sum divergent asymptotic series.

4.7.2.1 Padé Approximations

From a numerical point of view, the Padé approximants are important for computing functions outside the disk of convergence of the power series of the function, as well as inside the disk (for example, near the boundary of the disk). The Padé method can also be successfully applied for locating zeros and poles of the function.

Consider the power series

$$f(z) = c_0 + c_1 z + c_2 z^2 + \cdots, \tag{4.7.19}$$

with $c_0 \neq 0$. This series may be convergent or just a formal power series. We introduce a rational function $N_m^n(z)/D_m^n(z)$, where $N_m^n(z)$ and $D_m^n(z)$ are polynomials of maximal degree n and m, respectively. That is,

$$N_m^n(z) = a_0 + a_1 z + \cdots + a_n z^n, \qquad D_m^n(z) = b_0 + b_1 z + \cdots + b_m z^m. \tag{4.7.20}$$

We choose these polynomials such that the power series expansion of $N_m^n(z) - f(z)D_m^n(z)$ starts with a term $A_{n,m}z^{n+m+1}$. The ratio $N_m^n(z)/D_m^n(z)$, of which the polynomials $N_m^n(z)$ and $D_m^n(z)$ satisfy the conditions

$$\begin{aligned} &\text{degree } N_m^n(z) \leq n, \qquad \text{degree } D_m^n(z) \leq m, \\ &N_m^n(z) - f(z)D_m^n(z) = A_{n,m}z^{n+m+1} + \cdots, \end{aligned} \tag{4.7.21}$$

is called a Padé approximant of type (n, m) to the power series (4.7.19) (the function f). The ratio $N_m^n(z)/D_m^n(z)$ is denoted by $[n/m]_f$.

For each pair (n, m) at least one rational function exists that satisfies the conditions in (4.7.21), and this function can be found by solving the equations

$$\begin{cases} a_0 = c_0 b_0, \\ a_1 = c_1 b_0 + c_0 b_1, \\ \quad\vdots \\ a_n = c_n b_0 + c_{n-1} b_1 + \cdots + c_{n-m} b_m, \end{cases} \qquad \begin{cases} 0 = c_{n+1} b_0 + \cdots + c_{n-m+1} b_m, \\ \quad\vdots \\ 0 = c_{n+m} b_0 + \cdots + c_n b_m, \end{cases} \tag{4.7.22}$$

where $c_j = 0$ if $j < 0$. When $m = 0$ the system of equations at the right is empty. In this case $a_j = c_j (j = 0, 1, \ldots, n)$ and $b_0 = 1$, and the partial sums of (4.7.19) yield the Padé approximants of type $(n, 0)$. In general, first the set at the right-hand side of (4.7.22) is solved (a homogeneous set of m equations for the $m + 1$ values b_j), which has at least one nontrivial solution. We take a normalization, for example, by taking $b_0 = 1$ (see also the discussion in [8, p. 18]), and with this choice the last m equations give $b_1, \ldots, b_m$ as the solution of a system of m linear equations. The set on the left-hand side in (4.7.22) then yields $a_0, \ldots, a_n$.

The array of Padé approximants

$$\begin{array}{cccc} [0/0]_f & [0/1]_f & [0/2]_f & \cdots \\ [1/0]_f & [1/1]_f & [1/2]_f & \cdots \\ [2/0]_f & [2/1]_f & [2/2]_f & \cdots \\ \vdots & \vdots & \vdots & \ddots \end{array} \tag{4.7.23}$$

is called a Padé table. It is arranged here so that approximants with the same denominator degree are located in the same column. As remarked earlier, the first column corresponds to the partial sums of the power series in (4.7.19). The elements of the first row correspond to the partial sums of the power series of $1/f$.

In the literature special attention is paid to the diagonal elements $[n, n]_f$ of the table, with applications to orthogonal polynomials, quadrature formulas, moment problems, and other problems of classical analysis.

In applied mathematics and in theoretical physics, Padé approximants have become a useful tool for overcoming convergence problems with power series. The popularity of Padé approximants in theoretical physics is due to Baker [6], who also wrote a monograph on Padé approximants [7]. Of interest also is the monograph by Baker and Graves-Morris [8].

An extended bibliography on Padé approximants and related matters containing several thousand references was published by Brezinski in [14]. For an annotated bibliography focusing on computational aspects, see [126]. Luke gives many rational approximations of special functions, and usually these are Padé approximants; see [75, 76].

4.7.2.2 How to Compute the Padé Approximants

The approximants can be computed by Wynn's cross rule. Any five Padé approximants arranged in the Padé table as

$$\begin{array}{ccc} & N & \\ W & C & E \\ & S & \end{array}$$

satisfy Wynn's cross rule (see [128])

$$(N-C)^{-1} + (S-C)^{-1} = (W-C)^{-1} + (E-C)^{-1}. \tag{4.7.24}$$

Starting with the first column $[n/0]_f$, $n = 0, 1, 2, \ldots$, initializing the preceding column by $[n/-1]_f = \infty$, $n = 1, 2, \ldots$, (4.7.24) enables us to compute the lower triangular part of the table. Likewise, the upper triangular part follows from the first row $[0/n]_f$, $n = 0, 1, 2, \ldots$, by initializing $[-1/n]_f = 0$, $n = 1, 2, \ldots$.

The elements of the Padé table can also be computed by the epsilon algorithm of Wynn [127]. We consider the recursions

$$
\begin{aligned}
&\varepsilon_{-1}^{(n)} = 0, \qquad \varepsilon_{0}^{(n)} = s_n, \quad n = 0, 1, 2, \ldots, \\
&\varepsilon_{m+1}^{(n)} = \varepsilon_{m-1}^{(n+1)} + \frac{1}{\varepsilon_m^{(n+1)} - \varepsilon_m^{(n)}}, \qquad n, m = 0, 1, 2, \ldots.
\end{aligned}
\tag{4.7.25}
$$

If s_n is the nth partial sum of a power series f, then $\varepsilon_{2k}^{(n)}$ is the Padé approximant $[n+k/k]_f$ (cf. (4.7.23)). The elements $\varepsilon_{2k+1}^{(n)}$ are only auxiliary quantities which diverge if the whole transformation process converges and shouldn't be used for convergence tests or output. A recent review of the applications of the epsilon algorithm can be found in [63].

In applications one usually concentrates on obtaining diagonal elements $[n/n]_f$ and elements not far away from the diagonal; see [121], which also has an efficient modified algorithm for these elements.

4.7.2.3 Nonlinear Sequence Transformations

We discuss a few other sequence transformations that, in the case of power series, produce different rational approximants, and they can also be applied to other convergence acceleration problems.

Details on the history of sequence transformations and related topics, starting from the 17th century, can be found in [15]; see also [16]. For review papers, with many references to monographs devoted to this topic, we refer the reader to [65, 121]. See also Appendix A in [12], written by Dirk Laurie, with interesting observations and opinions about sequence transformations.

First we mention Levin's sequence transformation [71], which is defined by

$$
\mathcal{L}_k^{(n)}(s_n, \omega_n) = \frac{\sum_{j=0}^{k} (-1)^j \binom{k}{j} \frac{(n+j+1)^{k-1}}{(\zeta+n+k)^{k-1}} \frac{s_{n+j}}{\omega_{n+j}}}{\sum_{j=0}^{k} (-1)^j \binom{k}{j} \frac{n+j+1)^{k-1}}{(\zeta+n+k)^{k-1}} \frac{1}{\omega_{n+j}}},
\tag{4.7.26}
$$

where s_n are the partial sums of (4.7.18) and the quantities ω_n are remainder estimates. For example, we can simply take $\zeta = 1$ and

$$
\omega_n = s_{n+1} - s_n = a_{n+1},
\tag{4.7.27}
$$

but more explicit remainder estimates can be used.

Another transformation is due to Weniger [121], who replaced the powers $(n+j+1)^{k-1}$ in Levin's transformation by Pochhammer symbols $(n+j+1)_{k-1}$. That is, Weniger's transformation reads

$$
\mathcal{S}_k^{(n)}(s_n, \omega_n) = \frac{\sum_{j=0}^{k} (-1)^j \binom{k}{j} \frac{(\zeta+n+j)_{k-1}}{(\zeta+n+k)_{k-1}} \frac{s_{n+j}}{\omega_{n+j}}}{\sum_{j=0}^{k} (-1)^j \binom{k}{j} \frac{(\zeta+n+j)_{k-1}}{(\zeta+n+k)_{k-1}} \frac{1}{\omega_{n+j}}}.
\tag{4.7.28}
$$

Other sequence transformations can be found in [121] or in [17, Sect. 2.7]. The sequence transformations (4.7.26) and (4.7.28) differ from other sequence transformations because not only the elements of a sequence $\{s_n\}$ are required, but also explicit remainder estimates $\{\omega_n\}$. For special functions this information is usually available when divergent asymptotic expansions are considered. It was shown in several articles that the transformation (4.7.28) is apparently very effective, in particular if divergent asymptotic series are to be summed; see [10, 123, 125].

For transforming partial sums $f_n(z) = \sum_{k=0}^{n} \gamma_k z^k$ of a formal power series

$$f(z) = \sum_{k=0}^{\infty} \gamma_k z^k, \tag{4.7.29}$$

we can take the remainder estimates

$$\omega_n = \gamma_{n+1} z^{n+1}, \tag{4.7.30}$$

and we replace z by $1/z$ in the case of an asymptotic series.

With these modifications the transformations (4.7.26) and (4.7.28) become rational functions of the variable z. If the coefficients γ_n in (4.7.29) are all different from zero, these rational functions satisfy the asymptotic error estimates [124, Eqs. (4.28)–(4.29)]

$$\begin{aligned} f(z) - \mathcal{L}_k^{(n)}(f_n(z), \gamma_{n+1} z^{n+1}) &= \mathcal{O}(z^{k+n+2}), \quad z \to 0, \\ f(z) - \mathcal{S}_k^{(n)}(f_n(z), \gamma_{n+1} z^{n+1}) &= \mathcal{O}(z^{k+n+2}), \quad z \to 0. \end{aligned} \tag{4.7.31}$$

These estimates imply that all terms of the formal power series, which were used for construction of the rational approximants in this way, are reproduced exactly by a Taylor expansion around $z = 0$. Thus, the transformations $\mathcal{L}_k^{(n)}(f_n(z), \gamma_{n+1} z^{n+1})$ and $\mathcal{S}_k^{(n)}(f_n(z), \gamma_{n+1} z^{n+1})$ are formally very similar to the analogous estimate (4.7.21) satisfied by the Padé approximants $[n/m]_f(z) = N_m^n(z)/D_m^n(z)$.

4.7.2.4 Numerical Examples

Simple test problems, which nevertheless demonstrate convincingly the power of sequence transformations using explicit remainder estimates, are the integrals

$$\mathbf{E}^{(\nu)}(z) = \int_0^\infty \frac{e^{-t}\, dt}{1 + z t^\nu} \tag{4.7.32}$$

and their associated divergent asymptotic expansions

$$\mathbf{E}^{(\nu)}(z) \sim \sum_{k=0}^{\infty} (\nu k)! (-z)^k, \quad z \to 0. \tag{4.7.33}$$

For $\nu = 1$, $\mathbf{E}^{(\nu)}(z)$ is the exponential integral E_1 with argument $1/z$ according to $\mathbf{E}^{(1)}(z) = \mathrm{e}^{1/z} E_1(1/z)/z$. For $\nu = 2$ or $\nu = 3$, $\mathbf{E}^{(\nu)}(z)$ cannot be expressed in terms of known special functions.

In order to demonstrate the use of sequence transformations with explicit remainder estimates, both $\mathcal{S}_k^{(n)}(f_n(z), \gamma_{n+1} z^{n+1})$ and Padé approximants are applied to the partial sums

$$E_n^{(\nu)}(z) = \sum_{k=0}^{n} (\nu k)!(-z)^k, \quad 0 \le n \le 50, \tag{4.7.34}$$

of the asymptotic series (4.7.33) for $\nu = 1, 2, 3$. The Padé approximants were computed with the help of Wynn's epsilon algorithm (see Sect. 4.7.2.2). All calculations were done in Maple, and the integrals $\mathbf{E}^{(\nu)}(z)$ were computed to the desired precision with the help of numerical quadrature. For the remainder estimates we took $\omega_n = (\nu(n+1))!(-z)^{n+1}$.

The results for $\mathbf{E}^{(1)}(z)$ with $z = 1$ are

$$\begin{aligned} \mathbf{E}^{(1)}(1) &= 0.59634\,73623\,23194\,07434\,10785, \\ \mathcal{L}_0^{(50)}(E_0^{(1)}(1), \omega_{50}) &= 0.59634\,73623\,23194\,07434\,10759, \\ \mathcal{S}_0^{(50)}(E_0^{(1)}(1), \omega_{50}) &= 0.59634\,73623\,23194\,07434\,10785, \\ [25/24] &= 0.59634\,7322, \\ [25/25] &= 0.59634\,7387. \end{aligned} \tag{4.7.35}$$

The Padé approximants are not very efficient. Nevertheless, it seems that they are able to sum the divergent series (4.7.33) for $\nu = 1$.

The results for $\mathbf{E}^{(2)}(z)$ with $z = 1/10$ are

$$\begin{aligned} \mathbf{E}^{(2)}(1/10) &= 0.88425\,13061\,26979, \\ \mathcal{L}_0^{(50)}(E_0^{(2)}(1/10), \omega_{50}) &= 0.88425\,13061\,26980, \\ \mathcal{S}_0^{(50)}(E_0^{(2)}(1/10), \omega_{50}) &= 0.88425\,13061\,26985, \\ [25/24] &= 0.88409, \\ [25/25] &= 0.88437. \end{aligned} \tag{4.7.36}$$

Here, the Padé approximants are certainly not very useful since they can only extract an accuracy of three places.

The results for $\mathbf{E}^{(3)}(z)$ with $z = 1/100$ are

$$
\begin{aligned}
\mathbf{E}^{(3)}(1/100) &= 0.96206\,71061,\\
\mathcal{S}_0^{(50)}(E_0^{(3)}(1/100), \omega_{50}) &= 0.96206\,71055,\\
\mathcal{L}_0^{(50)}(E_0^{(3)}(1/100), \omega_{50}) &= 0.96206\,71057,\\
[25/24] &= 0.960,\\
[25/25] &= 0.964.
\end{aligned} \tag{4.7.37}
$$

In [62] it is shown that an asymptotic series, whose coefficients grow more rapidly than $(2n)!$, is not Padé summable since subsequences $[n + j/n]$ in the Padé table converge to different, j-dependent limits as $n \to \infty$. The Levin and Weniger transformations are apparently able to sum the asymptotic series (4.7.33) even for $\nu = 3$.

Other numerical examples of sequence transformations using explicit remainder estimates can be found in [10, 122, 123]. For a recent paper on transformations of hypergeometric series, see [22].

4.7.3 Other Quadrature Methods

In Sect. 4.5 we have described the use of the trapezoidal rule in the evaluation of special functions. In [54, Sects. 5.3, 9.6] several other methods are discussed, with as the main one Gaussian quadrature, and the relation with orthogonal polynomials.

Other methods are Romberg quadrature, which provides a scheme for computing successively refined rules with a higher degree of exactness. Fejér and Clenshaw–Curtis quadratures are interpolatory rules which behave quite similarly to Gauss–Legendre rules, but which are easier to compute and provide nested rules. Other nested rules, related to Gauss quadrature but harder to compute than the Clenshaw–Curtis rule, are Kronrod and Patterson quadratures. Specific methods for oscillatory integrands are also described, with special attention to Filon's method.

4.7.4 Numerical Inversion of Laplace Transforms

We consider the pair of Laplace transforms

$$
F(s) = \int_0^\infty e^{-st} f(t)\,dt, \qquad f(t) = \frac{1}{2\pi i}\int_{c-i\infty}^{c+i\infty} e^{st} F(s)\,ds, \tag{4.7.38}
$$

where f should be absolutely integrable on any finite interval $[0, a]$ and the number c is chosen such that all singularities of $F(s)$ are at the left of the vertical line $\Re s = c$.

The inversion problem is to find $f(t)$ when $F(s)$ is given. To solve this problem numerically, an essential condition is whether function values of $F(s)$ are only available for real s or for complex values of s. The first case is quite difficult and requires completely different techniques compared with those for the second case. In this section we consider a method for the case that $F(s)$ is available as an analytic function in part of the complex s-plane. We describe a method based on the deformation of the contour of integration.

We give an example in which an optimal representation of the integral is obtained by deforming the contour of integration and by using a proper value of c in the complex integral in (4.7.38). After selecting this new contour, the trapezoidal rule can be used for numerical quadrature. As explained in Sect. 4.5 this method may be very efficient for evaluating a class of integrals with analytic integrands.

We use the Laplace transform pair (see [1, Eq. (29.3.83)])

$$F(s) = \frac{1}{s}e^{-k\sqrt{s}} = \int_0^\infty e^{-st}\,\mathrm{erfc}\,\frac{k}{2\sqrt{t}}\,dt,$$

$$\mathrm{erfc}\,\frac{k}{2\sqrt{t}} = \frac{1}{2\pi i}\int_{c-i\infty}^{c+i\infty} e^{st-k\sqrt{s}}\,\frac{ds}{s}, \tag{4.7.39}$$

where in this case $c > 0$. We take $k = 2\lambda$ and $t = 1$, which gives

$$\mathrm{erfc}\,\lambda = \frac{1}{2\pi i}\int_{c-i\infty}^{c+i\infty} e^{s-2\lambda\sqrt{s}}\,\frac{ds}{s}, \tag{4.7.40}$$

and we assume that $\lambda > 0$. When λ is large the integral becomes exponentially small, and straightforward application of a quadrature rule is useless.

With the transformation $s = \lambda^2 t$, (4.7.40) becomes

$$\mathrm{erfc}\,\lambda = \frac{1}{2\pi i}\int_{c-i\infty}^{c+i\infty} e^{\lambda^2(t-2\sqrt{t})}\,\frac{dt}{t}. \tag{4.7.41}$$

When we take $c = 1$ the path runs through the saddle point at $t = 1$, where the exponential function of the integrand has the value $e^{-\lambda^2}$, which corresponds to the main term in the asymptotic estimate

$$\mathrm{erfc}\,\lambda \sim \frac{e^{-\lambda^2}}{\sqrt{\pi}\lambda}, \qquad \lambda \to \infty. \tag{4.7.42}$$

Because the convergence at $\pm i\infty$ along the vertical through $t = 1$ is rather poor, the next step is to deform the contour into a new contour that terminates in the left half-plane, with $\Re t \to -\infty$.

In fact many contours are suitable, but there is only one contour through $t = 1$ on which no oscillations occur. That contour, the steepest descent path, is given by $\Im(t - 2\sqrt{t}) = 0$, or in polar coordinates $t = re^{i\theta}$ we have $r = \sec^2(\frac{1}{2}\theta)$. See Fig. 4.2.

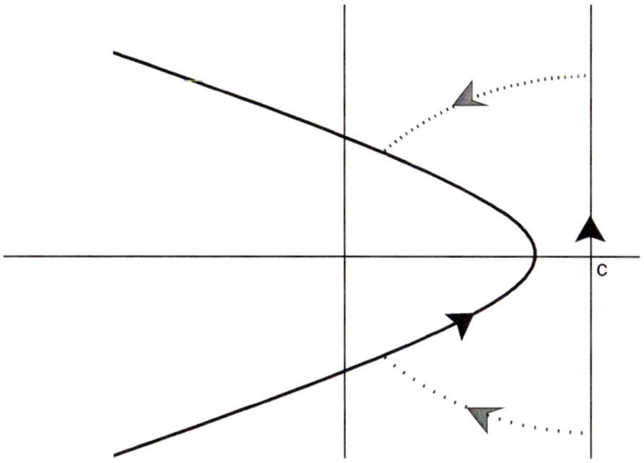

Fig. 4.2 The new contour of integration for (4.7.41) has the shape of a parabola

Table 4.4 Composite trapezoidal rule for the integral in (4.7.43) with $\lambda = 10$

h	$\operatorname{erfc}\lambda$	n
0.25	$0.20949\,49432\,96679_{10}{-44}$	5
0.20	$0.20886\,11645\,34559_{10}{-44}$	6
0.15	$0.20884\,87588\,72946_{10}{-44}$	8
0.10	$0.20884\,87583\,76254_{10}{-44}$	11

This gives, by integrating with respect to $\theta \in [-\pi, \pi]$,

$$\operatorname{erfc}\lambda = \frac{e^{-\lambda^2}}{2\pi}\int_{-\pi}^{\pi} e^{-\lambda^2 \tan^2(\frac{1}{2}\theta)}\,d\theta. \tag{4.7.43}$$

As discussed in Sect. 4.5.1 the trapezoidal rule is exceptionally accurate in this case.

Table 4.4 gives the results of applying the composite trapezoidal rule with step size h; n indicates the number of function values in the rule that are larger than 10^{-15} (we exploit the fact that the integrand is even). All digits shown in the approximation in the final row are correct.

When $F(s)$ in (4.7.38) has singularities or poles, a straightforward and optimal choice of the path of integration, as in the above example, might not be easy to find. In these cases, or when less information is available on the function $F(s)$, a less optimal contour may be chosen.

For example, we can take a parabola or a hyperbola that terminates in the left half-plane at $-\infty$. When we write $s = u + iv$, and consider the parabola defined by $u = p - qv^2$, and integrate with respect to v. When we choose p and q properly all singularities of $F(s)$ may remain inside the contour (unless $F(s)$ has an infinite number of singularities up to $\pm i\infty$).

Further details can be found in [84, 91, 107]. Recent investigations are discussed in [92, 114, 119, 120].

4.7.5 Computing Zeros of Special Functions

The zeros of special functions appear in a great number of applications in mathematics, physics, and engineering, from the computation of Gauss quadrature rules [60] in the case of orthogonal polynomials to many applications in which boundary value problems for second order ordinary differential equations arise.

In some sense, the computation of the zeros of special functions has nothing special: to compute the roots of an equation $y(x) = 0$, with $y(x)$ special or not, we can apply well-known methods (bisection, secant, Newton–Raphson, or whatever) once we know how to compute the function $y(x)$ (and in the case of Newton–Raphson also its derivative) accurately enough.

However, as is generally the case with nonlinear equations, some information on the location of the zeros is desirable, particularly when applying rapidly convergent (but often unpredictable) methods like Newton–Raphson or higher order methods.

The zeros of special functions usually appear nicely arranged, forming clear patterns from which a priori information can be found. For instance, the zeros of Jacobi polynomials $P_n^{(\alpha,\beta)}(x)$ are all real and in the interval $(-1, 1)$ for $\alpha > -1$ and $\beta > -1$, and they satisfy other regularity properties (such as, for instance, interlacing with the zeros of the derivative and with the zeros of contiguous orders). As α and/or β become smaller than -1, some or all of the n zeros escape from the real axis, forming a regular pattern in the complex plane (see Fig. 4.3).

These regular patterns formed by the zeros is a common feature of "classical" special functions and beyond [23]. The regularity in the distribution of zeros helps in the design of specific algorithms with good convergence properties. In addition, for many special functions, accurate a priori approximations are available. This a priori information, when wisely applied, will save computation time and avoid divergent (or even chaotic) algorithms. In a fortunate situation, as in the case of Bessel functions, asymptotic approximations provide accurate enough starting values for higher order Newton–Raphson methods; see [110].

In [54, Chap. 7], a variety of methods for computing zeros of special functions is discussed, starting with bisection and the fixed point method, including Newton-Raphson. In general, for computing zeros of special functions it is a wise idea to use some of their properties and to design specific methods.

Next to methods that take advantage of information about asymptotic approximations for the zeros of special functions, there are methods for which it is not necessary to compute values of these functions themselves in order to obtain their zeros. This is the case for the classical orthogonal polynomials, the zeros of which are the exact eigenvalues of real tridiagonal symmetric matrices with very simple entries; this method is usually named the Golub-Welsch algorithm [60]. The recurrence relation of the special functions plays a crucial role because the matrix is built from the coefficients of the recursion. Also, there are other functions, minimal solutions of three-term recurrence relations (the Bessel function $J_\nu(x)$ is among them) for which the problem of computing zeros is not exactly an eigenvalue problem for a (finite) matrix, but it can be approximated by it [61, 67].

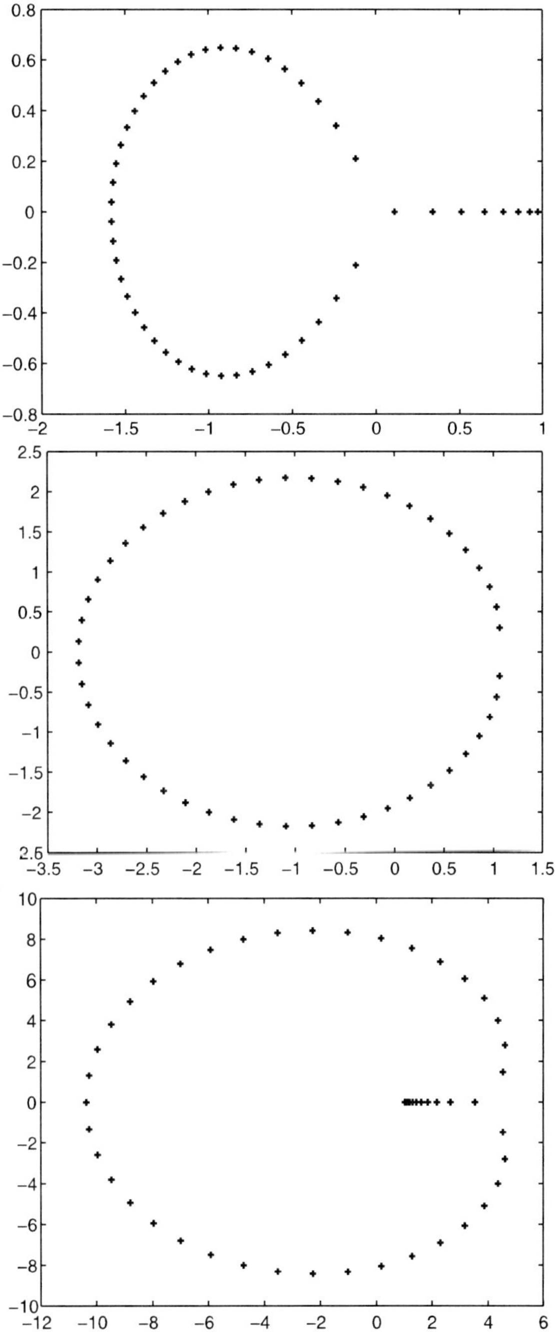

Fig. 4.3 Zeros of the Jacobi polynomials $P_{50}^{(2,-42.5)}(x)$ (*left*), $P_{50}^{(2,-52)}(x)$ (*center*), and $P_{50}^{(2,-63.5)}(x)$ (*right*) in the complex plane

Another type of methods, which are global in the sense that, similarly as matrix methods, don't require a priori estimations of the zeros for ensuring convergence, are the fixed point methods of [45, 100]; these methods use first order differential systems, which are satisfied by a large number of special functions (hypergeometric functions among them). More recently, a fixed point method of fourth order was obtained in [99], which can be used to compute the zeros of any solution of any second order equation $y''(x) + A(x)y(x) = 0$ in an interval where $A(x)$ is continuous. It is shown that, when $A(x) > 0$, the fixed point iteration

$$T(x) = x - \frac{1}{\sqrt{A(x)}} \arctan\bigl(\sqrt{A(x)}\, y(x)/y'(x)\bigr) \qquad (4.7.44)$$

has order four and the sequence $x_{n+1} = T(x_n)$ converges for any x_0 under mild conditions on $A(x)$. A scheme is given which guarantees convergence for any continuous $A(x)$ and allows for computing all the zeros in an interval. Typically 3 or 4 iterations per zero are required for 100 digits accuracy.

Additional methods and details on obtaining information on the zeros by using asymptotic expansions of the special functions, with examples such as Airy functions, Scorer functions, error functions, parabolic cylinder functions, Bessel functions, and Laguerre polynomials $L_n^{(\alpha)}(x)$ with large values of α are given in [54, Chap. 7].

4.7.6 Uniform Asymptotic Expansions

The asymptotic expansions considered in Sect. 4.2.1 are simple in the sense that they hold for large values of one variable, in fact the argument of the special function. There are many powerful expansions available that hold also for other large parameters.

For example, the expansion for the incomplete gamma function in (4.7.12) holds when z is large, but it becomes useless when a is also large. Also, the expansion in (4.7.9) converges for all a and z with trivial exceptions: $a \neq 0, -1, -2, \ldots$. But for computations it becomes useless when z is much larger than a.

There is a nice alternate asymptotic representations for these functions which can be used when a and/or z are large, and which in particular holds when $a \sim z$, a transition point in the behavior for large a and z. In this representation the complementary error function

$$\operatorname{erfc} z = \frac{2}{\sqrt{\pi}} \int_z^{\infty} e^{-t^2}\, dt \qquad (4.7.45)$$

plays the role of the transition form very small to very large values, or for the scaled functions

$$P(a, z) = \frac{\gamma(a, z)}{\Gamma(a)}, \qquad Q(a, z) = \frac{\Gamma(a, z)}{\Gamma(a)} \qquad (4.7.46)$$

from values close to 0 to values close to 1.

We have the following representations:

$$Q(a,z) = \frac{1}{2}\,\mathrm{erfc}(\eta\sqrt{a/2}) + R_a(\eta),$$
$$P(a,z) = \frac{1}{2}\,\mathrm{erfc}(-\eta\sqrt{a/2}) - R_a(\eta), \tag{4.7.47}$$

where

$$\frac{1}{2}\eta^2 = \lambda - 1 - \ln\lambda, \quad \lambda = \frac{z}{a}, \tag{4.7.48}$$

and

$$R_a(\eta) = \frac{e^{-\frac{1}{2}a\eta^2}}{\sqrt{2\pi a}} S_a(\eta), \qquad S_a(\eta) \sim \sum_{n=0}^{\infty} \frac{C_n(\eta)}{a^n}, \tag{4.7.49}$$

as $a \to \infty$.

The relation between η and λ in (4.7.48) becomes clear when we expand

$$\lambda - 1 - \ln\lambda = \frac{1}{2}(\lambda-1)^2 - \frac{1}{3}(\lambda-1)^3 + \frac{1}{4}(\lambda-1)^4 + \cdots, \tag{4.7.50}$$

and in fact the relation in (4.7.48) can also be written as

$$\eta = (\lambda - 1)\sqrt{\frac{2(\lambda - 1 - \ln\lambda)}{(\lambda-1)^2}}, \tag{4.7.51}$$

where the sign of the square root is positive for $\lambda > 0$. For complex values we use analytic continuation. An expansion for small values of $|\lambda - 1|$ reads

$$\eta = (\lambda - 1) - \frac{1}{3}(\lambda-1)^2 + \frac{7}{36}(\lambda-1)^3 + \cdots, \tag{4.7.52}$$

and, upon inverting this expansion,

$$\lambda = 1 + \eta + \frac{1}{3}\eta^2 + \frac{1}{36}\eta^3 + \cdots. \tag{4.7.53}$$

The asymptotic expansion for $S_a(\eta)$ in (4.7.49) holds uniformly with respect to $z \ge 0$. Both a and z may be complex. Note that the symmetry relation $P(a,z) + Q(a,z) = 1$ is preserved in the representations in (4.7.47) because $\mathrm{erfc}\, z + \mathrm{erfc}(-z) = 2$.

The first coefficients for $S_a(\eta)$ are

$$C_0 = \frac{1}{\lambda - 1} - \frac{1}{\eta}, \qquad C_1(\eta) = \frac{1}{\eta^3} - \frac{1}{(\lambda-1)^3} - \frac{1}{(\lambda-1)^2} - \frac{1}{12(\lambda-1)}. \tag{4.7.54}$$

These coefficients, and all higher ones, are regular at the transition point $a = z$, or $\lambda = 1$, or $\eta = 0$. For numerical applications Taylor expansions can be used, as explained in [54, Sect. 8.3].

For Bessel functions we have a similar problem in the design of efficient algorithms. All Bessel functions can be expanded in terms of Airy functions, and these expansions are in particular useful in the neighborhood of the turning point $z = \nu$. For example, the Bessel function $J_\nu(z)$ is oscillatory for $z > \nu$ and monotonic for $z < \nu$. Airy functions have a similar turning point behavior, as follows from their differential equation $w'' - zw = 0$.

The coefficients in the asymptotic series are regular at the turning point $z = \nu$, but for numerical evaluations we need expansions of the used coefficients in the neighborhood of the turning point. For more details we refer to [54, Sect. 8.4] and [112].

Airy-type expansions are used in software for Bessel functions in [5, 51, 70], and for parabolic cylinder functions in [56, 57, 95].

4.7.7 *Taylor Expansion Methods for Ordinary Differential Equations*

The special functions of mathematical physics usually arise as special solutions of ordinary linear differential equations, which follow from certain forms of the wave equation. Separation of the variables and the use of domains such as spheres, circles, cylinders, and so on, are the standard ways of introducing Bessel functions, Legendre functions, and confluent hypergeometric functions (also called Whittaker functions). For an introduction to this topic, see [111, Chap. 10].

In numerical mathematics, computing solutions of ordinary linear differential equations is a vast research area, with popular methods such as, for example, Runge–Kutta methods. These techniques are usually not used for computing special functions, mainly because so many other efficient methods are available for these functions. However, when the differential equation has coefficients in terms of analytic functions, as is the case for the equations of special functions, a method based on Taylor expansions may be considered as an alternative method, in particular for solving the equation in the complex plane.

In [54, Sect. 9.5] the basic steps are given for the Taylor expansion method, in particular for linear second order equations of the form

$$\frac{d^2w}{dz^2} + f(z)\frac{dw}{dz} + g(z)w = h(z), \tag{4.7.55}$$

where f, g, and h are analytic functions in a domain $D \subseteq \mathbb{C}$. For applications to special functions f, g, and h are often simple rational functions, and usually the equation is homogeneous.

For further information and examples, see [86] and [73]. For an application to compute solutions in the complex plane of the Airy differential equation, see [35] with a Fortran computer program in [34]. In [56] Taylor methods are used for computing the parabolic cylinder function $W(a, x)$.

4.7.8 Computing Symmetric Elliptic Integrals

Legendre's standard elliptic integrals are the incomplete elliptic integral of the first kind,

$$K(k)=\int_0^{\phi}\frac{d\theta}{\sqrt{1-k^2\sin^2\theta}}, \tag{4.7.56}$$

the incomplete integral of the second kind,

$$E(k)=\int_0^{\phi}\sqrt{1-k^2\sin^2\theta}\,d\theta, \tag{4.7.57}$$

and the incomplete elliptic integral of the third kind,

$$\Pi(n;\phi,k)=\int_0^{\phi}\frac{1}{1-n\sin^2\theta}\frac{d\theta}{\sqrt{1-k^2\sin^2\theta}}. \tag{4.7.58}$$

It is assumed here that $k\in[0,1]$, $\phi\in[0,\frac{1}{2}\pi]$, although the functions can also be defined for complex values of the parameters. Also, n is real, and if $n>1$, the integral of the third kind should be interpreted as a Cauchy principal value integral. When $\phi=\frac{1}{2}\pi$ the integrals are called complete elliptic integrals.

The computational problem for the elliptic integrals has received much attention in the literature, and the algorithms are usually based on successive Landen transformations or Gauss transformations, or by infinite series.

By considering a new set of integrals it is possible to compute the elliptic integrals, also by using successive transformations, by very efficient algorithms. The integrals are introduced in [18]. For example we have

$$R_F(x,y,z)=\frac{1}{2}\int_0^{\infty}\frac{dt}{\sqrt{(t+x)(t+y)(t+z)}}. \tag{4.7.59}$$

This function is symmetric and homogeneous of degree $-\frac{1}{2}$ in x,y,z and is normalized so that $R_F(x,x,x)=x^{-\frac{1}{2}}$.

Many elementary functions can be expressed in terms of these integrals. For example,

$$R_F(x,y,y)=\frac{1}{2}\int_0^{\infty}\frac{dt}{\sqrt{(t+x)}\,(t+y)}, \tag{4.7.60}$$

which is a logarithm if $0<y<x$ and an inverse circular function if $0\le x<y$.

The three standard elliptic integrals in (4.7.56)–(4.7.58) can be written in terms of symmetric integrals. For example, we have

$$F(\phi,k)=\sin\phi\,R_F\bigl(\cos^2\phi,1-k^2\sin^2\phi,1\bigr). \tag{4.7.61}$$

For further details we refer to [54, Sect. 11.4] and to the work of B.C. Carlson, who wrote very efficient algorithms for the computation of Legendre's standard elliptic integrals, also for complex parameters [19–21].

Other methods [82], based on series expansions [31], can be also considered and may be faster for some parameter values.

4.7.9 Best Rational Approximations

In the theory of best rational approximation (which includes the best polynomial of approximation) the goal is to find a rational function that approximates a function f on a finite real interval as best as possible. The rational approximations can be written in the form

$$\frac{N_m^n(x)}{D_m^n(x)} = \frac{a_0 + a_1 x + \cdots + a_n x^n}{b_0 + b_1 x + \cdots + b_m x^m}. \tag{4.7.62}$$

The characterization of the best approximation to a function f may be given in terms of oscillations of the error curve. Let $R = N/D$ be an irreducible rational function of the form (4.7.62). A necessary and sufficient condition that R be the best approximation to f is that the error function $R(x) - f(x)$ exhibits at least $2 + \max\{m + \partial N, n + \partial D\}$ points of alternation. (Here ∂P denotes the degree of the polynomial P.) For the proof see [79].

For the elementary and well-known higher transcendental functions the polynomials in the best rational approximations are not explicitly known, and the coefficients of these polynomials should be computed by an algorithm. This algorithm is not as simple as the one for computing coefficients in Chebyshev series (see Sect. 4.4) or Padé approximants (which can be based on solving a set of linear equations). For best approximation the *second algorithm of Remes* can be used [90, p. 176], and for a Fortran program see [69].

For many elementary and special functions best rational approximations have been computed. See [64] for many tables (and an explanation of the Remes algorithm). For several other special functions we refer to the survey [27]. Computer algebra packages, such as Maple, also have programs for computing best rational approximants.

For flexible algorithms for special functions we prefer the method based on Chebyshev polynomials. Chebyshev series (4.4.1) usually converge rapidly (for example, for functions in C^∞ on $[-1, 1]$), we obtain a very good first approximation to the polynomial $p_n(x)$ of best approximation for $[-1, 1]$ if we truncate (4.4.1) at its $(n + 1)$th term. This is because

$$f(x) - \sum_{k=0}^{n}{}' c_k T_k(x) \doteq c_{n+1} T_{n+1}(x), \tag{4.7.63}$$

and the right-hand side enjoys exactly those properties concerning its maxima and minima that are required for the polynomial of best approximation. In practice the gain in replacing a truncated Chebyshev series expansion by the corresponding minimax polynomial approximation is hardly worthwhile; see [88].

Inspection of the size of the coefficients c_k gives a good idea about the applicability for a certain choice of n, and for a new choice of n the computations are easy to modify. In best approximations for each choice of n (or of n and m in rational approximations), new coefficients have to be computed by using a complicated algorithm. In addition, representations of the polynomials in best approximations may be quite unstable.

4.8 Recent Software and Publications on Methods for Computing Special Functions

4.8.1 A Selection of Recent Software for Special Functions

Software packages such as Mathematica, Maple, and Matlab have many excellent algorithms in multi-length arithmetic. For large scale and high performance computing these packages are not the optimal platforms. Also, there are many published books with software for special functions, some with supplied sources of the algorithms. We mention [9, 83, 89, 113, 117, 129]. Many software collections of special functions are available on the web, for instance the Cephes math library[5] and in more general repositories[6,7].

For an extensive survey of the available software for special functions we refer to [74]. The latest update of this project appeared in December 2000. In this section we give a selection of software for special functions published in the period 2000–2009 (disclaimer: we provide references, but we don't claim that all the software listed is equally reliable).

Books describing published software

1. Cuyt et al. [29], a Handbook describing numerical methods for evaluating special functions by using continued fractions. Many tables are given based on Maple programs published elsewhere by the group. All kinds of functions are considered: from gamma to Gauss, confluent, generalized and basic hypergeometric functions.
2. Gautschi [40] describes routines for generating recursion coefficients of orthogonal polynomials as well as routines dealing with applications.
3. Gil et al. [54] describes software for Airy and Scorer functions, associated Legendre functions of integer and half-integer degrees (including toroidal harmonics), Bessel functions (including modified Bessel functions with purely imaginary orders), parabolic cylinder functions, and a module for computing zeros of Bessel functions.

---

[5] http://www.moshier.net.

[6] http://www.netlib.org/.

[7] http://gams.nist.gov/Classes.html.

Gamma, error, Fadeeva, Voigt and related functions

1. Smith [105]: Fortran 90 software for floating-point multiple precision arithmetic, gamma and related functions.
2. Linhart et al. [72]: the logarithm of the normal distribution.
3. Shippony and Read [104]: Voigt function (a special case of the plasma dispersion function or the complex error function).

Bessel functions

1. Kodama [70]: all kinds of cylindrical functions of complex order and nonnegative argument.
2. Gil et al. [52]: modified Bessel functions $I_{ia}(x)$ and $K_{ia}(x)$ for real a and positive x.
3. Van Deun and Cools [115]: infinite range integrals of an arbitrary product of Bessel functions.
4. Talman [108]: spherical Bessel transforms.

Airy and related functions

1. Gil et al. [48, 49]: complex Airy and Scorer functions.
2. Fabijonas [34]: complex Airy functions.

Parabolic cylinder functions

1. Gil et al. [57]: functions $U(a, x)$ and $V(a, x)$ for real a and x.

Coulomb wave functions

1. Michel [80]: functions $F_\ell(\eta, \rho)$ and $G_\ell(\eta, \rho)$ for complex parameters
2. Seaton [97]: functions $F_\ell(\eta, \rho)$ and $G_\ell(\eta, \rho)$.
3. Seaton [98]: Numerov integrations of Coulomb functions.
4. Noble [85]: negative energy Coulomb (Whittaker) functions.

Legendre functions

1. Gil and Segura [43, 44]: toroidal harmonics.
2. Inghoff et al. [68]: Maple procedures for the coupling of angular momenta.

Hypergeometric functions

1. Michel and Stoitsov [81]: Gauss hypergeometric function with all its parameters complex.
2. Colavecchia and Gasaneo [28]: Appell's F_1 function.
3. Huber and Maître [66]: expanding hypergeometric functions about half-integer parameters.

Mathieu functions

1. Alhargan [3, 4]: Mathieu functions and characteristic numbers.
2. Erricolo [33]: expansion coefficients of Mathieu functions using Blanch's algorithm.

4.8.2 Recent Literature on the Computation of Special Functions

From our website[8] a list will be soon available with recent literature from the last ten years (2000–2009). The list can be viewed as an addition to the bibliography of Lozier and Olver [74], and contains references to software and papers describing methods for computing special functions. Some of the references are mentioned in earlier sections. The following topics can be found in the list.

1. General aspects in books and papers: continued fractions, recurrence relations, Hadamard-type expansions, infinite products.
2. Gamma function, Barnes multiple gamma function, incomplete gamma functions, beta distribution, error functions, exponential integrals.
3. Bessel functions, integrals of Bessel functions, series of K-Bessel function.
4. Airy functions, Airy-type integrals: oscillatory cuspoid integrals with odd and even polynomial phase functions, Pearcy integral.
5. Hypergeometric functions: Gauss, confluent, Coulomb, Weber parabolic, Appell.
6. Legendre functions: toroidal, conical, spherical harmonics.
7. Orthogonal polynomials, Gauss quadrature.
8. q-functions.
9. Mathieu functions.
10. Spheroidal wave functions.
11. Polylogarithms.
12. Mittag-Leffler function, Wright function.
13. Elliptic integrals, elliptic functions.
14. Riemann zeta function, Riemann theta function.
15. Bose–Einstein, Fermi–Dirac integrals.
16. Hubbell rectangular source integrals, Lambert's W-function, leaky aquifer function.
17. Multicenter integrals, Slater orbitals, other integrals from physics.
18. Zeros of special functions.
19. Multiprecision implementation of elementary and special functions.

Acknowledgements We thank the referees for their helpful comments and Dr. Ernst Joachim Weniger for providing us with notes that we used for writing Sect. 4.7.2. We acknowledge financial support from *Ministerio de Educación y Ciencia*, project MTM2006–09050. NMT acknowledges financial support from *Gobierno of Navarra, Res. 07/05/2008*.

References

1. Abramowitz, M., Stegun, I.A.: Handbook of mathematical functions with formulas, graphs, and mathematical tables. National Bureau of Standards Applied Mathematics Series, vol. 55. US Printing Office (1964)

[8] http://functions.unican.es.

2. Airey, J.R.: The "converging factor" in asymptotic series and the calculation of Bessel, Laguerre and other functions. Philos. Mag. **24**, 521–552 (1937)
3. Alhargan, F.A.: Algorithm 804: subroutines for the computation of Mathieu functions of integer orders. ACM Trans. Math. Softw. **26**(3), 408–414 (2000)
4. Alhargan, F.A.: Algorithm 855: subroutines for the computation of Mathieu characteristic numbers and their general orders. ACM Trans. Math. Softw. **32**(3), 472–484 (2006)
5. Amos, D.E.: Algorithm 644: a portable package for Bessel functions of a complex argument and nonnegative order. ACM Trans. Math. Softw. **12**(3), 265–273 (1986)
6. Baker, G.A. Jr.: The theory and application of the Padé approximant method. In: Advances in Theoretical Physics, vol. 1, pp. 1–58. Academic Press, New York (1965)
7. Baker, G.A. Jr.: Essentials of Padé Approximants. Academic Press, New York/London (1975). [A subsidiary of Harcourt Brace Jovanovich, Publishers]
8. Baker, G.A. Jr., Graves-Morris, P.: Padé Approximants, 2nd edn. Encyclopedia of Mathematics and Its Applications, vol. 59. Cambridge University Press, Cambridge (1996)
9. Baker, L.: C Mathematical Function Handbook. Programming Tools For Engineers and Scientists. McGraw-Hill, New York (1992)
10. Bhattacharya, R., Roy, D., Bhowmick, S.: Rational interpolation using Levin-Weniger transforms. Comput. Phys. Commun. **101**(3), 213–222 (1997)
11. Bickley, W.G., Comrie, L.J., Miller, J.C.P., Sadler, D.H., Thompson, A.J.: Bessel Functions. Part II. Functions of Positive Integer Order. British Association for the Advancement of Science, Mathematical Tables, vol. X. University Press, Cambridge (1952)
12. Bornemann, F., Laurie, D., Wagon, S., Waldvogel, J.: The SIAM 100-Digit Challenge. SIAM, Philadelphia (2004). A study in high-accuracy numerical computing, With a foreword by David H. Bailey
13. Boyd, J.P.: The devil's invention: Asymptotic, superasymptotic and hyperasymptotic series. Acta Appl. Math. **56**(1), 1–98 (1999)
14. Brezinski, C.: A Bibliography on Continued Fractions, Padé Approximation, Sequence Transformation and Related Subjects. Prensas Universitarias de Zaragoza, Zaragoza (1991)
15. Brezinski, C.: History of Continued Fractions and Padé Approximants. Springer, Berlin (1991)
16. Brezinski, C.: Convergence acceleration during the 20th century. J. Comput. Appl. Math. **122**(1–2), 1–21 (2000). Numerical analysis 2000, Vol. II: Interpolation and extrapolation
17. Brezinski, C., Redivo-Zaglia, M.: Extrapolation Methods. Theory and Practice. Studies in Computational Mathematics, vol. 2. North-Holland, Amsterdam (1991)
18. Carlson, B.C.: Special Functions of Applied Mathematics. Academic Press, New York (1977). [Harcourt Brace Jovanovich Publishers]
19. Carlson, B.C.: Computing elliptic integrals by duplication. Numer. Math. **33**(1), 1–16 (1979)
20. Carlson, B.C.: Numerical computation of real or complex elliptic integrals. Numer. Algorithms **10**(1–2), 13–26 (1995). Special functions (Torino, 1993)
21. Carlson, B.C., FitzSimons, J.: Reduction theorems for elliptic integrals with the square root of two quadratic factors. J. Comput. Appl. Math. **118**(1–2), 71–85 (2000)
22. Chatterjee, S., Roy, D.: A class of new transforms tailored for the hypergeometric series. Comput. Phys. Commun. **179**(8), 555–561 (2008)
23. Clarkson, P.A., Mansfield, E.L.: The second Painlevé equation, its hierarchy and associated special polynomials. Nonlinearity **16**(3), R1–R26 (2003)
24. Clenshaw, C.W.: A note on the summation of Chebyshev series. Math. Tables Aids Comput. **9**(51), 118–120 (1955)
25. Clenshaw, C.W.: The numerical solution of linear differential equations in Chebyshev series. Proc. Camb. Philos. Soc. **53**, 134–149 (1957)
26. Clenshaw, C.W.: Chebyshev Series for Mathematical Functions. National Physical Laboratory Mathematical Tables, vol. 5. Her Majesty's Stationery Office, London (1962). Department of Scientific and Industrial Research
27. Cody, W.J.: A survey of practical rational and polynomial approximation of functions. SIAM Rev. **12**(3), 400–423 (1970)

28. Colavecchia, F.D., Gasaneo, G.: f1: a code to compute Appell's F_1 hypergeometric function. Comput. Phys. Commun. **157**(1), 32–38 (2004)
29. Cuyt, A., Petersen, V.B., Verdonk, B., Waadeland, H., Jones, W.B.: Handbook of Continued Fractions for Special Functions. Springer, New York (2008). With contributions by Franky Backeljauw and Catherine Bonan-Hamada, Verified numerical output by Stefan Becuwe and Cuyt
30. Deaño, A., Segura, J., Temme, N.M.: Identifying minimal and dominant solutions for Kummer recursions. Math. Comput. **77**(264), 2277–2293 (2008)
31. DiDonato, A.R., Hershey, A.V.: New formulas for computing incomplete elliptic integrals of the first and second kind. J. Assoc. Comput. Mach. **6**, 515–526 (1959)
32. Dingle, R.B.: Asymptotic expansions and converging factors. I. General theory and basic converging factors. Proc. R. Soc. Lond. Ser. A **244**, 456–475 (1958)
33. Erricolo, D.: Algorithm 861: Fortran 90 subroutines for computing the expansion coefficients of Mathieu functions using Blanch's algorithm. ACM Trans. Math. Softw. **32**(4), 622–634 (2006)
34. Fabijonas, B.R.: Algorithm 838: Airy functions. ACM Trans. Math. Softw. **30**(4), 491–501 (2004)
35. Fabijonas, B.R., Lozier, D.W., Olver, F.W.J.: Computation of complex Airy functions and their zeros using asymptotics and the differential equation. ACM Trans. Math. Softw. **30**(4), 471–490 (2004)
36. Forrey, R.C.: Computing the hypergeometric function. J. Comput. Phys. **137**(1), 79–100 (1997)
37. Gautschi, W.: Computational aspects of three-term recurrence relations. SIAM Rev. **9**(1), 24–82 (1967)
38. Gautschi, W.: A computational procedure for incomplete gamma functions. ACM Trans. Math. Softw. **5**(4), 466–481 (1979)
39. Gautschi, W.: Computation of Bessel and Airy functions and of related Gaussian quadrature formulae. BIT **42**(1), 110–118 (2002)
40. Gautschi, W.: Orthogonal Polynomials: Computation and Approximation. Numerical Mathematics and Scientific Computation. Oxford University Press, New York (2004)
41. Gil, A., Segura, J.: Evaluation of Legendre functions of argument greater than one. Comput. Phys. Commun. **105**(2–3), 273–283 (1997)
42. Gil, A., Segura, J.: A code to evaluate prolate and oblate spheroidal harmonics. Comput. Phys. Commun. **108**(2–3), 267–278 (1998)
43. Gil, A., Segura, J.: Evaluation of toroidal harmonics. Comput. Phys. Commun. **124**, 104–122 (2000)
44. Gil, A., Segura, J.: DTORH3 2.0: A new version of a computer program for the evaluation of toroidal harmonics. Comput. Phys. Commun. **139**(2), 186–191 (2001)
45. Gil, A., Segura, J.: Computing the zeros and turning points of solutions of second order homogeneous linear ODEs. SIAM J. Numer. Anal. **41**(3), 827–855 (2003)
46. Gil, A., Segura, J., Temme, N.M.: Computing toroidal functions for wide ranges of the parameters. J. Comput. Phys. **161**(1), 204–217 (2000)
47. Gil, A., Segura, J., Temme, N.M.: On nonoscillating integrals for computing inhomogeneous Airy functions. Math. Comput. **70**(235), 1183–1194 (2001)
48. Gil, A., Segura, J., Temme, N.M.: Algorithm 819: AIZ, BIZ: two Fortran 77 routines for the computation of complex Airy functions. ACM Trans. Math. Softw. **28**(3), 325–336 (2002)
49. Gil, A., Segura, J., Temme, N.M.: Algorithm 822: GIZ, HIZ: two Fortran 77 routines for the computation of complex Scorer functions. ACM Trans. Math. Softw. **28**(4), 436–447 (2002)
50. Gil, A., Segura, J., Temme, N.M.: Computing complex Airy functions by numerical quadrature. Numer. Algorithms **30**(1), 11–23 (2002)
51. Gil, A., Segura, J., Temme, N.M.: Algorithm 831: Modified Bessel functions of imaginary order and positive argument. ACM Trans. Math. Softw. **30**(2), 159–164 (2004)
52. Gil, A., Segura, J., Temme, N.M.: Computing solutions of the modified Bessel differential equation for imaginary orders and positive arguments. ACM Trans. Math. Softw. **30**(2), 145–158 (2004)

53. Gil, A., Segura, J., Temme, N.M.: The ABC of hyper recursions. J. Comput. Appl. Math. **190**(1–2), 270–286 (2006)
54. Gil, A., Segura, J., Temme, N.M.: Numerical Methods for Special Functions. SIAM, Philadelphia (2007)
55. Gil, A., Segura, J., Temme, N.M.: Computing the conical function $P^{\mu}_{-1/2+i\tau}(x)$. SIAM J. Sci. Comput. **31**(3), 1716–1741 (2009)
56. Gil, A., Segura, J., Temme, N.M.: Fast and accurate computation of the Weber parabolic cylinder function $w(a,x)$ (2009). Submitted to IMA J. Numer. Anal.
57. Gil, A., Segura, J., Temme, N.M.: Algorithm 850: Real parabolic cylinder functions $U(a,x)$, $V(a,x)$. ACM Trans. Math. Softw. **32**(1), 102–112 (2006)
58. Gil, A., Segura, J., Temme, N.M.: Computing the real parabolic cylinder functions $U(a,x)$, $V(a,x)$. ACM Trans. Math. Softw. **32**(1), 70–101 (2006)
59. Gil, A., Segura, J., Temme, N.M.: Numerically satisfactory solutions of hypergeometric recursions. Math. Comput. **76**(259), 1449–1468 (2007)
60. Golub, G.H., Welsch, J.H.: Calculation of Gauss quadrature rules. Math. Comput. **23**(106), 221–230 (1969). Loose microfiche suppl. A1–A10
61. Grad, J., Zakrajšek, E.: Method for evaluation of zeros of Bessel functions. J. Inst. Math. Appl. **11**, 57–72 (1973)
62. Graffi, S., Grecchi, V.: Borel summability and indeterminacy of the Stieltjes moment problem: Application to the anharmonic oscillators. J. Math. Phys. **19**(5), 1002–1006 (1978)
63. Graves-Morris, P.R., Roberts, D.E., Salam, A.: The epsilon algorithm and related topics. J. Comput. Appl. Math. **122**(1–2), 51–80 (2000). Numerical analysis 2000, vol. II: Interpolation and extrapolation
64. Hart, J.F., Cheney, E.W., Lawson, C.L., Maehly, H.J., Mesztenyi, C.K., Rice, J.R., Thacher, H.C. Jr., Witzgall, C.: Computer Approximations. SIAM Ser. in Appl. Math. Wiley, New York (1968)
65. Homeier, H.H.H.: Scalar Levin-type sequence transformations. J. Comput. Appl. Math. **122**(1–2), 81–147 (2000). Numerical analysis 2000, Vol. II: Interpolation and extrapolation
66. Huber, T., Maître, D.: HypExp 2, expanding hypergeometric functions about half-integer parameters. Comput. Phys. Commun. **178**(10), 755–776 (2008)
67. Ikebe, Y.: The zeros of regular Coulomb wave functions and of their derivatives. Math. Comput. **29**, 878–887 (1975)
68. Inghoff, T., Fritzsche, S., Fricke, B.: Maple procedures for the coupling of angular momenta. IV: Spherical harmonics. Comput. Phys. Commun. **139**(3), 297–313 (2001)
69. Johnson, J.H., Blair, J.M.: REMES2—a Fortran program to calculate rational minimax approximations to a given function. Technical Report AECL-4210, Atomic Energy of Canada Limited. Chalk River Nuclear Laboratories, Chalk River, Ontario (1973)
70. Kodama, M.: Algorithm 877: A subroutine package for cylindrical functions of complex order and nonnegative argument. ACM Trans. Math. Softw. **34**(4), Art. 22, 21 (2008)
71. Levin, D.: Development of non-linear transformations of improving convergence of sequences. Int. J. Comput. Math. **3**, 371–388 (1973)
72. Linhart, J.M.: Algorithm 885: Computing the logarithm of the normal distribution. ACM Trans. Math. Softw. **35**(3), Art. 20 (2008)
73. Lozier, D.W., Olver, F.W.J.: Airy and Bessel functions by parallel integration of ODEs. In: Sincovec, R.F., Keyes, D.E., Leuze, M.R., Petzold, L.R., Reed, D.A. (eds.) Parallel Processing for Scientific Computing. Proceedings of the Sixth SIAM Conference, vol. II, pp. 530–538. SIAM, Philadelphia (1993)
74. Lozier, D.W., Olver, F.W.J.: Numerical evaluation of special functions. In: Mathematics of Computation 1943–1993: A Half-century of Computational Mathematics, Vancouver, BC, 1993. Proc. Sympos. Appl. Math., vol. 48, pp. 79–125. Am. Math. Soc., Providence (1994). Updates are available at http://math.nist.gov/mcsd/Reports/2001/nesf/
75. Luke, Y.L.: The Special Functions and Their Approximations II. Mathematics in Science and Engineering, vol. 53. Academic Press, New York (1969)
76. Luke, Y.L.: Mathematical Functions and Their Approximations. Academic Press, New York (1975)

77. MacLeod, A.J.: An instability problem in Chebyshev expansions for special functions. ACM SigNum Newslett. **28**(2), 2–7 (1993)
78. Maino, G., Menapace, E., Ventura, A.: Computation of parabolic cylinder functions by means of a Tricomi expansion. J. Comput. Phys. **40**(2), 294–304 (1981)
79. Meinardus, G.: Approximation of Functions: Theory and Numerical Methods. Springer Tracts in Natural Philosophy, vol. 13. Springer, New York (1967). Expanded translation from the German edition. Translated by Larry L. Schumaker
80. Michel, N.: Precise Coulomb wave functions for a wide range of complex l, η and z. Comput. Phys. Commun. **176**, 232–249 (2007)
81. Michel, N., Stoitsov, M.V.: Fast computation of the Gauss hypergeometric function with all its parameters complex with application to the Pöschl-Teller-Ginocchio potential wave functions. Comput. Phys. Commun. **178**, 535–551 (2008)
82. Morris, A.H. Jr.: NSWC library of mathematical subroutines. Naval Surface Warfare Center, Dahlgren Division, Dahlgren, VA (1993)
83. Lloyd Baluk Moshier, S.: Methods and Programs for Mathematical Functions. Ellis Horwood Series: Mathematics and Its Applications. Ellis Horwood, Chichester (1989)
84. Murli, A., Rizzardi, M.: Algorithm 682: Talbot's method for the Laplace inversion problem. ACM Trans. Math. Softw. **16**(2), 158–168 (1990)
85. Noble, C.J.: Evaluation of negative energy Coulomb (Whittaker) functions. Comput. Phys. Commun. **159**(1), 55–62 (2004)
86. Olde Daalhuis, A.B., Olver, F.W.J.: On the asymptotic and numerical solution of linear ordinary differential equations. SIAM Rev. **40**(3), 463–495 (1998)
87. Paris, R.B., Wood, A.D.: Stokes phenomenon demystified. Bull. Inst. Math. Appl. **31**(1–2), 21–28 (1995)
88. Powell, M.J.D.: On the maximum errors of polynomial approximations defined by interpolation and by least squares criteria. Comput. J. **9**(4), 404–407 (1967)
89. Press, W.H., Teukolsky, S.A., Vetterling, W.T., Flannery, B.P.: Numerical Recipes in C++. Cambridge University Press, Cambridge (2002). The art of scientific computing, 2nd edition, updated for C++
90. Rice, J.R.: The Approximation of Functions. Vol. I: Linear Theory. Addison-Wesley, Reading (1964)
91. Rizzardi, M.: A modification of Talbot's method for the simultaneous approximation of several values of the inverse Laplace transform. ACM Trans. Math. Softw. **21**(4), 347–371 (1995)
92. Schmelzer, T., Trefethen, L.N.: Computing the gamma function using contour integrals and rational approximations. SIAM J. Numer. Anal. **45**(2), 558–571 (2007) (electronic)
93. Schonfelder, J.L.: Chebyshev expansions for the error and related functions. Math. Comput. **32**(144), 1232–1240 (1978)
94. Schulten, Z., Anderson, D.G.M., Gordon, R.G.: An algorithm for the evaluation of the complex Airy functions. J. Comput. Phys. **31**(1), 60–75 (1979)
95. Schulten, Z., Gordon, R.G., Anderson, D.G.M.: A numerical algorithm for the evaluation of Weber parabolic cylinder functions $U(a, x)$, $V(a, x)$, and $W(a, \pm x)$. J. Comput. Phys. **42**(2), 213–237 (1981)
96. Seaton, M.J.: Coulomb functions for attractive and repulsive potentials and for positive and negative energies. Comput. Phys. Commun. **146**(2), 225–249 (2002)
97. Seaton, M.J.: FGH, a code for the calculation of Coulomb radial wave functions from series expansions. Comput. Phys. Commun. **146**(2), 250–253 (2002)
98. Seaton, M.J.: NUMER, a code for Numerov integrations of Coulomb functions. Comput. Phys. Commun. **146**(2), 254–260 (2002)
99. Segura, J.: Reliable computation of the zeros of solutions of second order linear ODEs with a fourth order method. SIAM J. Numer. Anal. **48**(2), 452–469 (2010)
100. Segura, J.: The zeros of special functions from a fixed point method. SIAM J. Numer. Anal. **40**(1), 114–133 (2002)
101. Segura, J., de Córdoba, P.F., Ratis, Yu.L.: A code to evaluate modified Bessel functions based on the continued fraction method. Comput. Phys. Commun. **105**(2–3), 263–272 (1997)

102. Segura, J., Gil, A.: Parabolic cylinder functions of integer and half-integer orders for nonnegative arguments. Comput. Phys. Commun. **115**(1), 69–86 (1998)
103. Segura, J., Temme, N.M.: Numerically satisfactory solutions of Kummer recurrence relations. Numer. Math. **111**(1), 109–119 (2008)
104. Shippony, Z., Read, W.G.: A correction to a highly accurate Voigt function algorithm. JQSRT **78**(2), 255 (2003)
105. Smith, D.M.: Algorithm 814: Fortran 90 software for floating-point multiple precision arithmetic, gamma and related functions. ACM Trans. Math. Softw. **27**(4), 377–387 (2001)
106. Stieltjes, T.-J.: Recherches sur quelques séries semi-convergentes. Ann. Sci. École Norm. Sup. (3) **3**, 201–258 (1886)
107. Talbot, A.: The accurate numerical inversion of Laplace transforms. J. Inst. Math. Appl. **23**(1), 97–120 (1979)
108. Talman, J.D.: NumSBT: A subroutine for calculating spherical Bessel transforms numerically. Comput. Phys. Commun. **180**(2), 332–338 (2009)
109. Temme, N.M.: On the numerical evaluation of the modified Bessel function of the third kind. J. Comput. Phys. **19**(3), 324–337 (1975)
110. Temme, N.M.: An algorithm with ALGOL 60 program for the computation of the zeros of ordinary Bessel functions and those of their derivatives. J. Comput. Phys. **32**, 270–279 (1979)
111. Temme, N.M.: Special Functions. Wiley, New York (1996). An introduction to the classical functions of mathematical physics
112. Temme, N.M.: Numerical algorithms for uniform Airy-type asymptotic expansions. Numer. Algorithms **15**(2), 207–225 (1997)
113. Thompson, W.J.: An Atlas for Computing Mathematical Functions: An Illustrated Guide for Practitioners, with Programs in Fortran 90 and Mathematica. Wiley, New York (1997)
114. Trefethen, L.N., Weideman, J.A.C., Schmelzer, T.: Talbot quadrature and rational approximations. Technical report, Oxford University Computing Laboratory Numerical Analysis Group (2005)
115. Van Deun, J., Cools, R.: Algorithm 858: Computing infinite range integrals of an arbitrary product of Bessel functions. ACM Trans. Math. Softw. **32**(4), 580–596 (2006)
116. Van Loan, C.: Computational Frameworks for the Fast Fourier Transform. Frontiers in Applied Mathematics, vol. 10. SIAM, Philadelphia (1992)
117. Wang, Z.X., Guo, D.R.: Special Functions. World Scientific, Teaneck (1989). Translated from the Chinese by Guo and X.J. Xia
118. Watson, G.N.: A Treatise on the Theory of Bessel Functions, 2nd edn. Cambridge Mathematical Library. Cambridge University Press, Cambridge (1944)
119. Weideman, J.A.C.: Optimizing Talbot's contours for the inversion of the Laplace transform. Technical Report NA 05/05, Oxford U. Computing Lab. (2005)
120. Weideman, J.A.C., Trefethen, L.N.: Parabolic and hyperbolic contours for computing the Bromwich integral. Math. Comput. **76**(259), 1341–1356 (2007)
121. Weniger, E.J.: Nonlinear sequence transformations for the acceleration of convergence and the summation of divergent series. Comput. Phys. Rep. **10**(5,6), 189–371 (1989)
122. Weniger, E.J.: On the summation of some divergent hypergeometric series and related perturbation expansions. J. Comput. Appl. Math. **32**(1–2), 291–300 (1990). Extrapolation and rational approximation (Luminy, 1989)
123. Weniger, E.J., Čížek, J.: Rational approximations for the modified Bessel function of the second kind. Comput. Phys. Commun. **59**(3), 471–493 (1990)
124. Weniger, E.J., Čížek, J., Vinette, F.: The summation of the ordinary and renormalized perturbation series for the ground state energy of the quartic, sextic, and octic anharmonic oscillators using nonlinear sequence transformations. J. Math. Phys. **34**(2), 571–609 (1993)
125. Weniger, E.J.: Computation of the Whittaker function of the second kind by summing its divergent asymptotic series with the help of nonlinear sequence transformations. Comput. Phys. **10**, 496–503 (1996)

126. Wuytack, L.: Commented bibliography on techniques for computing Padé approximants. In: Padé approximation and its applications, Proc. Conf., Univ. Antwerp, Antwerp, 1979. Lecture Notes in Math., vol. 765, pp. 375–392. Springer, Berlin (1979)
127. Wynn, P.: On a device for computing the $e_m(S_n)$ tranformation. Math. Tables Aids Comput. **10**, 91–96 (1956)
128. Wynn, P.: Upon systems of recursions which obtain among the quotients of the Padé table. Numer. Math. **8**(3), 264–269 (1966)
129. Zhang, S., Jin, J.: Computation of Special Functions. Wiley, New York (1996)

Chapter 5
Melt Spinning: Optimal Control and Stability Issues

Thomas Götz and Shyam S.N. Perera

Abstract A mathematical model describing the melt spinning process of polymer fibers is considered. Newtonian and non-Newtonian models are used to describe the rheology of the polymeric material. Two key questions related to the industrial application of melt spinning are considered: the optimization and the stability of the process. Concerning the optimization question, the extrusion velocity of the polymer at the spinneret as well as the velocity and temperature of the quench air serve as control variables. A constrained optimization problem is derived and the first-order optimality system is set up to obtain the adjoint equations. Numerical solutions are carried out using a steepest descent algorithm. Concerning the stability with respect to variations of the velocity and temperature of the quench air, a linear stability analysis is carried out. The critical draw ratio, indicating the onset of instabilities, is computed numerically solving the eigenvalue problem for the linearized equations.

Keywords Melt spinning · Non-Newtonian fluids · Optimal control · First-order optimality system · Linear stability analysis

Mathematics Subject Classification (2000) 49K15 · 93C15

5.1 Introduction

Many kinds of synthetic textile fibers, like Nylon, Polyester, etc. are manufactured by a so-called melt spinning process. In this process, the molten polymer is extruded

T. Götz (✉)
Department of Mathematics, TU Kaiserslautern, 67663 Kaiserslautern, Germany
e-mail: goetz@mathematik.uni-kl.de

S.S.N. Perera
Department of Mathematics, University of Colombo, Colombo 03, Sri Lanka
e-mail: ssnp@maths.cmb.ac.lk

T.E. Simos (ed.), *Recent Advances in Computational and Applied Mathematics*,
DOI 10.1007/978-90-481-9981-5_5,

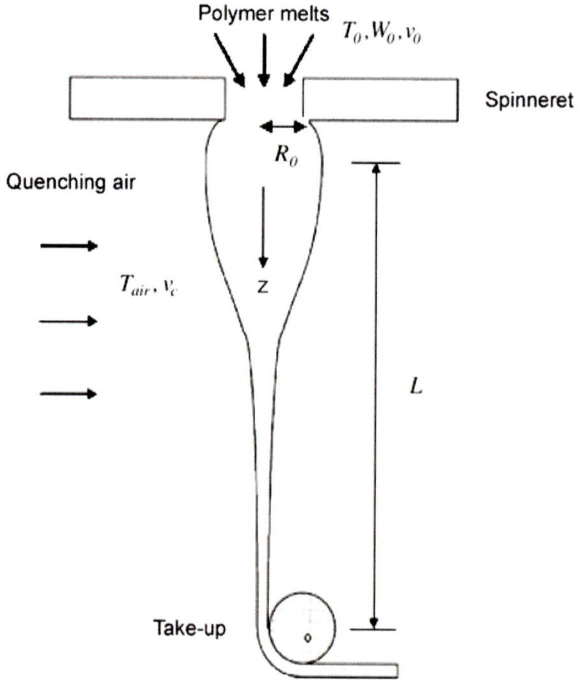

Fig. 5.1 Sketch of the melt spinning process

through a die, called the spinneret, to create a slender, cylindrical jet of viscous polymer, the fiber. Far away from the spinneret, the fiber is wrapped around a drum, which pulls it away at a pre-determined take-up speed, see Fig. 5.1. The take-up speed is much higher than the extrusion speed; in industrial processes the take-up speed is about 50 m/s and the extrusion speed is about 10 m/s, see [2, 10]. The ratio between the take-up speed v_L and the extrusion speed v_0 is called draw-ratio $d = v_L/v_0 > 1$. Hence the filament is stretched considerably in length and therefore it decreases in diameter. The ambient atmosphere temperature is below the polymer solidification temperature such that the polymer is cooled and solidifies before the take-up. In industrial processes a whole bundle of hundreds of single filaments is extruded and spun in parallel, however for the analysis we consider a single filament.

The dynamics of melt spinning processes has been studied by many research groups throughout the world during the last decades starting with early works of Kase and Matsuo [8] and Ziabicki [18]. In later works more and more sophisticated descriptions including crystallization kinetics and viscoelastic behavior were developed by several authors in order to achieve a better understanding of the fiber formation process. Up to now it is possible to use the basic models with more or less modifications in different technological aspects of the melt spinning process. The outcome of the melt spinning process depends significantly on the boundary conditions, e.g. the draw ratio, the ambient temperature, the quench air velocity and temperature. The questions of optimizing the fiber production or investigating the stability of the melt spinning process with respect to these external variables has not yet been treated in the literature.

The aim of this paper is twofold. First, we want to optimize the melt spinning process with respect to the final temperature, the quench air velocity and temperature. To model the fiber spinning process, we consider both a Newtonian model for the viscosity and a non-Newtonian model, where the viscosity is temperature-dependent. We formulate the optimal control problem as a constrained minimization problem, see [6], and derive formally the corresponding first-order optimality system via the Lagrange functional. For the numerical computation of the optimal control variables we present a steepest descent algorithm using the adjoint variables.

Second, we want to investigate the stability of the process. The onset of the so-called draw resonance instability occurring in the melt spinning process as well as related processes like fiber spinning, film casting and tubular film blowing, have been investigated by many research groups theoretically as well as experimentally during the last four decades, see [3–5]. The draw resonance phenomenon was first experimentally observed and named as such in the early 1960s. Draw resonance is of importance, both theoretically as well as practically, since it is closely related to the quality of the final product. Its theoretical analysis involves a fundamental understanding of the nonlinear dynamics of the fiber formation process. Earlier studies [3–5, 11] aimed to understand the physics behind draw resonance. In this paper, we try to analyse the stability of the melt spinning process with respect to the quench air velocity and temperature. The stability analysis of the melt spinning process is based on a linear approach.

The paper is organized as follows. In Sect. 5.2, we shortly derived the underlying stationary model. In Sect. 5.3 we introduce the constrained optimization problem and derive the according first-order optimality system. A steepest descent algorithm is proposed to solve the optimization problem numerically. Numerical results of the optimization are shown in Sect. 5.4. Section 5.5 deals with the linear stability analysis for the isothermal Newtonian and the temperature-dependent non-Newtonian case. The numerical results are presented in Sect. 5.6 and concluding remarks can be found in Sect. 5.7.

5.2 Governing Equations for Melt Spinning

5.2.1 Melt Spinning Model

Considering the conservation laws for mass, momentum and energy of a viscous polymer jet, one can obtain by averaging over the cross-section of the slender fiber, the following set of equations, see [10–12]:

$$\frac{\partial A}{\partial t} + \frac{\partial}{\partial z}(Av) = 0, \tag{5.2.1a}$$

$$\rho A\left(\frac{\partial v}{\partial t} + v\frac{\partial v}{\partial z}\right) = \frac{\partial}{\partial z}(A\tau) - \sqrt{\pi A}\, C_d\, \rho_{\text{air}}\, v^2 + \rho\, A\, g, \tag{5.2.1b}$$

$$\rho C_p\left(\frac{\partial T}{\partial t}+v\frac{\partial T}{\partial z}\right)=-\frac{2\alpha\sqrt{\pi}}{\sqrt{A}}(T-T_\infty). \tag{5.2.1c}$$

In the mass balance (5.2.1a), z denotes the coordinate along the spinline, t is the time, A denotes the cross-sectional area of the fiber and v is the velocity of the fiber along the spinline. The density ρ of the polymer is assumed to be constant. In the momentum balance (5.2.1b), the axial stress τ is related via the constitutive equation

$$\tau=3\eta\frac{dv}{dz}$$

to the viscosity η [1]. By g we denote the gravitational acceleration and C_d is the air drag coefficient. In the energy equation (5.2.1c), T and C_p denote the temperature and the heat capacity of the polymer, T_∞ is the temperature of the quench air and α denotes the heat transfer coefficient between the fiber and the quench air.

According to [10], we assume the following relations for the air drag and the heat transfer coefficient

$$C_d=0.37\,\mathrm{Re}_{\mathrm{air}}^{-0.61}$$

and

$$\alpha=\frac{0.21}{\sqrt{A_0}}\kappa\,\mathrm{Re}_{\mathrm{air}}^{\frac{1}{3}}\left[1+\frac{64v_c^2}{v^2}\right]^{\frac{1}{6}}$$

depending on the Reynolds-number of the quench air flow

$$\mathrm{Re}_{\mathrm{air}}=\frac{2v\rho_{\mathrm{air}}}{\eta_{\mathrm{air}}}\sqrt{\frac{A}{\pi}}.$$

Here $A_0=\pi R_0^2$ denotes the cross-sectional area of the spinneret, $\rho_{\mathrm{air}}, \eta_{\mathrm{air}}$ and κ represent the density, viscosity and heat conductivity of the air and v_c is the velocity of the quench air.

In the Newtonian model, the viscosity $\eta=\eta_0$ of the polymer is constant, whereas in the non-Newtonian case, we consider an Arrhenius-type temperature dependence [10]

$$\eta=\eta_0\exp\left[\frac{E_a}{R_G}\left(\frac{1}{T}-\frac{1}{T_0}\right)\right],$$

where $\eta_0>0$ is the zero shear viscosity at the initial temperature T_0, E_a denotes the activation energy and R_G equals to the gas constant.

The system (5.2.1a)–(5.2.1c) is subject to the boundary conditions

$$A=A_0,\quad v=v_0\quad\text{and}\quad T=T_0\quad\text{at } z=0\quad\text{for all } t, \tag{5.2.1d}$$

$$v=v_L\quad\text{at } z=L\quad\text{for all } t, \tag{5.2.1e}$$

where L denotes the length of the spinline.

In the momentum balance (5.2.1b), the air drag and gravity effects are neglected for the sake of brevity and clarity. However, including these effects into the model would not change the general outline of our study.

5.2.2 Dimensionless Form

Introducing the dimensionless quantities

$$t^* = \frac{t v_0}{L}, \quad v^* = \frac{v}{v_0}, \quad z^* = \frac{z}{L}, \quad T^* = \frac{T}{T_0}, \quad A^* = \frac{A}{A_0} \quad \text{and} \quad \tau^* = \frac{\tau L}{\eta_0 v_0},$$

and dropping the star, the system (5.2.1a)–(5.2.1e) can be re-formulated in dimensionless form

$$\frac{\partial A}{\partial t} + \frac{\partial}{\partial z}(Av) = 0, \tag{5.2.2a}$$

$$\frac{\partial v}{\partial t} + v\frac{\partial v}{\partial z} = \frac{3}{\mathrm{Re}\, A}\frac{\partial}{\partial z}\left(\eta A \frac{\partial v}{\partial z}\right) - C\frac{v^2}{\sqrt{A}} + \frac{1}{\mathrm{Fr}}, \tag{5.2.2b}$$

$$\frac{\partial T}{\partial t} + v\frac{\partial T}{\partial z} = -\gamma \frac{T - T_\infty}{\sqrt{A}}, \tag{5.2.2c}$$

where $\mathrm{Re} = \frac{\rho L v_0}{\eta_0}$ is the Reynolds number, $\mathrm{Fr} = \frac{v_0^2}{gL}$ stands for the Froude number, $C = \frac{\pi C_d \rho_{\mathrm{air}} L}{\sqrt{A_0}\rho}$ and $\gamma = \frac{2\alpha L}{\rho C_p v_0 R_0}$ are the scaled air drag and heat transfer coefficients.

The viscosity is given by

$$\eta = \begin{cases} 1 & \text{for the Newtonian model,} \\ \exp\left[\frac{E_a}{R_G T_0}\left(\frac{1}{T} - 1\right)\right] & \text{for the non-Newtonian model.} \end{cases} \tag{5.2.2d}$$

The boundary conditions read as

$$A(0) = 1, \quad v(0) = 1 \quad \text{and} \quad T(0) = 1 \quad \text{for all } t, \tag{5.2.2e}$$

$$v(1) = d \quad \text{for all } t, \tag{5.2.2f}$$

where $d = v_L/v_0 > 1$ denotes the draw-ratio.

Now, we can easily read off the steady state equations of melt spinning. The continuity equation (5.2.2a) reduces to $A = 1/v$ and writing the second order momentum equation (5.2.2b) as two first order equations, we finally arrive at

$$\frac{dv}{dz} = \frac{\tau}{3\eta}, \tag{5.2.3a}$$

$$\frac{d\tau}{dz} = \mathrm{Re}\frac{v\tau}{3\eta} + \frac{\tau^2}{3\eta v} - \frac{1}{\mathrm{Fr}} + C v^{3/2}, \tag{5.2.3b}$$

$$\frac{dT}{dz} = -\gamma \frac{T - T_\infty}{\sqrt{v}}, \tag{5.2.3c}$$

subject to the boundary conditions

$$v(0) = 1 \quad \text{and} \quad T(0) = 1, \tag{5.2.3d}$$

$$v(1) = d. \tag{5.2.3e}$$

In the sequel we will consider two main questions related to the model (5.2.3a)–(5.2.3e)

(1) How to choose the process conditions, i.e. the external air velocity v_c and temperature T_∞, such that the fiber is solid at the take-up? Moreover, we wish to maximize the mass flux, i.e. the inflow velocity v_0. In Sects. 5.3 and 5.4 this optimization problem is treated using optimal control techniques. Some numerical simulations illustrate our findings.
(2) Is the system (5.2.3a)–(5.2.3e) stable with respect to small perturbations of the process parameters? In Sect. 5.5 we apply a linear stability analysis to investigate this question. Again, some numerical computations highlight the results, see Sect. 5.6.

5.3 Optimal Control of the Melt Spinning Process

We want to control the temperature profile of the fiber, such that the final temperature $T(1)$ is below the solidification point $T_s^* = T_s/T_0$. On the other hand we want to maximize the outflow, i.e. maximize v_0. The air temperature T_∞ and the air velocity v_c can be influenced and hence serve as control variables. Therefore, we consider the following cost functional

$$\begin{aligned} J = J(y,u) &= J_1 + J_2 + J_3 + J_4 \\ &= -\omega_1 u_3 + \omega_2 (y_3(1) - T_s^*) \\ &\quad + \frac{\omega_3}{2}\int_0^1 (u_2(z) - T_{\text{ref}})^2 dz + \frac{\omega_4}{2}\int_0^1 u_1(z)^2 dz, \end{aligned} \tag{5.3.1}$$

where $y = (v, \tau, T) \in Y$ denotes the vector of state variables and $u = (v_c, T_\infty, v_0) \in U$ are the controls. The weighting coefficients $\omega_i > 0$, $i = 1, \dots, 4$ allow to adjust the cost functional to different scenarios. The larger the coefficient ω_1, the more importance is given to the maximization of the outflow; ω_2 and ω_3 measure the influence of the temperature constraints and the larger ω_4, the more weight is given to a low and hence "cheap" cooling air velocity. The actual choice of the numerical values of the weighting coefficients $\omega_1, \dots, \omega_4$ depends of the demands of the applicant.

Summarizing, we consider the following constrained optimization problem

$$\text{minimize } J(y,u) \text{ with respect to } u, \quad \text{subject to (5.2.3a)–(5.2.3e)}. \tag{5.3.2}$$

In the sequel, we will address this problem using the calculus of adjoint variables.

5.3.1 First-Order Optimality System

In this section we introduce the Lagrangian associated to the constrained minimization problem (5.3.2) and derive the system of first-order optimality conditions.

Let $Y = C^1([0,1];\mathbb{R}^3)$ be the state space consisting of triples of differentiable functions $y = (v, \tau, T)$ denoting velocity, stress and temperature of the fiber. Further, let $U = C^1([0,1];\mathbb{R}^2) \times \mathbb{R}$ be the control space consisting of a pair $(u_1, u_2) = (v_c, T_\infty)$ of differentiable functions, i.e. air velocity and temperature, and a scalar $u_3 = v_0$ interpreted as the inflow velocity.

We define the operator $e = (e_v, e_\tau, e_T) : Y \times U \to Y^*$ via the weak formulation of the state system (5.2.3a)–(5.2.3e):

$$\langle e(y,u), \xi\rangle_{Y,Y^*} = 0, \quad \forall \xi \in Y^*,$$

where $\langle \cdot, \cdot\rangle_{Y,Y^*}$ denotes the duality pairing between Y and its dual space Y^*. Now, the minimization problem (5.3.2) reads as

$$\text{minimize } J(y,u) \text{ with respect to } u \in U, \quad \text{subject to } e(y,u) = 0.$$

Introducing the Lagrangian $\mathcal{L} : Y \times U \times Y^* \to \mathbb{R}$ defined as

$$\mathcal{L}(y,u,\xi) = J(y,u) + \langle e(y,u), \xi\rangle_{Y,Y^*},$$

the first-order optimality system reads as

$$\nabla_{y,u,\xi}\mathcal{L}(y,u,\xi) = 0.$$

Considering the variation of $\mathcal{L}$ with respect to the adjoint variable ξ, we recover the state system

$$e(y,u) = 0$$

or in the classical form

$$\frac{dy}{dz} = f(y,u), \quad \text{with } v(0) = 1,\ v(1) = d,\ T(0) = 1, \tag{5.3.3}$$

where

$$f(y,u) = \begin{pmatrix} \frac{\tau}{3\eta} \\ \mathrm{Re}\,\frac{v\tau}{3\eta} + \frac{\tau^2}{3\eta} - \frac{1}{\mathrm{Fr}} + Cv^{3/2} \\ -\gamma\frac{T - T_\infty}{\sqrt{v}} \end{pmatrix}.$$

Second, taking variations of $\mathcal{L}$ with respect to the state variable y we get the adjoint system

$$J_y(y,u) + e_y^*(y,u)\xi = 0$$

or in classical form

$$-\frac{d\xi}{dz} = F(y, u, \xi), \quad \text{with } \xi_q(0) = 0,\ \xi_q(1) = 0,\ \xi_T(1) = -\omega_2, \tag{5.3.4}$$

where

$$F(y, u, \xi) = \left(\frac{\partial f}{\partial y}\right)^{\top} \xi.$$

Finally, considering variations of $\mathcal{L}$ with respect to the control variable u in a direction of δu we get the optimality condition

$$\langle J_u(y, u), \delta u\rangle + \langle e_u(y, u)\delta u, \xi\rangle = 0. \tag{5.3.5}$$

In the optimum, this holds for all $\delta u \in U$.

5.3.2 Decent Algorithm

To solve the nonlinear first-order optimality system consisting of (5.3.3), (5.3.4) and (5.3.5), we propose an iterative steepest-descent method.

1. *Set* $k = 0$ *and choose initial control* $u^{(0)} \in U$.
2. *Given the control* $u^{(k)}$. *Solve the state system* (5.3.3) *with a shooting method to obtain* $y^{(k+1)}$.
3. *Solve the adjoint system* (5.3.4) *with a shooting method to obtain* $\xi^{(k+1)}$.
4. *Compute the gradient* $g^{(k+1)}$ *of the cost functional.*
5. *Update the control* $u^{(k+1)} = u^{(k)} - \beta g^{(k+1)}$ *for a step size* $\beta > 0$.
6. *Compute the cost functional* $J^{(k+1)} = J(y^{(k+1)}, u^{(k+1)})$.
7. *If* $|g^{(k+1)}| \geq \delta$, *goto* 2.

Here, $\delta > 0$ is some prescribed relative tolerance for the termination of the optimization procedure. In each iteration step, we need to solve two boundary value problems, i.e. the state system (5.3.3) and the adjoint system (5.3.4) in the steps 2 and 3 of the algorithm. Both systems are solved using a shooting method based on a Newton-iteration.

The main steps of the shooting method for solving, e.g. the state system (5.3.3) are the following. Choose some initial guess s for $y_2(0)$ and denote by $y(z; s)$ the solution of the initial value problem

$$\frac{dy}{dz} = f(y, u), \quad \text{with } y_1(0) = 1,\ y_2(0) = s,\ y_3(0) = 1. \tag{5.3.6}$$

Now we introduce new dependent variables

$$x(z; s) = \frac{\partial y}{\partial s}$$

and define a second system as follows

$$\frac{\partial x}{\partial z} = \left(\frac{\partial f}{\partial y}\right) x, \quad \text{with } x_1(0; s) = 0,\ x_2(0; s) = 1,\ x_3(0; s) = 0. \tag{5.3.7}$$

The solution of $y(z; s)$ of the initial value problem (5.3.6) coincides with the solution $y(z)$ of the boundary value state system (5.3.3) provided that the value s can be found such that

$$\phi(s) = y_1(1; s) - d = 0.$$

Using the system (5.3.7), $\phi'(s)$ can be computed as

$$\phi'(s) = x_1(1; s).$$

Applying Newton's method generates a sequence $(s_n)_{n\in\mathbb{N}}$

$$s_{n+1} = s_n - \frac{\phi(s_n)}{\phi'(s_n)} \quad \text{for a given initial guess } s_0.$$

If the initial guess s_0 is a sufficiently good approximation to the required root of $\phi(s) = 0$, the theory of the Newton-iteration method ensures that the sequence $(s_n)_{n\in\mathbb{N}}$ converges to the desired root s.

A crucial ingredient for the convergence of the decent algorithm is the choice of the step size β in the direction of the gradient, see step 5 of the algorithm. Clearly, the best choice would be the result of a line search

$$\beta^* = \operatorname*{argmin}_{\beta>0} J(u_k - \beta g_k).$$

However this is numerically quite expensive although it is a one dimensional minimization problem. Instead of the exact line search method, we use an approximation based on a quadratic polynomial method [9] in order to find β^* such that we minimize $J(u_k - \beta g_k)$. We construct a quadratic polynomial $p(\beta)$ to approximate $J(u_k - \beta g_k)$ using following data points

$$p(0) = J(u_k), \qquad p(1) = J(u_k - g_k), \qquad p'(0) = -\nabla J(u_k)^{\mathrm{T}} g_k < 0.$$

Now, the quadratic approximation to our cost functional reads as

$$p(\beta) = p(0) + p'(0)\beta + (p(1) - p(0) - p'(0))\beta^2$$

and its global minimum is located at

$$\beta^* = \frac{-p'(0)}{2(p(1) - p(0) - p'(0))} \in (0, 1).$$

Table 5.1 Processing conditions and parameter values

Parameter	Value	Unit
ρ	0.98	$\mathrm{g/cm^3}$
C_p	0.46	cal/(g°C)
R_0	0.01	cm
T_0	300	°C
v_0	16.667	m/s
v_L	50	m/s
L	1	m
T_∞	24	°C
T_{ref}	20	°C
v_c	0.4	m/s
ρ_{air}	1.189	$\mathrm{kg/m^3}$
η_{air}	1.819×10^{-5}	pa s
R_G	1.9859	cal/(K mol)
E_a	13500	cal/mol
κ	0.0257	W/(m K)

5.4 Optimal Control Results

Both the state and the adjoint system (5.3.3) and (5.3.4) were solved using the MATLAB routine `ode23tb`. This routine uses an implicit method with backward differentiation to solve stiff differential equations. As weighting coefficients for the cost functional (5.3.1) we use $(\omega_1, \dots, \omega_4) = (1, 1, 1.5, 2.5)$. The relevant simulation conditions and parameter values are shown in Table 5.1. These values are typical for Nylon-66.

5.4.1 Newtonian Model

Figure 5.2 shows the spinline velocity and temperature as well as the cooling air velocity and temperature profiles before and after optimization for the Newtonian model. Some of the intermediate profiles are also included in the graphs. The corresponding cost functional is shown on the left in Fig. 5.3.

It can be seen that in the optimum, the final temperature is below 50°C. The extrusion velocity drops from 16.67 m/s to 12.65 m/s. The optimal air temperature is more or less close to 20°C, which we considered as a reference temperature. In this case, the optimal air velocity is quite large near the spinneret and drops after the first 30 cm almost to zero. Figures 5.4 and 5.5 visualize the optimal air velocity and air temperature profiles in Newtonian model for different weighting coefficients.

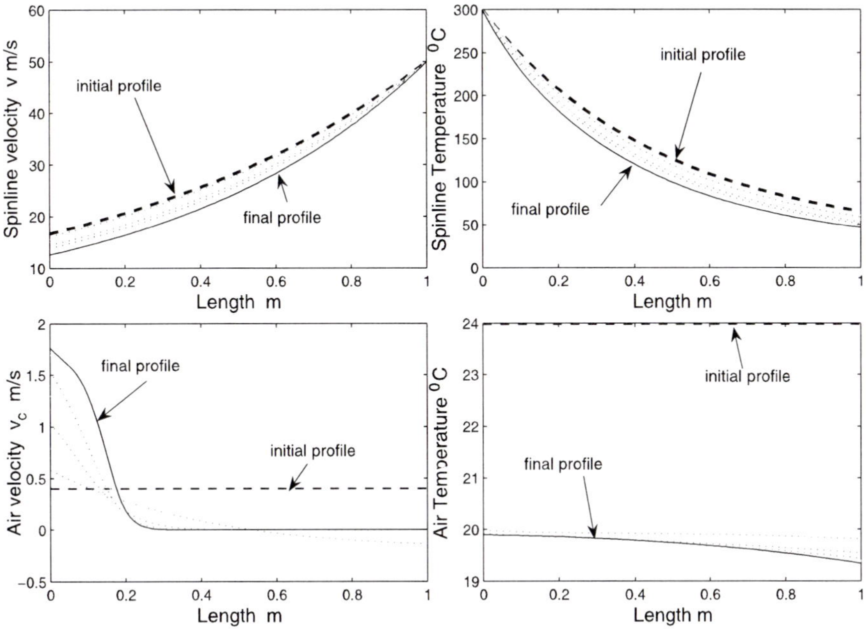

Fig. 5.2 Graphs for the spinline velocity (*top-left*) and temperature (*top-right*) as well as for the air velocity (*bottom-left*) and temperature (*bottom-right*). The *dashed curve* is the initial profile, while *dotted curves* correspond to intermediate profiles. The final, optimized profile is shown with the *solid line*

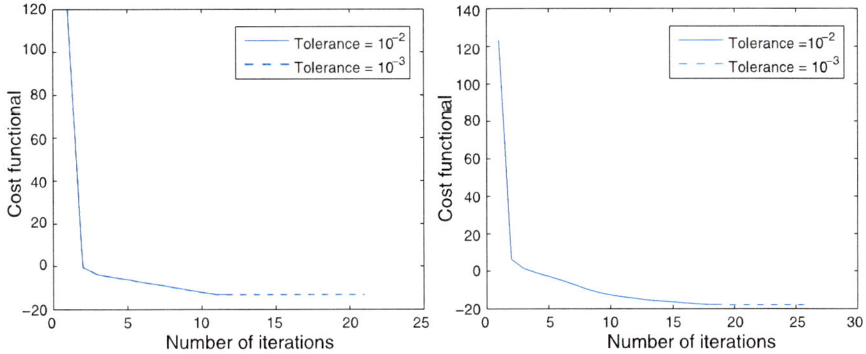

Fig. 5.3 Reduction of the cost functional for the Newtonian (*left*) and non-Newtonian model (*right*)

5.4.2 Non-Newtonian Model

Figure 5.6 visualizes the spinline velocity and temperature as well as the cooling air velocity and temperature before and after optimization for the non-Newtonian model. The reduction of the cost functional is shown on the right in Fig. 5.3.

Fig. 5.4 Optimal air velocity in the Newtonian case for different weighting coefficients

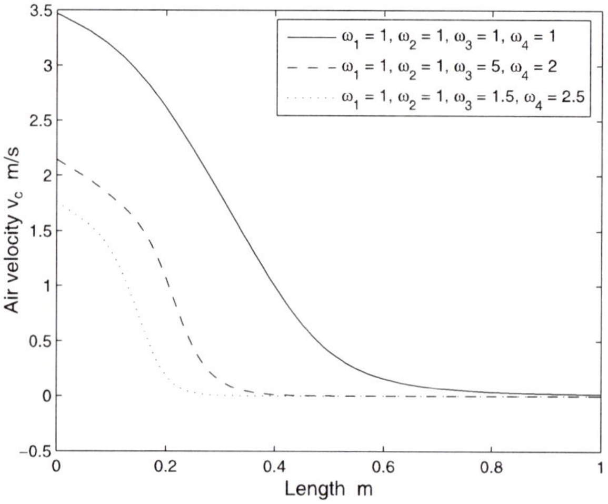

Fig. 5.5 Optimal air temperature in the Newtonian case for different weighting coefficients

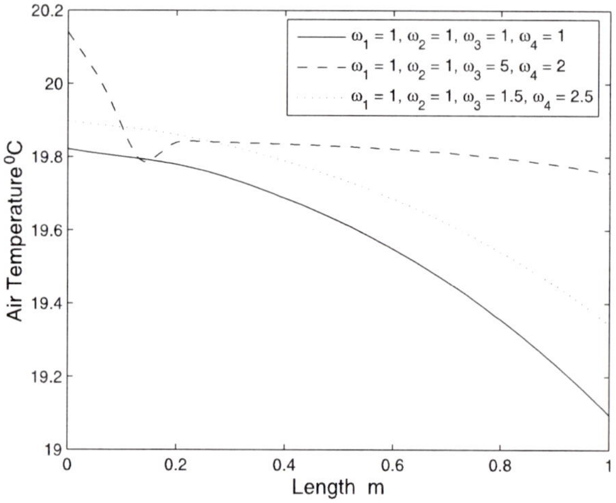

As for the Newtonian model, the optimized final temperature is well below 50°C and the optimal air temperature profile is more or less close to 20°C. The extrusion velocity drops from 16.67 m/s to 10.42 m/s. In the optimal state, the air velocity reaches a peak value near the spinneret exit point and just after this point it almost close to zero.

Figures 5.7 and 5.8 show the optimized air velocity and temperature profiles in the non-Newtonian case for different cost coefficients.

5.5 Linear Stability Analysis

Now, we turn our attention the second question treated in this paper. Are the solutions of the steady state system (5.2.3a)–(5.2.3e) stable with respect to small pertur-

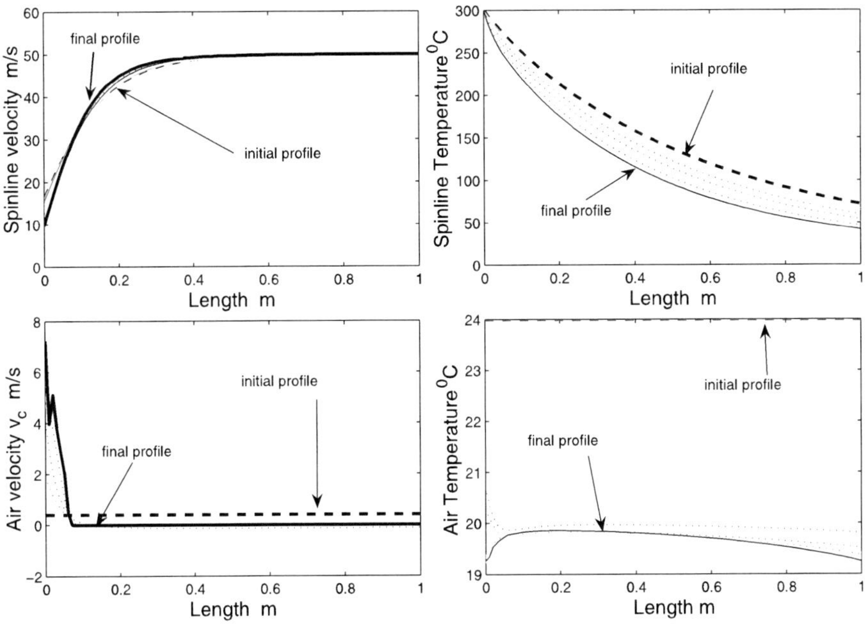

Fig. 5.6 Graphs for the spinline velocity (*top-left*) and temperature (*top-right*) as well as for the air velocity (*bottom-left*) and temperature (*bottom-right*). The *dashed curve* is the initial profile, while *dotted curves* correspond to intermediate profiles. The final, optimized profile is shown with the *solid line*

Fig. 5.7 Optimal air velocity in the non-Newtonian model for different weighting coefficients

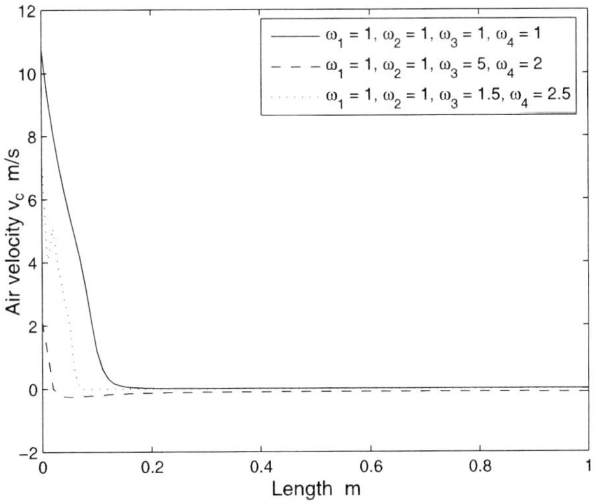

bations of the process conditions? For the sake of simplicity of notation, we assume $\mathrm{Fr} = 0$ and $C = 0$, i.e.—we neglect the influence of gravity and air drag. Including those effects in the analysis, however, is straightforward.

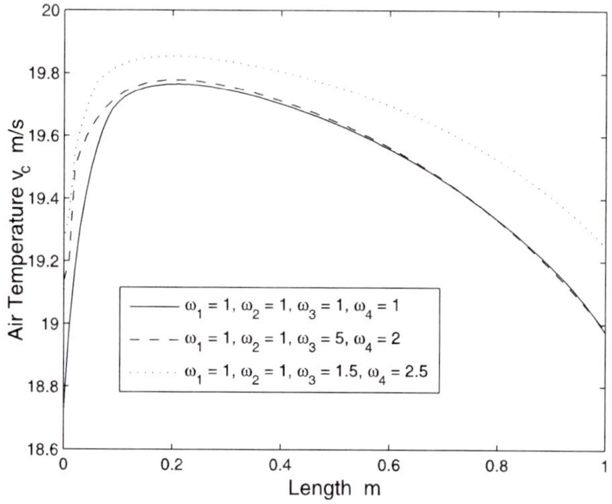

Fig. 5.8 Optimal air temperature in the non-Newtonian model for different weighting coefficients

Let $v_s(z)$ and $T_s(z)$ denote the solutions of the steady state system (5.2.3a)–(5.2.3e). Then $A_s(z) = 1/v_s(z)$ solves the stationary version of the mass balance (5.2.2a). Now, we consider small perturbations ϕ, φ and θ of the state variables A, v and T around their steady states. A linearization leads to

$$\begin{aligned} A(t,z) &= A_s(z) + \phi(z)\exp(\Omega t), \\ v(t,z) &= v_s(z) + \varphi(z)\exp(\Omega t), \\ T(t,z) &= T_s(z) + \theta(z)\exp(\Omega t), \end{aligned} \tag{5.5.1}$$

where $\Omega \in \mathbb{C}$ is a complex eigenvalue that accounts for the growth rate of the perturbation. Substituting (5.5.1) into the transient system (5.2.2a)–(5.2.2f) and neglecting higher order terms yields the following linearized equations.

In the case of a isothermal Newtonian flow, i.e. $\eta \equiv 1$, we get

$$\Omega\phi = -\left[(v_s\phi)' + \left(\frac{\varphi}{v_s}\right)'\right], \tag{5.5.2a}$$

$$\Omega\varphi = -(v_s\varphi)' + \frac{3}{\text{Re}}\left[v_s'(v_s\phi)' + v_s\left(\frac{\varphi'}{v_s}\right)'\right], \tag{5.5.2b}$$

subject to the boundary conditions

$$0 = \phi(0) = \varphi(0) = \varphi(1). \tag{5.5.2c}$$

In the non-Newtonian case, we get

$$\Omega\phi = -\left[(v_s\phi)' + \left(\frac{\varphi}{v_s}\right)'\right], \tag{5.5.3a}$$

$$\Omega\varphi = -(v_s\varphi)' + \frac{3\eta}{\mathrm{Re}}\left[v_s'(v_s\phi)' + v_s\left(\frac{\varphi'}{v_s}\right)'\right] - \frac{3E_a\eta\theta}{R_G\,\mathrm{Re}\,T_0T_s^2}\left[v_s\left(\frac{v_s'}{v_s}\right)' + T_s'\varphi'\right]$$
$$+\frac{3E_a\eta}{R_G\,\mathrm{Re}\,T_0T_s}v_s'\left[T_s'\theta\left(\frac{E_a}{R_G\,\mathrm{Re}\,T_0T_s^3} + \frac{2}{T_s^2}\right) - \theta'\right], \tag{5.5.3b}$$

$$\Omega\theta = \frac{1}{2}C_0(T_s - T_\infty)v_s^{\frac{11}{6}}\Psi^{\frac{1}{6}}\phi - C_0v_s^{\frac{5}{6}}\Psi^{\frac{1}{6}}\theta - v_s\theta'$$
$$+\left(\frac{64}{3}C_0v_c^2v_s^{-\frac{13}{6}}(T_s - T_\infty)\Psi^{-\frac{5}{6}} - \frac{1}{3}C_0v_s^{-\frac{1}{6}}(T_s - T_\infty)\Psi^{\frac{1}{6}} - T_s'\right)\varphi, \tag{5.5.3c}$$

subject to the boundary conditions

$$0 = \phi(0) = \varphi(0) = \theta(0) = \varphi(1), \tag{5.5.3d}$$

where

$$\Psi = \left[1 + 64\left(\frac{v_c}{v_s}\right)^2\right] \quad \text{and} \quad C_0 = \frac{0.42L\kappa}{\rho C_p v_0 R_0^2}\left(\frac{2R_0v_0\rho_{\mathrm{air}}}{\eta_{\mathrm{air}}}\right)^{\frac{1}{3}}.$$

5.5.1 *Numerical Solution of the Eigenvalue Problem*

Now we discretize the linearized system (5.5.2a)–(5.5.2c) or (5.5.3a)–(5.5.3d) respectively, using a finite-difference scheme on a uniform grid $z_i = ih$, $i = 0, \ldots, n$ with grid size $h = 1/n$. The derivatives are approximated using centered differences

$$\frac{dy}{dz} \approx \frac{y_{i+1} - y_{i-1}}{2h} \quad \text{and} \quad \frac{d^2y}{dz^2} \approx \frac{y_{i+1} \quad 2y_i + y_{i-1}}{h^2},$$

where $i = 1, \ldots, n-1$ ranges through the interior points. At the endpoint $i = n$ we apply a backward difference formula

$$\frac{dy}{dz} \approx \frac{y_n - y_{n-1}}{h},$$

In each point z_i we have two unknowns, namely ϕ_i, φ_i, in the isothermal Newtonian case and three unknowns ϕ_i , φ_i and θ_i in the non-Newtonian case. Due to the boundary conditions $\phi_0 = \varphi_0 = \varphi_n = 0$ and $\theta_0 = 0$, we end up with $m = 2n - 1$ degrees of freedom in the isothermal Newtonian model and $m = 3n - 1$ in the non-Newtonian model.

Plugging the difference approximations into the linearized systems (5.5.2a)–(5.5.2c) or (5.5.3a)–(5.5.3d) and re-arranging the terms, we arrive at the algebraic eigenvalue problem

$$\Omega Y = MY, \tag{5.5.4}$$

Table 5.2 Critical draw ratio in the isothermal Newtonian case

d_c	Literature source
20.218	This study
20.20	Pearson & Shah [14]
20.22	van der Hout [17]
20.218	Jung, Song, Hyun [7]

where $Y = [\phi_1, \varphi_1, \ldots \phi_{n-1}, \varphi_{n-1}, \phi_n]^T \in \mathbb{R}^m$ for the isothermal Newtonian case and $Y = [\phi_1, \varphi_1, \theta_1, \ldots, \phi_{n-1}, \varphi_{n-1}, \theta_{n-1}, \phi_n, \theta_n]^T$ in the non-Newtonian case and $M \in \mathbb{R}^{m \times m}$ arises from the discretization.

The stability is determined by the real parts of the eigenvalues Ω. Positive real parts indicate instability due to unbounded growth of the state variables with time. The critical draw-ratio d_c is defined to be the minimal draw-ratio, where instability, i.e. an eigenvalue Ω with $\mathrm{Re}(\Omega) > 0$ occurs. In order to determine the critical draw-ratio d_c, we have to compute the eigenvalues of the system (5.5.4). To do so, we first solve the stationary state system (5.2.3a)–(5.2.3e) for a given draw ratio d and given parameters v_c and T_∞. In this step, we use the MATLAB-routine `ode23tb`. This routine uses an implicit method with backward differentiation to solve stiff differential equations. It is an implementation of `TR-BDF2` [15], an implicit two stage Runge–Kutta formula where the first stage is a trapezoidal rule step and the second stage is a backward differentiation formula of order two. The computed stationary solution A_s, v_s and T_s is then plugged into the eigenvalue problem (5.5.4). The eigenvalue problem is solved using the MATLAB-routine `eigs` based on the `ARPACK`-library for large sparse matrices. This algorithm is based upon an algorithmic variant of the Arnoldi process called the implicitly restarted Arnoldi method, see [13, 16]. As a result, we finally obtain the eigenvalues Ω as functions of d.

5.6 Stability Results

Table 5.2 compares the computed critical draw ratio d_c in the isothermal Newtonian case with values reported in literature. An excellent agreement is found.

Table 5.3 reports the critical draw ratio for the non-Newtonian model depending on the velocity of the quench air. These simulations are carried out using a fixed air temperature of 24°C. With increasing air velocity, the onset of instability is shifted to higher draw ratios. Hence, the stability of the spinning process can be improved by increasing the quench air velocity. Figure 5.9 visualizes the relation between the air velocity v_c and the critical draw ratio in the non-Newtonian case. The simulation results suggest an almost linear relation between the air velocity v_c and the critical draw ratio. A regression analysis yields—as a rule of thumb—

$$d_c = 3.4 v_c + 27, \tag{5.6.1}$$

where the air velocity v_c is measured in $\mathrm{m/s}$.

Table 5.3 Critical draw ratio for the non-Newtonian case depending on the air velocity (air temperature 24°C fixed)

Air velocity [m/s]	Critical draw ratio
0	30.50
2.5	36.12
5.0	45.27
7.7	52.49
10.0	60.04
15.0	76.94
20.0	93.31
25.0	111.44
30.0	135.17

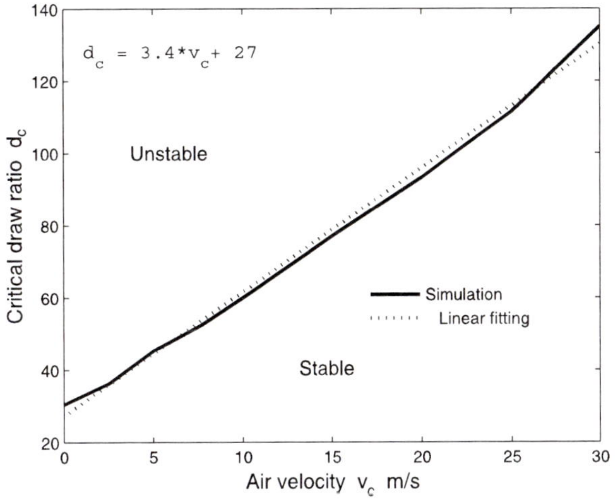

Fig. 5.9 Plot of the critical draw ratio vs. the quench air velocity for the non-Newtonian model

Figure 5.10 shows the critical draw ratio vs. the temperature of the quench air at a fixed air velocity of 0 m/s. The critical draw ratio decreases with a cooler quench air; the according numerical values are reported in Table 5.4. Again a linear regression was carried out. Measuring the air temperature T_∞ in °C, one finds $d_c = 0.21T_\infty + 25$ indicating a weak dependence of the critical draw ratio on the air temperature.

5.7 Conclusions

We considered two key questions related to the industrial application of the melt spinning process: the optimization and the stability of the process. The mathematical model describing the melt spinning process of polymer fibers uses Newtonian and non-Newtonian models for the rheology of the polymeric material.

Table 5.4 Critical draw ratio depending on the air temperature (air velocity 0 m/s)

Air temperature [°C]	Critical draw ratio
10	27.40
15	28.20
20	29.40
24	30.50
30	31.60
35	32.50

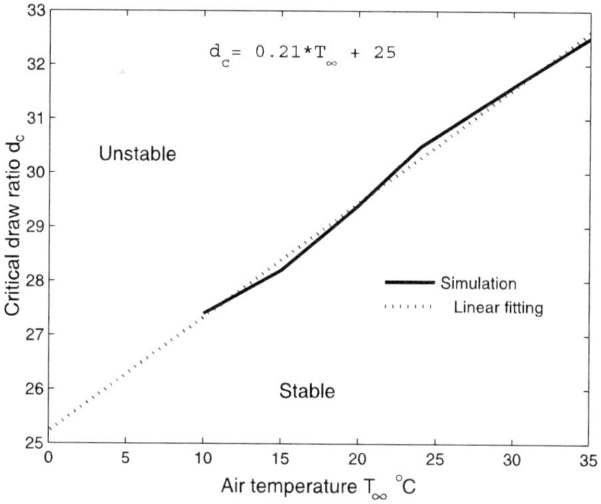

Fig. 5.10 Critical draw ratio vs. air temperature

Concerning the optimization question, the aim was to maximize the outflow, minimize the air velocity and air temperature and to get the final spinline temperature below the fiber solidification temperature. Defining an appropriate cost functional, we converted the problem into a constrained optimization problem and derived the first-order optimality system. Based on the adjoint variables, we propose a steepest descent algorithm with suitable step size control to solve the problem numerically. Simulations show the applicability of our approach and yield the desired results. After the optimization, the final fiber temperature is well below the solidification temperature. The velocity of the quench air was reduced as much as possible.

Regarding the stability of the process, instabilities may arise, if some process variables exceeded a certain critical value. Furthermore, instabilities lead to irregular fibers or induces breakage of the individual filaments along the spinline. Hence, a stability analysis of the process with respect to process parameters is needed. We analysed the stability of the melt spinning process with respect to the quench air velocity and the temperature. From this results the following points can be concluded.

In the Newtonian regime, the stability domain is independent of the quench air velocity and air temperature. The critical draw ratio is found to be $d_c \approx 20.218$; in accordance with the results of other research groups.

In the non-Newtonian case, the stability domain strongly depends on the quench air velocity. A linear regression reveals the approximate relation $d_c = 3.4v_c + 27$, hence the process stability can be improved by increasing the air velocity. The quench air temperature only has a minor influence on the process stability. However, there is a weak trend towards improved stability with higher air temperature.

Acknowledgements The research of S.S.N. Perera was supported by a grant from the DAAD (German Academic Exchange Services).

References

1. Bird, R.B., Amstrong, R.C., Hassager, O.: Dynamics of Polymeric Liquids, 2nd edn. Fluid Mechanics, vol. 1. Wiley, New York (1987)
2. Brünig, H., Roland, H., Blechschmidt, D.: High filament velocities in the underpressure spunbonding nonwoven process. In: IFJ, pp. 129–134, December (1997)
3. Gelder, D.: The stability of fiber drawing processes. Ind. Eng. Chem. Fundam. **10**, 534–543 (1978)
4. Hyun, J.C.: Theory of draw resonance: Part I. Newtonian fluids. AIChE J. **24**(3), 418–422 (1978)
5. Hyun, J.C.: Theory of draw resonance: Part II. Power-law and Maxwell fluids. AIChE J. **24**(3), 423–426 (1978)
6. Ito, K., Ravindran, S.S.: Optimal control of thermally convected fluid flows. SIAM J. Sci. Comput. **19**(6), 1847–1869 (1998)
7. Jung, H.W., Song, H.S., Hyun, J.C.: Draw resonance and kinematic waves in viscoelastic isothermal spinning. AIChE J. **46**(10), 2106–2110 (2000)
8. Kase, S., Matsuo, T.: Studies on melt spinning, fundamental equations on the dynamics of melt spinning. J. Polym. Sci., Part A **3**, 2541–2554 (1965)
9. Kelley, C.T.: Iterative Methods for Optimization. SIAM, Philadelphia (1999)
10. Langtangen, H.P.: Derivation of a mathematical model for fiber spinning. Department of Mathematics, Mechanics Division, University of Oslo, December (1997)
11. Lee, J.S., Jung, H.W., Hyun, J.C., Seriven, LE: Simple indicator of draw resonance instability in melt spinning processes. AIChE J. **51**(10), 2869–2874 (2005)
12. Lee, J.S., Shin, D.M., Jung, H.W., Hyun, J.C.: Transient solution of the dynamics in low-speed fiber spinning process accompanied by flow-induced crystallization. J. Non-Newtonian Fluid Mech. **130**, 110–116 (2005)
13. Lehoucq, R.B., Sorensen, D.C., Yang, C.: ARPACK users' guide: Solution of large scale eigenvalue problems with implicitly restarted Arnoldi methods. Department of computational and Applied mathematics. Rice University, Houston, Texas (1997)
14. Pearson, J.R.A., Shah, Y.T.: On the stability of isothermal and nonisothermal fiber spinning of power law fluids. Ind. Eng. Chem. Fund. **13**(2), 134–138 (1974)
15. Shampine, L.F., Reichelt, M.W.: The MATLAB ODE Suite. SIAM J. Sci. Comput. **18**, 1–22 (1997)
16. Sorensen, D.C.: Implicitly restarted Arnoldi/Lanczos methods for large scale eigenvalue calculations. Department of Computational and Applied Mathematics, Rice University, Houston, Texas (1995)
17. van der Hout, R.: Draw resonance in isothermal fibre spinning of Newtonian and power-law fluids. Eur. J. Appl. Math. **11**, 129–136 (2000)
18. Ziabicki, A.: Fundamentals of Fiber Formation. Wiley, New York (1976)

Chapter 6
On Orthonormal Polynomial Solutions of the Riesz System in $\mathbb{R}^3$

K. Gürlebeck and J. Morais

Abstract The main goal of the paper is to deal with a special orthogonal system of polynomial solutions of the Riesz system in $\mathbb{R}^3$. The restriction to the sphere of this system is analogous to the complex case of the Fourier exponential functions $\{e^{in\theta}\}_{n\geq 0}$ on the unit circle and has the additional property that also the scalar parts of the polynomials form an orthogonal system. The properties of the system will be applied to the explicit calculation of conjugate harmonic functions with a certain regularity.

Keywords Quaternionic analysis · Homogeneous monogenic polynomials · Riesz system · Harmonic conjugates

Mathematics Subject Classification (2000) 30G35 · 31B05 · 31B35

6.1 Introduction

Quaternionic analysis is nowadays a comfortable tool to solve boundary value problems of mathematical physics given in three or four dimensions. The reader who wishes a comprehensive account of such theory and its more recent achievements can profitably consult [28–31, 33, 34, 43–45] and elsewhere. Related to concrete boundary value problems, we need special systems of functions (well-adapted to series expansions and special geometries) that can be used theoretically and/or numerically in stable procedures. A starting point for these considerations relies on

K. Gürlebeck (✉) · J. Morais
Institut für Mathematik/Physik, Bauhaus-Universität Weimar, Coudraystr. 13B, 99421 Weimar, Germany
e-mail: klaus.guerlebeck@uni-weimar.de

J. Morais
e-mail: jmorais@mat.ua.pt

T.E. Simos (ed.), *Recent Advances in Computational and Applied Mathematics*,
DOI 10.1007/978-90-481-9981-5_6,

the concept of null solutions of higher-dimensional Cauchy-Riemann or Dirac systems, known as monogenic functions. The present work is intended as a study on the best approximation of a given class of monogenic functions in the quaternionic analysis setting. For a general orientation the reader is suggested to read some of the works [9, 11, 12] and [39], of which the first three deal entirely with the four-dimensional case and the last the three-dimensional case.

In view of many applications to boundary value problems and for simplicity one is mainly interested in the approximation of functions defined in domains of $\mathbb{R}^3$ and with values again in $\mathbb{R}^3$. One possible approach is then, roughly speaking, the identification of vectors from $\mathbb{R}^3$ with the so-called reduced quaternions. Unfortunately, such a structure is not closed under the quaternionic multiplication. For those reasons, Sect. 6.3 studies the approximation of monogenic functions in the linear space of square integrable functions over $\mathbb{R}$. This is also supported by the fact that the operators of some boundary value problems are not $\mathbb{H}$-linear but are nevertheless efficiently treated by means of quaternionic analysis tools (e.g., Lamé system [30], Stokes system [29]). Therefore, we aim to approximate reduced quaternion-valued monogenic functions in terms of well-defined homogeneous monogenic polynomials. To pave the way, we choose a convenient system of homogeneous monogenic polynomials that should replace the holomorphic function systems (polynomials) used in the complex case. We wish to deal with a system that has a simple structure, in the sense that the underlying functions can be explicitly calculated and the numerical costs are only slightly growing. To define some differential operators on the basis functions and to extend them to more general functions via Fourier expansions, one has forcibly to consider complete orthonormal systems. This gives then via continuous extension a basis to approximate monogenic functions or solutions of more general differential equations by series expansions in terms of homogeneous monogenic polynomials. The latter is also necessary to ensure the numerical stability of the considered approximations.

The remarks above cover only a part of the problem. Analogously to the one-dimensional case, the following facts have also to be considered:

1. Homogeneous monogenic polynomials of different order are orthogonal;
2. The scalar parts of the basis polynomials are orthogonal to each other;
3. All (hypercomplex-)derivatives of the basis elements deliver again basis elements, one degree lower.

The general problem of approximating a monogenic function by monogenic polynomials started basically with early works of Fueter [21, 22]. This was managed by means of the notion of hypercomplex variables. Later, in [6] and [36] it is shown that a monogenic function can be developed locally as a Taylor series in terms of homogeneous monogenic polynomials based on those variables. The resulting homogeneous monogenic polynomials are the so-called Fueter polynomials. In this line of reasoning Leutwiler in 2001 (see [35]), based on these polynomials, constructed a basis in the real-linear Hilbert space of reduced quaternion-valued homogeneous monogenic polynomials in $\mathbb{R}^3$. His results were recently generalized to arbitrary dimensions in the framework of a Clifford algebra by Delanghe in 2007 [16]. The approach followed from both authors relies on the notion of harmonic conjugates. However, a drawback remains in the fact that in general the Fueter polynomials and their associated scalar parts are not orthogonal with respect to the real-inner product (see [39] Chap. 2 for a special approach). A naive approach is to apply the Gram-Schmidt procedure for normalization of these polynomials. Unfortunately, this orthonormalization process is not easy to handle and it is numerically highly unstable. For this reason, research has been directed to the construction of a more suitable basis.

A different effort in this direction was done by Ryan in 1986 [41]. Therein, the author constructed a complete orthonormal system of homogeneous monogenic polynomials for even dimensions. However, his system is not appropriate for our case since we consider only functions defined in domains of the Euclidean space $\mathbb{R}^3$, of odd dimension. In this context we also mention the works by Brackx, Delanghe and Sommen in [6] and the first author in [23]. The authors have constructed shifted systems of Cauchy kernels as rational basis systems to approximate a monogenic function. Although the constructed systems are complete and give a simple structured basis, they are not orthogonal and do not have the property of having (hypercomplex-)derivatives within the same basis. Also, in these systems the construction of a series expansion is not possible and derivatives of the basis functions are not even finite linear combinations of the original basis functions. Other intensive works in the 90-ties were done by Abul-Ez and Constales (see e.g. [1–3]). The authors have constructed so-called special monogenic polynomials in the framework of Clifford analysis, as an extension of the basic sets of polynomials of one complex variable that appeared in the thirties in the work of Whittaker and resumed later in his book [49]. The completeness of the set in the space of homogeneous monogenic polynomials is not considered. Recently, Falcão and Malonek [18–20] have also constructed a set of special homogeneous monogenic polynomials involving only products of a hypercomplex variable and its hypercomplex conjugate. It is proved that the obtained set is an Appell set of monogenic polynomials with respect to the (hypercomplex-)derivative. Since the authors have just constructed one polynomial for each degree, the latter is not enough to form a basis for the space of square integrable monogenic functions. In a different setting, there are several other works as for example those by Cnops [13], Brackx, De Schepper and Sommen [8], De Bie and Sommen [14], related to Clifford-Hermite and Gegenbauer polynomials.

An important effort was done recently by Cação, worked out in her thesis [9] and in follow-up papers [11, 12]. Cação et al. have constructed a complete orthonormal system of homogeneous monogenic polynomials in the unit ball of $\mathbb{R}^3$ and taking values in the full quaternions ($\mathbb{R}^3 \to \mathbb{R}^4$ case). In [10] it is proved that the rate of convergence for a monogenic function approximated by this system of homogeneous monogenic polynomials has the same quality as in the similar case of a harmonic function approximated by homogeneous harmonic polynomials (for comparison see [32]). In particular one part of the resulting polynomials, contrary to the sets described above, carry the property of having monogenic (hypercomplex-)derivatives within the same basis one degree lower, like in the complex case. Partially motivated by the strategy adopted by the previous authors, in [39] the second author constructed a basis ($\mathbb{R}^3 \to \mathbb{R}^3$ case) such that property 2 of the previous list has been also regarded. In this paper we mainly follow the notations as in [39]; cf. also [27].

The paper is organized as follows. After some preliminaries, in Sect. 6.3 we start by presenting the orthogonal polynomial system exploited in [39]. The usage of a system of spherical harmonics in $\mathbb{R}^3$ (considered e.g. in [42]) for the construction of our system allows to use some well-known results like its orthogonality on the unit sphere. As a consequence of the interrelation between spherical harmonics and Legendre polynomials, the constructed homogeneous monogenic polynomials are related to the Legendre polynomials as well. Surprisingly, the set formed by the scalar parts of the basis elements are multiples of the used spherical harmonics. This reproducing property is in no way self-evident and it establishes a special relationship between quaternionic analysis and harmonic analysis. By the nature of the given approach, which is genuinely constructive, this relation is also fundamental in the study of the notion of conjugate harmonicity. As an illustration, we are able to set up under some particular asymptotic conditions on the Fourier coefficients of a scalar-valued function U, a (sufficient) condition that ensures the existence of a vector-valued function V conjugate to U such that $U + V$ is square integrable and reduced quaternion-valued monogenic. It is also described an explicit formula for the construction of V.

6.2 Basic Notions and Terminology

Consider a holomorphic function $f(z) = u(x, y) + iv(x, y)$ defined in a domain $\Omega \subset \mathbb{C}$. As is well-known, its real and imaginary parts are real-valued harmonic functions in Ω, satisfying the so-called Cauchy-Riemann system

$$\begin{cases} \frac{\partial u}{\partial x} = \frac{\partial v}{\partial y}, \\ \frac{\partial u}{\partial y} = -\frac{\partial v}{\partial x}. \end{cases}$$

As in the case of two variables, we may characterize the analogue of the Cauchy-Riemann system in an open domain of the Euclidean space $\mathbb{R}^3$. More precisely, consider a vector-valued function $f^* = (u_0, u_1, u_2)$ whose components $u_i = u_i(x_0, x_1, x_2)$ are real functions of real variables x_0, x_1, x_2 for which

$$\begin{cases} \sum_{i=0}^{2} \partial_{x_i} u_i = 0, \\ \partial_{x_j} u_i - \partial_{x_i} u_j = 0 \quad (i \neq j, 0 \leq i, j \leq 2) \end{cases}$$

or, equivalently, in a more compact form:

$$\begin{cases} \operatorname{div} f^* = 0, \\ \operatorname{rot} f^* = 0. \end{cases} \tag{6.2.1}$$

This 3-tuple f^* is said to be a system of conjugate harmonic functions in the sense of Stein-Weiß [46, 47] and system (6.2.1) is called the Riesz system [40].

The system (6.2.1) can be obtained naturally using the quaternionic algebra. Let $\mathbb{H} := \{\mathbf{a} = a_0 + a_1\mathbf{e}_1 + a_2\mathbf{e}_2 + a_3\mathbf{e}_3, a_i \in \mathbb{R}, i = 0, 1, 2, 3\}$ be the algebra of the real quaternions, where the imaginary units $\mathbf{e}_i$ $(i = 1, 2, 3)$ are subject to the multiplication rules

$$\mathbf{e}_1^2 = \mathbf{e}_2^2 = \mathbf{e}_3^2 = -1,$$
$$\mathbf{e}_1\mathbf{e}_2 = \mathbf{e}_3 = -\mathbf{e}_2\mathbf{e}_1, \quad \mathbf{e}_2\mathbf{e}_3 = \mathbf{e}_1 = -\mathbf{e}_3\mathbf{e}_2, \quad \mathbf{e}_3\mathbf{e}_1 = \mathbf{e}_2 = -\mathbf{e}_1\mathbf{e}_3.$$

For the standard basis system of the Hamiltonian quaternions one uses the original notation $\{1, \mathbf{i}, \mathbf{j}, \mathbf{k}\}$. In this paper we use a more general notation, $\{1, \mathbf{e}_1, \mathbf{e}_2, \mathbf{e}_3\}$, more suited for a future extension.

The real number $\mathbf{Sc}(\mathbf{a}) := a_0$ and $\mathbf{Vec}(\mathbf{a}) := a_1\mathbf{e}_1 + a_2\mathbf{e}_2 + a_3\mathbf{e}_3$ are the scalar and vector parts of $\mathbf{a}$, respectively. Analogously to the complex case, the conjugate of $\mathbf{a}$ is the quaternion $\overline{\mathbf{a}} = a_0 - a_1\mathbf{e}_1 - a_2\mathbf{e}_2 - a_3\mathbf{e}_3$. The norm of $\mathbf{a}$ is given by $|\mathbf{a}| = \sqrt{\mathbf{a}\overline{\mathbf{a}}}$, and coincides with its corresponding Euclidean norm, as a vector in $\mathbb{R}^4$.

We consider the subset

$$\mathcal{A} := \operatorname{span}_{\mathbb{R}}\{1, \mathbf{e}_1, \mathbf{e}_2\}$$

of $\mathbb{H}$. Its elements are usually called reduced quaternions. The real vector space $\mathbb{R}^3$ is to be embedded in $\mathcal{A}$ via the identification

$$x := (x_0, x_1, x_2) \in \mathbb{R}^3 \leftrightarrow \mathbf{x} := x_0 + x_1\mathbf{e}_1 + x_2\mathbf{e}_2 \in \mathcal{A}.$$

We denote by $\underline{x}$ the vectorial part of the reduced quaternion $\mathbf{x}$, that is, $\underline{x} := x_1\mathbf{e}_1 + x_2\mathbf{e}_2$. Also, we emphasize that $\mathcal{A}$ is a real vectorial subspace, but not a sub-algebra, of $\mathbb{H}$.

Let Ω be an open subset of $\mathbb{R}^3$ with a piecewise smooth boundary. A reduced quaternion-valued function or, briefly, an $\mathcal{A}$-valued function is a mapping $\mathbf{f} : \Omega \longrightarrow \mathcal{A}$ such that

$$\mathbf{f}(x) = [\mathbf{f}(x)]_0 + \sum_{i=1}^{2} [\mathbf{f}(x)]_i \mathbf{e}_i, \quad x \in \Omega,$$

where the coordinate-functions $[\mathbf{f}]_i$ $(i = 0, 1, 2)$ are real-valued functions defined in Ω. Properties such as continuity, differentiability or integrability are ascribed coordinate-wise. We will work with the real-linear Hilbert space of square integrable $\mathcal{A}$-valued functions defined in Ω, that we denote by $L_2(\Omega; \mathcal{A}; \mathbb{R})$. The real-valued inner product is defined by

$$\langle \mathbf{f}, \mathbf{g} \rangle_{L_2(\Omega;\mathcal{A};\mathbb{R})} = \int_\Omega \mathbf{Sc}(\overline{\mathbf{f}}\mathbf{g})\, dV, \tag{6.2.2}$$

where dV denotes the Lebesgue measure on Ω.

One possibility to generalize complex holomorphy is offered by following the Riemann approach. In this context for continuously real-differentiable functions $\mathbf{f} : \Omega \longrightarrow \mathcal{A}$, we consider the (reduced) quaternionic operator

$$D = \frac{\partial}{\partial x_0} + \mathbf{e}_1 \frac{\partial}{\partial x_1} + \mathbf{e}_2 \frac{\partial}{\partial x_2} \tag{6.2.3}$$

thet is called generalized Cauchy-Riemann operator on $\mathbb{R}^3$, as it corresponds to the 3-dimensional extension of the classical Cauchy-Riemann operator $\partial_{\overline{z}}$. In the same way, we define the conjugate quaternionic Cauchy-Riemann operator

$$\overline{D} = \frac{\partial}{\partial x_0} - \mathbf{e}_1 \frac{\partial}{\partial x_1} - \mathbf{e}_2 \frac{\partial}{\partial x_2}, \tag{6.2.4}$$

which is a generalization of the operator ∂_z.

Definition 6.1 (Monogenicity) A continuously real-differentiable $\mathcal{A}$-valued function $\mathbf{f}$ is called monogenic in Ω if $D\mathbf{f} = 0$ in Ω.

Remark 6.1 It is not necessary to distinguish between left and right monogenic in the case of $\mathcal{A}$-valued functions because $D\mathbf{f} = 0$ implies $\mathbf{f}D = 0$ and vice versa.

The generalized Cauchy-Riemann operator (6.2.3) and its conjugate (6.2.4) factorize the 3-dimensional Laplace operator, in the sense that $\Delta_3 \mathbf{f} = D\overline{D}\mathbf{f} = \overline{D}D\mathbf{f}$, whenever $\mathbf{f} \in C^2$, which implies that any monogenic function is also a harmonic function. This means that quaternionic analysis can be seen as a refinement of harmonic analysis.

According to [24, 37, 48], we use $(\frac{1}{2}\overline{D})\mathbf{f}$ as the *hypercomplex derivative* of a monogenic function $\mathbf{f}$. An $\mathcal{A}$-valued monogenic function with an identically vanishing hypercomplex derivative is called *hyperholomorphic constant* (see again [24]). It is immediately clear that such a function depends only on the variables x_1 and x_2.

Reconsider now a function $\mathbf{f} : \Omega \subset \mathbb{R}^3 \longrightarrow \mathcal{A}$ and write the Riesz system (6.2.1) explicitly as

$$(\mathrm{R}) \quad \begin{cases} \frac{\partial [\mathbf{f}]_0}{\partial x_0} - \frac{\partial [\mathbf{f}]_1}{\partial x_1} - \frac{\partial [\mathbf{f}]_2}{\partial x_2} = 0, \\ \frac{\partial [\mathbf{f}]_0}{\partial x_1} + \frac{\partial [\mathbf{f}]_1}{\partial x_0} = 0, \\ \frac{\partial [\mathbf{f}]_0}{\partial x_2} + \frac{\partial [\mathbf{f}]_2}{\partial x_0} = 0, \\ \frac{\partial [\mathbf{f}]_1}{\partial x_2} - \frac{\partial [\mathbf{f}]_2}{\partial x_1} = 0. \end{cases}$$

Using the generalized Cauchy-Riemann operator this reads as $D\mathbf{f} = \mathbf{f}D = 0$.

Following [35], the solutions of the system (R) are called (R)-solutions. The subspace of polynomial (R)- solutions of degree n is denoted by $\mathcal{M}^+(\Omega; \mathcal{A}; n)$. In [35], it is shown that the space $\mathcal{M}^+(\Omega; \mathcal{A}; n)$ has dimension $2n+3$. Later, this result was generalized to arbitrary higher dimensions in the framework of a Clifford algebra by Delanghe in [17]. We denote further by $\mathcal{M}^+(\Omega; \mathcal{A}) := L_2(\Omega; \mathcal{A}; \mathbb{R}) \cap \ker D$ the space of square integrable $\mathcal{A}$-valued monogenic functions defined in Ω. A basis of $\mathcal{M}^+(\mathbb{R}^3; \mathcal{A}; n)$, based on the well-known Fueter polynomials, was constructed by Leutwiler in [35]. His results were recently generalized to arbitrary dimensions by Delanghe in [16]. Both papers rely on the notion of harmonic conjugates. The main difficulty in carrying out such constructions is that only a certain number of these polynomials and their associated scalar parts are orthogonal to each other with respect to the real-inner product (6.2.2). Also, their norms are not easy to handle and the calculations are numerically highly unstable. For an account of such arguments, see [39], Chap. 2.

6.3 A Basis of $\mathcal{M}^+(B; \mathcal{A}; n)$

In ([9], Chap. 3) and [11], special $\mathbb{R}$-linear and $\mathbb{H}$-linear complete orthonormal systems of $\mathbb{H}$-valued homogeneous monogenic polynomials defined in the unit ball B of $\mathbb{R}^3$ are explicitly constructed ($\mathbb{R}^3 \to \mathbb{R}^4$ case). The main idea of such constructions is the already referred factorization of the Laplace operator. Partially motivated by these results (see [39], Chap. 1) we deal from now on with constructive aspects of approximation theory by presenting a convenient system of polynomial solutions of the Riesz system ($\mathbb{R}^3 \to \mathbb{R}^3$ case). The restriction to the sphere of this system can be viewed as analogue to the complex case of the Fourier exponential functions $\{e^{in\theta}\}_{n\geq 0}$ on the unit circle and constitutes a refinement of the well-known spherical harmonics.

The strategy adopted (the idea is taken from [9]) is the following: we start by considering the set of homogeneous harmonic polynomials

$$\left\{r^{n+1}U_{n+1}^0, r^{n+1}U_{n+1}^m, r^{n+1}V_{n+1}^m, m = 1, \ldots, n+1\right\}_{n\in\mathbb{N}_0}, \tag{6.3.1}$$

formed by the extensions in the ball of an orthogonal basis of spherical harmonics in $\mathbb{R}^3$ considered e.g. in [42]. The application of the operator $\frac{1}{2}\overline{D}$ to the homogeneous harmonic polynomials in (6.3.1) leads to the following set of homogeneous monogenic polynomials:

$$\left\{\mathbf{X}_n^{0,\dagger}, \mathbf{X}_n^{m,\dagger}, \mathbf{Y}_n^{m,\dagger} : m = 1, \ldots, n+1\right\}, \tag{6.3.2}$$

with the notation

$$\mathbf{X}_n^{0,\dagger} := r^n \mathbf{X}_n^0, \qquad \mathbf{X}_n^{m,\dagger} := r^n \mathbf{X}_n^m, \qquad \mathbf{Y}_n^{m,\dagger} := r^n \mathbf{Y}_n^m.$$

In [9] it is proved the following result:

Lemma 6.1 *For each n, the set*

$$\left\{\mathbf{X}_n^{0,\dagger}, \mathbf{X}_n^{m,\dagger}, \mathbf{Y}_n^{m,\dagger} : m = 1, \ldots, n+1\right\}$$

is orthogonal with respect to the real-inner product (6.2.2).

The consideration of the orthogonal system of harmonic spherical functions $\{U_n^0, U_n^m, V_n^m : m = 1, \ldots, n\}$ for the construction of our system allows to use properly some well-known results like its orthogonality on the unit sphere. Of course, as a consequence of the interrelation between spherical harmonics and Legendre polynomials (resp. Legendre functions), the constructed homogeneous monogenic polynomials are related to the Legendre polynomials (resp. Legendre functions) as well. A detailed study of Legendre polynomials and associated Legendre functions together with their relations with the polynomials (6.3.2) shows more interesting properties of the basis polynomials but they are not needed in the present article and so will not be discussed.

We now come to the aim of this section. In order to establish our main results, we first state the following properties of the basis polynomials.

Theorem 6.1 *For a fixed n, the homogeneous monogenic polynomials*

$$\left\{\mathbf{X}_n^{l,\dagger}, \mathbf{Y}_n^{m,\dagger}, \mathbf{X}_n^{n+1,\dagger}, \mathbf{Y}_n^{n+1,\dagger} : l = 0, \ldots, n, m = 1, \ldots, n\right\},$$

satisfy the properties:

1. *The following relations hold*:

$$\mathbf{Sc}(\mathbf{X}_n^{l,\dagger}) := \frac{(n+l+1)}{2} U_n^{l,\dagger},$$

$$\mathbf{Sc}(\mathbf{Y}_n^{m,\dagger}) := \frac{(n+m+1)}{2} V_n^{m,\dagger};$$

2. *The polynomials* $\mathbf{X}_n^{n+1,\dagger}$ *and* $\mathbf{Y}_n^{n+1,\dagger}$ *are hyperholomorphic constants.*

Proof The proof of the Affirmation 1 may be found in [25] and [39] by direct inspection of the scalar parts of the basis polynomials and their relations with the Legendre polynomials and associated functions. Affirmation 2 is already known from [9] because our constructed system is a subsystem of the polynomials which were constructed in [9]. □

The surprising observation in Affirmation 1 is that the scalar parts of the $\mathcal{A}$-valued homogeneous monogenic polynomials, which were obtained by applying the $\overline{D}$ operator to scalar-valued harmonic polynomials, are strongly related to the original polynomials. This means nothing else than the scalar parts being orthogonal to each other. In this sense, it reflects one of the most noteworthy properties of the system and it shows an immediate relationship with the classical complex function theory in the plane, where the real parts of the complex variables z^n are also orthogonal to each other. We remark that such a behavior is very different in the case of the Fueter polynomials (see [39], Chap. 2 for a special approach).

Now we want to orthonormalize the basis polynomials (6.3.2), and for that we proceed by presenting explicit formulae of their norms (see [9]) and of their corresponding scalar parts (see [25] and [26]).

Proposition 6.1 *For $n \in \mathbb{N}_0$, the norms of the homogeneous monogenic polynomials $\mathbf{X}_n^{0,\dagger}$, $\mathbf{X}_n^{m,\dagger}$ and $\mathbf{Y}_n^{m,\dagger}$ $(m = 1, \ldots, n)$ and their associated scalar parts are given by*

$$\|\mathbf{X}_n^{0,\dagger}\|_{L_2(B;\mathcal{A};\mathbb{R})} = \sqrt{\frac{\pi(n+1)}{2n+3}},$$

$$\|\mathbf{X}_n^{m,\dagger}\|_{L_2(B;\mathcal{A};\mathbb{R})} = \|\mathbf{Y}_n^{m,\dagger}\|_{L_2(B;\mathcal{A};\mathbb{R})} = \sqrt{\frac{\pi}{2}\frac{(n+1)}{2n+3}\frac{(n+1+m)!}{(n+1-m)!}},$$

$$\|\mathbf{X}_n^{n+1,\dagger}\|_{L_2(B;\mathcal{A};\mathbb{R})} = \|\mathbf{Y}_n^{n+1,\dagger}\|_{L_2(B;\mathcal{A};\mathbb{R})} = \sqrt{\frac{\pi}{2}\frac{(n+1)(2n+2)!}{2n+3}},$$

$$\|\mathbf{Sc}(\mathbf{X}_n^{0,\dagger})\|_{L_2(B)} = \frac{(n+1)}{\sqrt{2n+3}}\sqrt{\frac{\pi}{2n+1}},$$

$$\|\mathbf{Sc}(\mathbf{X}_n^{m,\dagger})\|_{L_2(B)} = \|\mathbf{Sc}(\mathbf{Y}_n^{m,\dagger})\|_{L_2(B)} = \frac{(n+1+m)}{\sqrt{2n+3}}\sqrt{\frac{\pi}{2}\frac{1}{(2n+1)}\frac{(n+m)!}{(n-m)!}}.$$

From now on we shall denote by $\mathbf{X}_n^{0,\dagger,*}$, $\mathbf{X}_n^{m,\dagger,*}$, $\mathbf{Y}_n^{m,\dagger,*}$ $(m = 1, \ldots, n+1)$ the new normalized basis functions $\mathbf{X}_n^{0,\dagger}$, $\mathbf{X}_n^{m,\dagger}$, $\mathbf{Y}_n^{m,\dagger}$ in $L_2(B;\mathcal{A};\mathbb{R})$. We begin by formulating a theorem from [39], p. 96:

Theorem 6.2 *For each n, the normalized set of $2n+3$ homogeneous monogenic polynomials*

$$\left\{\mathbf{X}_n^{0,\dagger,*}, \mathbf{X}_n^{m,\dagger,*}, \mathbf{Y}_n^{m,\dagger,*} : m = 1, \ldots, n+1\right\} \tag{6.3.3}$$

forms an orthonormal basis in the subspace $\mathcal{M}^+(B;\mathcal{A};n)$ with respect to the real-inner product (6.2.2). Consequently,

$$\left\{\mathbf{X}_n^{0,\dagger,*}, \mathbf{X}_n^{m,\dagger,*}, \mathbf{Y}_n^{m,\dagger,*}, m = 1, \ldots, n+1; n = 0, 1, \ldots\right\}$$

is an orthonormal basis in $\mathcal{M}^+(B;\mathcal{A})$.

The above consideration makes it possible to define the Fourier expansion of a square integrable $\mathcal{A}$-valued monogenic function. Next we formulate the result:

Theorem 6.3 *Let $\mathbf{f}$ be a square integrable $\mathcal{A}$-valued monogenic function. The function $\mathbf{f}$ can then be represented with the orthonormal system (6.3.3):*

$$\mathbf{f} = \sum_{n=0}^{\infty}\left[\mathbf{X}_n^{0,\dagger,*}a_n^0 + \sum_{m=1}^{n+1}\left(\mathbf{X}_n^{m,\dagger,*}a_n^m + \mathbf{Y}_n^{m,\dagger,*}b_n^m\right)\right], \tag{6.3.4}$$

where for each $n \in \mathbb{N}_0$, $a_n^0, a_n^m, b_n^m \in \mathbb{R}$ $(m = 1, \ldots, n+1)$ *are the associated Fourier coefficients.*

Remark 6.2 Since by construction the system (6.3.3) forms an orthonormal basis with respect to the real-inner product (6.2.2), we stress that the coefficients a_n^0, a_n^m and b_n^m $(m = 1, \ldots, n+1)$ are real constants.

Based on representation (6.3.4), in [26] we have proved that each $\mathcal{A}$-valued monogenic function can be decomposed in an orthogonal sum of a monogenic "main part" of the function ($\mathbf{g}$) and a hyperholomorphic constant ($\mathbf{h}$).

Lemma 6.2 (Decomposition theorem) *A function* $\mathbf{f} \in \mathcal{M}^+(B; \mathcal{A})$ *can be decomposed into*

$$\mathbf{f} := \mathbf{f}(0) + \mathbf{g} + \mathbf{h}, \tag{6.3.5}$$

where the functions $\mathbf{g}$ *and* $\mathbf{h}$ *have the Fourier series*

$$\mathbf{g}(x) = \sum_{n=1}^{\infty} \left(\mathbf{X}_n^{0,\dagger,*}(x) a_n^0 + \sum_{m=1}^{n} \left[\mathbf{X}_n^{m,\dagger,*}(x) a_n^m + \mathbf{Y}_n^{m,\dagger,*}(x) b_n^m \right] \right),$$

$$\mathbf{h}(\underline{x}) = \sum_{n=1}^{\infty} \left[\mathbf{X}_n^{n+1,\dagger,*}(\underline{x}) a_n^{n+1} + \mathbf{Y}_n^{n+1,\dagger,*}(\underline{x}) b_n^{n+1} \right].$$

Obviously, using Parseval's identity, $\mathbf{f}$ may be characterized by its coefficients in the following way:

Theorem 6.4 *The function* $\mathbf{f}$ *is a square integrable* $\mathcal{A}$*-valued monogenic function iff*

$$\sum_{n=0}^{\infty} \left((a_n^0)^2 + \sum_{m=1}^{n+1} \left[(a_n^m)^2 + (b_n^m)^2 \right] \right) < \infty. \tag{6.3.6}$$

6.4 Generation of $\mathcal{A}$-Valued Monogenic Functions

Despite the fact that Quaternionic analysis offers a possibility to generalize some of the most important features of classical complex analysis, the monogenic functions do not enjoy all the properties of the holomorphic functions in one complex variable. In fact, because of the non-commutativity of the quaternionic multiplication the product of two monogenic functions is seldom a monogenic function. Therefore, the construction of elementary monogenic functions is a challenging problem. It is already known that there are a lot of techniques, which generate monogenic functions (a list of those techniques can be found e.g. in [15]). In this section we consider the problem of deriving $\mathcal{A}$-valued monogenic functions with prescribed asymptotic properties, in particular, asymptotic properties related to the Fourier coefficients of a scalar-valued harmonic function. This approach is related to the generation of monogenic functions by means of conjugate harmonic functions.

Given a harmonic function U in a domain Ω of $\mathbb{R}^4$ with a specific geometric property, the problem of finding a harmonic conjugate V, generalizing the well-known case of the complex plane to the case of quaternion-valued monogenic functions, was introduced by Sudbery in [48]. The author proposed an algorithm for the calculation of quaternion-valued monogenic functions. However, since our particular interest in this contribution is the study of monogenic functions with values in the reduced quaternions, the described procedure is not well-adapted to our case. Later and independently, Xu in [50] considered the problem of conjugate harmonics in the framework of Clifford analysis. In [52] and [51] the construction of conjugate harmonics to the Poisson kernel in the open unit ball and the upper half space respectively are obtained in this setting. The extension and completeness of these results were obtained in [7] and [5] by Brackx, Delanghe and Sommen. Therein an algorithm is constructed for the calculation of a harmonic conjugate V to a given harmonic function U in Ω. By the nature of the given construction, the authors observed that such function V is not unique. While these results establish a general form of V, its precise description is given only up to a solution of a Poisson equation. Also, it is not discussed if the functions U and V belong to certain function spaces. For this reason, our research has been focused on the construction of an explicit algorithm respecting the underlying function spaces.

Next, we briefly introduce the notion of harmonic conjugates:

Definition 6.2 (Conjugate harmonic functions, see [7]) Let U be a harmonic function defined in an open subset Ω of $\mathbb{R}^3$. A vector-valued harmonic function V in Ω is called conjugate harmonic to U if $\mathbf{f} := U + V$ is monogenic in Ω. The pair $(U; V)$ is called a pair of conjugate harmonic functions in Ω.

In the sequel, assume U be a square integrable harmonic function defined in an open subset Ω of $\mathbb{R}^3$. In contrast to the approach presented in [5, 7], the general theory developed in the previous section gives the possibility to express explicitly the general form of a square integrable vector-valued harmonic function V conjugate to the scalar-valued U. This special case of conjugate harmonicity was first introduced by Moisil in [38] and taken up again by Stein and Weiß in [47].

First result in our line of consideration was already proved in [39].

Theorem 6.5 *Let U be harmonic and square integrable in $B \subset \mathbb{R}^3$ with respect to the orthonormal system*

$$\left\{ \frac{\mathbf{Sc}(\mathbf{X}_n^{0,\dagger})}{\|\mathbf{Sc}(\mathbf{X}_n^{0,\dagger})\|_{L_2(B)}}, \frac{\mathbf{Sc}(\mathbf{X}_n^{m,\dagger})}{\|\mathbf{Sc}(\mathbf{X}_n^{m,\dagger})\|_{L_2(B)}}, \frac{\mathbf{Sc}(\mathbf{Y}_n^{m,\dagger})}{\|\mathbf{Sc}(\mathbf{Y}_n^{m,\dagger})\|_{L_2(B)}} : m = 1, \ldots, n \right\} \quad (6.4.1)$$

given by

$$U = \sum_{n=0}^{\infty} \Bigg[\frac{\mathbf{Sc}(\mathbf{X}_n^{0,\dagger})}{\|\mathbf{Sc}(\mathbf{X}_n^{0,\dagger})\|_{L_2(B)}} a_n^0 + \sum_{m=1}^{n} \Bigg(\frac{\mathbf{Sc}(\mathbf{X}_n^{m,\dagger})}{\|\mathbf{Sc}(\mathbf{X}_n^{m,\dagger})\|_{L_2(B)}} a_n^m + \frac{\mathbf{Sc}(\mathbf{Y}_n^{m,\dagger})}{\|\mathbf{Sc}(\mathbf{Y}_n^{m,\dagger})\|_{L_2(B)}} b_n^m \Bigg) \Bigg], \tag{6.4.2}$$

where for each $n \in \mathbb{N}_0$, $a_n^0, a_n^m, b_n^m \in \mathbb{R}$ $(m = 1, \ldots, n)$ *are the associated Fourier coefficients. If the series*

$$\sum_{n=0}^{\infty} \Bigg[\frac{([\mathbf{X}_n^{0,\dagger}]_1 \mathbf{e}_1 + [\mathbf{X}_n^{0,\dagger}]_2 \mathbf{e}_2)}{\|\mathbf{Sc}(\mathbf{X}_n^{0,\dagger})\|_{L_2(B)}} a_n^0 + \sum_{m=1}^{n} \Bigg(\frac{([\mathbf{X}_n^{m,\dagger}]_1 \mathbf{e}_1 + [\mathbf{X}_n^{m,\dagger}]_2 \mathbf{e}_2)}{\|\mathbf{Sc}(\mathbf{X}_n^{m,\dagger})\|_{L_2(B)}} a_n^m + \frac{([\mathbf{Y}_n^{m,\dagger}]_1 \mathbf{e}_1 + [\mathbf{Y}_n^{m,\dagger}]_2 \mathbf{e}_2)}{\|\mathbf{Sc}(\mathbf{Y}_n^{m,\dagger})\|_{L_2(B)}} b_n^m \Bigg) \Bigg] \tag{6.4.3}$$

is convergent, then it defines a square integrable vector-valued harmonic function V *conjugate to* U.

Proof Let $U \in L_2(B)$ be a harmonic function. Consider the Fourier series of U with respect to the orthonormal system (6.4.1). Since the series (6.4.2) is convergent in L_2, it converges uniformly to U in each compact subset of B. Now, replacing the scalar part of each polynomial by the full polynomial we get formally

$$\begin{aligned} \mathbf{f} &= \sum_{n=0}^{\infty} \Bigg[\frac{\mathbf{X}_n^{0,\dagger}}{\|\mathbf{Sc}(\mathbf{X}_n^{0,\dagger})\|_{L_2(B)}} a_n^0 + \sum_{m=1}^{n} \Bigg(\frac{\mathbf{X}_n^{m,\dagger}}{\|\mathbf{Sc}(\mathbf{X}_n^{m,\dagger})\|_{L_2(B)}} a_n^m + \frac{\mathbf{Y}_n^{m,\dagger}}{\|\mathbf{Sc}(\mathbf{Y}_n^{m,\dagger})\|_{L_2(B)}} b_n^m \Bigg) \Bigg] \\ &:= U + V. \end{aligned} \tag{6.4.4}$$

□

Remark 6.3 If the previous series are finite sums then the functions U and V are polynomials. Then, it is clear that the partial expansion (6.4.3) makes always sense. In this special case our approach covers the results obtained in [7] and [5].

Remark 6.4 The vector-valued function V conjugate to the scalar-valued U is not unique. This is due to the fact that by adding any hyperholomorphic constant $\boldsymbol{\varphi}$ to V the resulting function $\tilde{V} := V + \boldsymbol{\varphi}$ is also harmonic conjugate to U.

Remark 6.5 By the direct construction of formula (6.4.4), we get only $2n + 1$ homogeneous monogenic polynomials (i.e., the monogenic "main part" of $\mathbf{f}$). However, since $\dim \mathcal{M}^+(\mathbb{R}^3; \mathcal{A}; n) = 2n+3$, adding two hyperholomorphic constants the necessary number of independent polynomials is achieved.

Remark 6.6 Originally the Fourier coefficients a_n^0, a_n^m, b_n^m $(m = 1, \dots, n)$ are defined as inner products of U and elements of the space $\mathcal{H}_n(\mathbb{R}^3)$. The new result is that one obtains, up to a factor, the same coefficients but now as inner products between $\mathbf{f}$ and elements of the space $\mathcal{M}^+(\mathbb{R}^3; \mathcal{A}; n)$.

The main point in the approach presented in [5, 7] as well as Sudbery's formula [48] is the construction of harmonic conjugates "function by function". No attention was paid to the question to which function spaces these conjugate harmonics and the whole monogenic function belong. In [4] this question was studied for conjugate harmonics via Sudbery's formula in the scale of Bergman spaces. As already mentioned this result is not applicable for $\mathcal{A}$-valued functions. If we want to make the previous theorem more precise then we need an a-priori criterion that ensures the convergence of the constructed series for mononogenic functions in $L_2(B; \mathcal{A}; \mathbb{R})$.

Theorem 6.6 *Let U be harmonic and square integrable in $B \subset \mathbb{R}^3$ with respect to the orthonormal system* (6.4.1)*. Additionally, we assume that the Fourier coefficients satisfy the condition*

$$\sum_{n=0}^{\infty} \left(\frac{2n+1}{n+1} (a_n^0)^2 + \sum_{m=1}^{n} \frac{(n+1)(2n+1)}{(n+1)^2 - m^2} [(a_n^m)^2 + (b_n^m)^2] \right) < \infty.$$

Then, the series in Theorem 6.5 *converges in $L_2(B)$.*

The proof runs analogously to the previous theorem. Of course, this criterion is not well-applicable in practice and it is still open to characterize which function space (for the functions U) is described by the condition of the theorem. If we suppose for the moment more smoothness of the given function U, then we can count with an exponential decay of the Fourier coefficients and we can formulate a simple sufficient condition to guarantee the convergence of the series expansion for V.

Theorem 6.7 (See [39]) *Let U be harmonic and square integrable in $B \subset \mathbb{R}^3$. If the absolute values of its Fourier coefficients a_n^0, a_n^m and b_n^m $(m = 1, \dots, n)$, with respect to the orthonormal system* (6.4.1) *in $L_2(B)$, are less than $\frac{c}{(n+1)^{1+\alpha}}$ $(\alpha > 1/2)$ with a positive constant c, then $\mathbf{f} := U + V \in \mathcal{M}^+(B; \mathcal{A})$.*

Proof Let $U \in L_2(B)$ be a harmonic function. Consider the Fourier series of U with respect to the orthonormal system (6.4.1). Now, as before, we replace the scalar part of each polynomial by the full polynomial and by introducing suitable correction factors we can rewrite the obtained series as a series expansion with respect to the normalized full polynomials. We get formally

$$\mathbf{f}=\sum_{n=0}^{\infty}\left[\mathbf{X}_n^{0,\dagger,*}\sqrt{\frac{2n+1}{n+1}}a_n^0+\sum_{m=1}^{n}\sqrt{\frac{(n+1)(2n+1)}{(n+1)^2-m^2}}\left(\mathbf{X}_n^{m,\dagger,*}a_n^m+\mathbf{Y}_n^{m,\dagger,*}b_n^m\right)\right].$$

On the right-hand side of the previous equality we recognize the Fourier expansion of the function $\mathbf{f}$ with respect to the orthonormal system (6.3.3) in $\mathcal{M}^+(B;\mathcal{A})$. Having in mind the conditions of the L_2-convergence of (6.4.2), our task now is to find out if the series

$$\sum_{n=0}^{\infty}\left(\frac{(2n+1)}{n+1}(a_n^0)^2+\sum_{m=1}^{n}\frac{(n+1)(2n+1)}{(n+1)^2-m^2}\left[(a_n^m)^2+(b_n^m)^2\right]\right) \tag{6.4.5}$$

is convergent. By assumption, there exists a constant c such that the Fourier coefficients a_n^0, a_n^m, b_n^m $(n\in\mathbb{N}_0, m=1,\ldots,n)$ satisfy

$$|a_n^0|,\ |a_n^m|,\ |b_n^m|<\frac{c}{(n+1)^{1+\alpha}},\qquad \alpha>1/2,\qquad m=1,\ldots,n.$$

Substituting in the expression (6.4.5) we get

$$\|\mathbf{f}\|^2_{L_2(B;\mathcal{A};\mathbb{R})}<\sum_{n=0}^{\infty}\frac{c^2}{(n+1)^{2(1+\alpha)}}\left(\frac{2n+1}{n+1}+2(n+1)n\right)\leq\sum_{n=0}^{\infty}\frac{2c^2}{(n+1)^{2\alpha}}.$$

The series on the right-hand side is convergent, because by assumption $\alpha>\frac{1}{2}$. Consequently, the series (6.4.5) is convergent. □

References

1. Abul-Ez, M.A., Constales, D.: Basic sets of polynomials in Clifford analysis. Complex Var. **14**(1–4), 177–185 (1990)
2. Abul-Ez, M.A., Constales, D.: Linear substitution for basic sets of polynomials in Clifford analysis. Portugaliae Math. **48**(2), 143–154 (1991)
3. Abul-Ez, M.A., Constales, D.: The square root base of polynomials in Clifford analysis. Arch. Math. **80**(5), 486–495 (2003)
4. Avetisyan, K., Gürlebeck, K., Sprößig, W.: Harmonic conjugates in weighted Bergman spaces of quaternion-valued functions. Comput. Methods Funct. Theory **9**(2), 593–608 (2009)
5. Brackx, F., Delanghe, R.: On harmonic potential fields and the structure of monogenic functions. Z. Anal. Anwend. **22**(2), 261–273 (2003)
6. Brackx, F., Delanghe, R., Sommen, F.: Clifford Analysis. Pitman, Boston/London/Melbourne (1982)
7. Brackx, F., Delanghe, R., Sommen, F.: On conjugate harmonic functions in Euclidean space. Math. Methods Appl. Sci. **25**(16–18), 1553–1562 (2002)
8. Brackx, F., De Schepper, N., Sommen, F.: Clifford algebra-valued orthogonal polynomials in Euclidean space. J. Approx. **137**(1), 108–122 (2005)
9. Cação, I.: Constructive Approximation by Monogenic Polynomials. Ph.D. thesis, Universidade de Aveiro, Departamento de Matemática, Dissertation (2004)

10. Cação, I., Gürlebeck, K., Malonek, H.: Special monogenic polynomials and L_2-approximation. Adv. Appl. Clifford Algebras **11**(S2), 47–60 (2001)
11. Cação, I., Gürlebeck, K., Bock, S.: On derivatives of spherical monogenics. Complex Var. Elliptic Equ. **51**(8–11), 847–869 (2006)
12. Cação, I., Gürlebeck, K., Bock, S.: Complete orthonormal systems of spherical monogenics—a constructive approach. In: Son, L.H., et al. (eds.) Methods of Complex and Clifford Analysis, pp. 241–260. SAS International, Delhi (2005)
13. Cnops, J.: Orthogonal functions associated with the Dirac operator. Ph.D. thesis, Ghent university (1989)
14. De Bie, H., Sommen, F.: Hermite and Gegenbauer polynomials in superspace using Clifford analysis. J. Phys. A, Math. Theor. **40**(34), 10441–10456 (2007)
15. Delanghe, R.: Clifford analysis: history and perspective. Comput. Methods Funct. Theory **1**(1), 107–153 (2001)
16. Delanghe, R.: On homogeneous polynomial solutions of the Riesz system and their harmonic potentials. Complex Var. Elliptic Equ. **52**(10–11), 1047–1062 (2007)
17. Delanghe, R.: On a class of inner spherical monogenics and their primitives. Adv. Appl. Clifford Algebras **18**(3–4), 557–566 (2008)
18. Falcão, M.I., Malonek, H.: Generalized exponentials through Appell sets in $\mathbb{R}^{n+1}$ and Bessel functions. In: AIP-Proceedings, pp. 738–741 (2007)
19. Falcão, M.I., Malonek, H.: Special monogenic polynomials—properties and applications. In: AIP-Proceedings, pp. 764–767 (2007)
20. Falcão, M., Cruz, J., Malonek, H.: Remarks on the generation of monogenic functions. In: Gürlebeck, K., Könke, C. (eds.) Proceedings 17th International Conference on the Applications of Computer Science and Mathematics in Architecture and Civil Engineering. Weimar (2006)
21. Fueter, R.: Analytische Funktionen einer Quaternionenvariablen. Comment. Math. Helv. **4**, 9–20 (1932)
22. Fueter, R.: Functions of a Hyper Complex Variable. Lecture notes written and supplemented by E. Bareiss, Math. Inst. Univ. Zürich, Fall Semester (1949)
23. Gürlebeck, K.: Interpolation and best approximation in spaces of monogenic functions. Wiss. Z. TU Karl-Marx-Stadt **30**, 38–40 (1988)
24. Gürlebeck, K., Malonek, H.: A hypercomplex derivative of monogenic functions in $\mathbb{R}^{n+1}$ and its applications. Complex Var. Elliptic Equ. **39**(3), 199–228 (1999)
25. Gürlebeck, K., Morais, J.: On mapping properties of monogenic functions. CUBO A Math. J. **11**(1), 73–100 (2009)
26. Gürlebeck, K., Morais, J.: Bohr type theorems for monogenic power series. Comput. Methods Funct. Theory **9**(2), 633–651 (2009)
27. Gürlebeck, K., Morais, J.: On local mapping properties of monogenic functions. In: Gürlebeck, K., Könke, C. (eds.) Proceedings 18th International Conference on the Applications of Computer Science and Mathematics in Architecture and Civil Engineering. Weimar (2009)
28. Gürlebeck, K., Sprössig, W.: Quaternionic Analysis and Elliptic Boundary Value Problems. Akademie Verlag, Berlin (1989)
29. Gürlebeck, K., Sprössig, W.: On the treatment of fluid problems by methods of Clifford analysis. Math. Comput. Simul. **44**(4), 401–413 (1997)
30. Gürlebeck, K., Sprössig, W.: Quaternionic Calculus for Engineers and Physicists. Wiley, Chichester (1997)
31. Gürlebeck, K., Sprössig, W.: On eigenvalue estimates of nonlinear Stokes eigenvalue problems. In: Micali, A., et al. (eds.) Clifford Algebras and Their Applications in Mathematical Physics, pp. 327–333. Kluwer Academic, Amsterdam (1992)
32. Kamzolov, A.: The best approximation of the classes of functions $W_p^\alpha(S^n)$ by polynomials in spherical harmonics. Math. Not. **32**(3), 622–626 (1982)
33. Kravchenko, V.: Applied Quaternionic Analysis. Research and Exposition in Mathematics, vol. 28. Heldermann, Lemgo (2003)

34. Kravchenko, V., Shapiro, M.: Integral Representations for Spatial Models of Mathematical Physics. Research Notes in Mathematics. Pitman Advanced Publishing Program, London (1996)
35. Leutwiler, H.: Quaternionic analysis in $\mathbb{R}^3$ versus its hyperbolic modification. In: Brackx, F., Chisholm, J.S.R., Soucek, V. (eds.) NATO Science Series II. Mathematics, Physics and Chemistry, vol. 25. Kluwer Academic, Dordrecht/Boston/London (2001)
36. Malonek, H.: Power series representation for monogenic functions in $\mathbb{R}^{m+1}$ based on a permutational product. Complex Var. Elliptic Equ. **15**(3), 181–191 (1990)
37. Mitelman, I., Shapiro, M.: Differentiation of the Martinelli-Bochner integrals and the notion of hyperderivability. Math. Nachr. **172**(1), 211–238 (1995)
38. Moisil, G.: Sur la généralisation des fonctions conjuguées. Rend. Acad. Naz. Lincei **14**, 401–408 (1931)
39. Morais, J.: Approximation by homogeneous polynomial solutions of the Riesz system in $\mathbb{R}^3$. Ph.D. thesis, Bauhaus-Universität Weimar (2009)
40. Riesz, M.: Clifford numbers and spinors. Inst. Phys. Sci. Techn. Lect. Ser., vol. 38. Maryland (1958)
41. Ryan, J.: Left regular polynomials in even dimensions, and tensor products of Clifford algebras. In: Chisholm, J.S.R., Common, A.K. (eds.) Clifford Algebras and Their Applications in Mathematical Physics, pp. 133–147. Reidel, Dordrecht (1986)
42. Sansone, G.: Orthogonal Functions. Pure and Applied Mathematics, vol. IX. Interscience Publishers, New York (1959)
43. Shapiro, M., Vasilevski, N.L.: Quaternionic ψ-hyperholomorphic functions, singular operators and boundary value problems I. Complex Var. Theory Appl. (1995)
44. Shapiro, M., Vasilevski, N.L.: Quaternionic ψ-hyperholomorphic functions, singular operators and boundary value problems II. Complex Var. Theory Appl. (1995)
45. Sprössig, W.: Boundary value problems treated with methods of Clifford analysis. Contemp. Math. **212**, 255–268 (1998)
46. Stein, E.M.: Singular Integrals and Differentiability Properties of Functions. Princeton University Press, Princeton (1970)
47. Stein, E.M., Weiß, G.: On the theory of harmonic functions of several variables. Part I: the theory of H^p spaces. Acta Math. **103**, 25–62 (1960)
48. Sudbery, A.: Quaternionic analysis. Math. Proc. Camb. Philos. Soc. **85**, 199–225 (1979)
49. Whittaker, J.M.: Sur les séries de base de polynômes quelconques. Avec la col laboration de C. Gattegno (Collection de monographies sur la theorie des fonctions). Paris: Gauthier-Villars (1949)
50. Xu, Z.: Boundary value problems and function theory for spin-invariant differential operators. Ph.D thesis. Gent (1989)
51. Xu, Z., Zhou, C.: On boundary value problems of Riemann-Hilbert type for monogenic functions in a half space of $\mathbb{R}^m$ $(m \geq 2)$. Complex Var. Theory Appl. **22**, 181–194 (1993)
52. Xu, Z., Chen, J., Zhang, W.: A harmonic conjugate of the Poisson kernel and a boundary value problem for monogenic functions in the unit ball of $\mathbb{R}^n$ $(n \geq 2)$. Simon Stevin **64**, 187–201 (1990)

Chapter 7
Brief Survey on the CP Methods for the Schrödinger Equation

L.Gr. Ixaru

Abstract The CP methods have some salient advantages over other methods, viz.: (i) the accuracy is uniform with respect to the energy E; (ii) there is an easy control of the error; (iii) the step widths are unusually big and the computation is fast; (iv) the form of the algorithm allows a direct evaluation of collateral quantities such as normalisation constant, Prüfer phase, or the derivative of the solution with respect to E; (v) the algorithm is of a form which allows using parallel computation.

Keywords Schrödinger equation · CP methods · η_m set of functions

Mathematics Subject Classification (2000) 81Q05

7.1 Introduction

The piecewise perturbation methods (PPM) are a class of numerical methods specially devised for the solution of the Schrödinger equation, and the CP methods form a subclass of these. The main element of attractivity for the methods in this subclass is that their algorithms are the easiest to construct and the fastest in runs.

Given the one-dimensional Schrödinger equation

$$\frac{d^2y}{dx^2} + [E - V(x)]y = 0, \quad x \in [a, b], \tag{7.1.1}$$

L.Gr. Ixaru (✉)
Department of Theoretical Physics, "Horia Hulubei" National Institute of Physics and Nuclear Engineering, P.O. Box MG-6, Bucharest, Romania
e-mail: ixaru@theory.nipne.ro

L.Gr. Ixaru
Academy of Romanian Scientists, 54 Splaiul Independentei, 050094, Bucharest, Romania

T.E. Simos (ed.), *Recent Advances in Computational and Applied Mathematics*,
DOI 10.1007/978-90-481-9981-5_7,

where the potential $V(x)$ is a well behaved function and the energy E is a free parameter, one can formulate either an initial value (IV) or a boundary value (BV) problem. The form of $V(x)$ is determined by the physical phenomenon under investigation, and a large variety of shapes are encountered. However, analytic solutions of (7.1.1), either for the IV or for the BV problem, are known only for a small number of expressions for the function $V(x)$, let such functions be denoted by $\bar{V}(x)$, such that any attempt to obtain an approximate solution in analytic form for a realistic problem must have the selection of $\bar{V}(x)$ which best fits the shape of the original $V(x)$ as its first step. In fact, this was the standard way of doing the things decades ago, when computers were at their infancy.

In the early stages the fit was global (that is a single fitted $\bar{V}(x)$ was used in place of $V(x)$ over the whole equation interval) but quite soon it became clear that the quality of the approximation increases if this fit is made *piecewise*, i.e., when the interval $I_{a,b} = [a,\ b]$ is first partitioned, $x_0 = a,\ x_1,\ x_2, \dots, x_{k_{max}} = b$, and a suitable $\bar{V}(x)$ is introduced on each step $I_k = [x_{k-1},\ x_k]$. If so is done, the solution on the whole $I_{a,b}$ requires an extra algebraic manipulation of the piecewise analytic solutions, a task which can be accomplished efficiently only on a computer, and this explains why the first systematic investigations along these lines are in the 60's. In [4, 6, 7] the piecewise $\bar{V}(x)$ was a constant, that is the average value of $V(x)$ over I_k, while in [5] it was a straightline segment which is tangent to $V(x)$. The versions corresponding to the two options are called CP (short for constant-based potential) and LP (line-based potential) methods, respectively. The two linear independent solutions are expressed by trigonometric or hyperbolic functions for the CP methods, and by the Airy functions for the LP methods.

7.2 The Algorithm of a CP Method

To fix the ideas we concentrate on a single step I_k denoted generically as $[X, X+h]$, and show how the solution can be propagated from one end to the other end. As said before, the first versions [4–7] contained only the solution from the reference potential $\bar{V}(X+\delta), \delta \in [0, h]$ and therefore their order was low, namely 2, see [8]. It has also became clear that the order can be increased only if corrections from the perturbation $\Delta V(\delta) = V(X+\delta) - \bar{V}(X+\delta)$ can be added into the algorithm but this raised another problem: are the formulae of these corrections simple enough for not charging the run time too much? The answer was given by Ixaru in [9]: if $\bar{V}(X+\delta)$ is a constant and $\Delta V(\delta)$ is a polynomial then a set of special functions can be introduced such that the perturbation corrections have simple analytic forms. In fact, all CP versions built up later on were based on this.

The starting point then technically consists in approximating the potential function by a polynomial, and a first problem is how the polynomial coefficients have to be chosen in order to ensure the best fit. The most important result is due to Pruess [24] and it says in essence that the best fit consists in taking a finite number of terms in the expansion of $V(x)$ over shifted Legendre polynomials of argument δ/h,

$$V(X+\delta) \approx V^p(X+\delta) = \sum_{n=0}^{N} V_n h^n P_n^*\left(\frac{\delta}{h}\right). \tag{7.2.1}$$

In [9] V^p is called a pilot potential and all subsequent operations use only this. As a matter of fact, taking V^p for V induces a residual error of $\mathcal{O}(h^{2N+2})$.

The expressions of several $P_n^*(\gamma)$ polynomials,$\gamma \in [0, 1]$, are as follows (see [1]):

$$P_0^*(\gamma) = 1, \qquad P_1^*(\gamma) = -1 + 2\gamma,$$
$$P_2^*(\gamma) = 1 - 6\gamma + 6\gamma^2, \qquad P_3^*(\gamma) = -1 + 12\gamma - 30\gamma^2 + 20\gamma^3.$$

The form (7.2.1) is separated into the constant reference potential $\bar{V}(X+\delta) = V_0$ and the perturbation $\Delta V(\delta) = V^p(X+\delta) - V_0$ such that the one step solution results by taking the exact solution from the reference potential and by adding to this as many corrections as possible from the perturbation ΔV.

Each member of the CPM family is identified by the degree N of the polynomial $\Delta V(\delta)$ and by the number of perturbation corrections Q retained in the algorithm, CPM$[N,\ Q]$ for short. The simplest version, that is when no correction is introduced, is identified either as CPM[0, 0] or directly as CPM(0).

How do N and Q influence the quality of the method? The main parameter for the quality of a numerical method is its *order*. Roughly speaking, if $\epsilon(h)$ and $\epsilon(h/2)$ are the errors produced by using the sufficiently small steps h and $h/2$ then the order p is that number such that

$$\epsilon(h/2) = \epsilon(h)/2^p.$$

Example For $p = 4$ the error at $h/2$ is by a factor $2^4 = 16$ smaller than that at h, but for $p = 6$ it is smaller by a factor $2^6 = 64$. The Numerov and standard Runge-Kutta methods are of order $p = 4$.

Intuition says that the order p must increase with N and Q and indeed this is confirmed by a theorem proved in [16]. In particular, for fixed N the maximal p is $2N+2$ because the residual error acts as a barrier. The dependence of p with respect to N and Q is listed in Table 7.1 for a number of cases. In that table the value of the minimal Q is given, with the meaning that any greater number of corrections does not influence anymore the value of the order. For example, versions CPM[4, 3] and CPM[4, 4] have the same order $p = 10$.

As for the versions for solving the one-channel equation (7.1.1) published in the literature, we find versions of orders between 2 (when no correction is added) and 6 in [9], up to 12 in [16], and 18 in [19]. A Fortran code for the CP version of [16] is in [17]. In all these the independent variable x, potential function $V(x)$ and energy E are assumed real. CPM versions when these are complex are also available, see [15].

Table 7.1 Link between order p and parameters N and Q of the version CPM[N, Q]

p	N	minimal Q
2	0	0
4	1	1
6	2	2
8	3	3
10	4	3
12	5	4

The case of systems of coupled-channel equations also enjoyed much attention. Again, the need of first having each element of the potential matrix approximated by a polynomial plays a central role in making the expressions of the corrections attractive from computational point of view. However the orders of the existing versions is comparatively lower than before. In fact, the orders are between 2 and 6 in [9, 11], and up to 10 in [21]. Other numerical approaches pertinent to the Schrödinger equation can be found in [25].

What about the computational effort? For the one-channel case the CPU time/step increases with p but moderately: for $p = 12$ it is only 2–3 times bigger than for $p = 2$. For the coupled-channel case the rate of increase depends on the number of channels.

As said, we look for a procedure which gives the values of y and y' at $X+h$ when the values at X are given (forward propagation) or, conversely, yields the values at X when the values at $X + h$ are given (backward propagation).

The values of y and y' at X and $X+h$ are connected via the so-called propagation (or transfer) matrix

$$\mathbf{P} = \begin{bmatrix} u(h) & v(h) \\ u'(h) & v'(h) \end{bmatrix}, \tag{7.2.2}$$

where $u(\delta)$ and $v(\delta)$, $\delta = x - X \in [0, h]$ are two linear independent solutions of (7.1.1) on $[X, X + h]$ which satisfy the following initial conditions

$$u(0) = 1, \quad u'(0) = 0, \qquad v(0) = 0, \quad v'(0) = 1. \tag{7.2.3}$$

Indeed, with the column vector

$$\mathbf{y}(x) = [y(x), y'(x)]^T \tag{7.2.4}$$

we have

$$\mathbf{y}(X + h) = \mathbf{P}\mathbf{y}(X), \qquad \mathbf{y}(X) = \mathbf{P}^{-1}\mathbf{y}(X + h) \tag{7.2.5}$$

and then the propagation of the solution in either of the two directions requires the generation of the elements of matrix $\mathbf{P}$, i.e. the values at $\delta = h$ of the two independent solutions $u(\delta)$ and $v(\delta)$ and of their first derivatives.

Functions $u(\delta)$ and $v(\delta)$ are constructed by perturbation. The procedure consists in taking the constant $\bar{V} = V_0$ as the *reference* potential and the polynomial $\Delta V(\delta)$ as a perturbation. Each of the two propagators, denoted generically as $p(\delta)$, is written as a perturbation series,

$$p(\delta) = p_0(\delta) + p_1(\delta) + p_2(\delta) + p_3(\delta) + \cdots, \tag{7.2.6}$$

where the zeroth-order term $p_0(\delta)$ is the solution of

$$p_0'' = [V_0 - E]p_0 \tag{7.2.7}$$

with $p_0(0) = 1$, $p_0'(0) = 0$ for u_0 and $p_0(0) = 0$, $p_0'(0) = 1$ for v_0. The correction $p_q, q = 1, 2, \ldots$ obeys the equation

$$p_q'' = [V_0 - E]p_q + \Delta V(\delta)p_{q-1}, \qquad p_q(0) = p_q'(0) = 0. \tag{7.2.8}$$

With $Z(\delta) = (V_0 - E)\delta^2$ and functions $\xi(Z), \eta_0(Z), \eta_1(Z), \ldots$, defined in the Appendix, the zeroth order propagators are

$$u_0(\delta) = \xi(Z(\delta)), \qquad v_0(\delta) = \delta\eta_0(Z(\delta)) \tag{7.2.9}$$

and the following iteration procedure exists to construct the corrections. Correction p_{q-1} is assumed as known and of such a form that the product $\Delta V(\delta)p_{q-1}$ reads

$$\Delta V(\delta)p_{q-1}(\delta) = Q(\delta)\xi(Z(\delta)) + \sum_{m=0}^{\infty} R_m(\delta)\delta^{2m+1}\eta_m(Z(\delta)). \tag{7.2.10}$$

Then $p_q(\delta)$ and $p_q'(\delta)$ are of the form

$$p_q(\delta) = \sum_{m=0}^{\infty} C_m(\delta)\delta^{2m+1}\eta_m(Z(\delta)), \tag{7.2.11}$$

$$\begin{aligned} p_q'(\delta) &= C_0(\delta)\xi(Z(\delta)) \\ &\quad + \sum_{m=0}^{\infty}(C_m(\delta) + \delta C_{m+1}(\delta))\delta^{2m+1}\eta_m(Z(\delta)), \end{aligned} \tag{7.2.12}$$

where $C_0(\delta), C_1(\delta), \ldots$ are given by quadrature (see again [9]):

$$C_0(\delta) = \frac{1}{2}\int_0^{\delta} Q(\delta_1)d\delta_1, \tag{7.2.13}$$

$$\begin{aligned} C_m(\delta) &= \frac{1}{2}\delta^{-m}\int_0^{\delta} \delta_1^{m-1}[R_{m-1}(\delta_1) \\ &\quad - C_{m-1}''(\delta_1)]d\delta_1, \quad m = 1, 2, \ldots. \end{aligned} \tag{7.2.14}$$

To calculate successive corrections for u, the starting functions in $\Delta V(\delta)p_0(\delta)$ are $Q(\delta)=\Delta V(\delta)$, $R_0(\delta)=R_1(\delta)=\cdots=0$, while for v they are $Q(\delta)=0$, $R_0(\delta)=\Delta V(\delta)$, $R_1(\delta)=R_2(\delta)=\cdots=0$. Integrals (7.2.13) and (7.2.14) have analytic forms. Each $C_m(\delta)$ is a polynomial and the series (7.2.11) and (7.2.12) are finite.

We give below the expression of $u(h)$ for the CPM version of order 12, [16]; the expressions of the other three matrix elements of the propagation matrix can be found in the same paper. With $V_0, V_1, V_2, \ldots, V_{10}$ defined in (7.2.1) and $\bar{V}_i = V_i h^{i+2}$, $i=1,2,\ldots,10$, $Z=(V_0-E)h^2$ and functions $\xi(Z)$, $\eta_m(Z)$, $m=0,1,2,\ldots$ as in the Appendix, this is:

$$\begin{aligned}
u(h) = \xi &+ \bar{V}_1[-576\eta_1 - 48\eta_2\bar{V}_1 + 24\eta_4\bar{V}_1^2 + \eta_5\bar{V}_1^3]/1152 \\
&+ \bar{V}_2[280\eta_3\bar{V}_1 + 14\eta_4\bar{V}_1^2 - 14(\eta_2+3\eta_3)\bar{V}_2 \\
&+ 7(\eta_4 - 31\eta_5)\bar{V}_1\bar{V}_2 + 2(\eta_4+10\eta_5)\bar{V}_2^2]/560 \\
&+ \bar{V}_3[-840\eta_1 + 4200\eta_2 + 420\eta_3\bar{V}_1 + 35(\eta_4 - 23\eta_5)\bar{V}_1^2 \\
&+ 420(2\eta_3 - 15\eta_4)\bar{V}_2 + 6(9\eta_4 - 50\eta_5)\bar{V}_1\bar{V}_2 \\
&- 30(\eta_2 - 4\eta_3 + 15\eta_4)\bar{V}_3]/1680 \\
&+ \bar{V}_4[18(2\eta_3 - 21\eta_4)\bar{V}_1 - 18\eta_5\bar{V}_1^2 + 54\eta_4\bar{V}_2 \\
&+ 36(\eta_3 - 18\eta_4 + 105\eta_5)\bar{V}_3 - (\eta_2 + 3\eta_3 - 75\eta_4 + 105\eta_5)\bar{V}_4]/72 \\
&+ \bar{V}_5[-2\eta_1 + 28\eta_2 - 126\eta_3 + (\eta_3 - 9\eta_4)\bar{V}_1 \\
&+ 2(\eta_3 - 21\eta_4 + 126\eta_5)\bar{V}_2 + (\eta_3 - 9\eta_4 + 15\eta_5)\bar{V}_3]/4 \\
&+ \bar{V}_6[2(\eta_3 - 27\eta_4 + 198\eta_5)\bar{V}_1 + 3(\eta_4 - 11\eta_5)\bar{V}_2]/4 \\
&+ \bar{V}_7[-2\eta_1 + 54\eta_2 - 594\eta_3 + 2574\eta_4 + (\eta_3 - 22\eta_4 + 143\eta_5)\bar{V}_1]/4 \\
&+ \bar{V}_9[-\eta_1 + 44\eta_2 - 858\eta_3 + 8580\eta_4 - 36465\eta_5]/2.
\end{aligned} \tag{7.2.15}$$

7.3 Advantages of the CPM Versions

A list of advantages is largely publicised in the literature (see, e.g., [9, 10, 22]), to mention only: (i) the accuracy is uniform with respect to E, a feature unparallelled by any other numerical method; (ii) there is an easy control of the error; (iii) the step widths are unusually big and the computation is fast; (iv) the form of the algorithm allows a direct evaluation of the normalisation constant, of the Prüfer phase and of the partial derivative with respect to E of the solution. Finally, (v), the algorithms are of a form which allows using parallel computation.

Related to point (i) above, it is instructive mentioning that another class of numerical methods is also often presented as much suited for application at big values of the energy. This is the class of methods whose algorithms are built up by exponential fitting, e.g., [2, 3, 13, 14, 18, 26–28] and references therein. Exponential fitting (ef) is a procedure for building up approximation formulae which are appropriate for linear combinations of trigonometric or hyperbolic functions sin, cos, with coefficients that are smooth enough to be well approximated by low degree polynomials.

The dependence of the error with respect to E has been examined in detail for various versions of the Numerov method, see [13, 14]. Four existing degrees of ef-tuning are available for this method, that is 0-classical version, 1, 2, and 3, and it has been proved that the error *increases* asymptotically as E^s where $s = 3, 2, 3/2$ and 1, respectively. As a matter of fact, there are reasons to believe that the same behaviour holds true for all other ef-based methods: each new higher degree of ef-tuning results in a reduction of the rate of increase or the error with E. However, the important fact is that the exponent s remains positive. For contrast, in the case of the CPM[N, Q] the exponent is negative. In fact the error now *decreases* as $E^{-1/2}$, as proved in [16]. Thus, though the ef-based methods certainly have some important merits, to mention only a vast area of applicability (approaching differential or integral equations, numerical differentiation, quadrature, interpolation etc.) their performance when solving the Schrödinger equation at big values of E is not one of these, in particular when this is compared to that of the CPM versions.

A number of advantages mentioned for the CPM versions is also shared by the LPM versions. The main difference is in the run time: the LP methods are about 15 times slower than their CP counterparts, see [20]. There are two main reasons for this. The first is that the zeroth order solutions are expressed by Airy functions and the existing subroutines for these, e.g. [23], are much slower than for the computation of the η functions. Second, the formulae of the perturbation corrections are no more so short and compact as they are in the CP methods. A way to improve the things was undertaken in [12]. It consisted in building up a procedure for the computation of the Airy propagators by a CPM version, and by converting the expressions of the perturbation corrections in the standard form accepted for the CPM versions, that is with the η functions. In this way the computational effort was drastically reduced: the new form of the LP method is only by a factor 1.5 slower than that of the CPM version of the same order.

7.4 A Numerical Illustration

We consider the Woods-Saxon potential defined by

$$V(x) = v_0 w(x)\left(1 - \frac{1 - w(x)}{a_0}\right) \tag{7.4.1}$$

with $w(x) = \{1 + \exp[(x - x_0)/a_0]\}^{-1}$, $v_0 = -50$, $x_0 = 7$, $a_0 = 0.6$, $x \in [0, x_f = 15]$ (see [9]). For this potential we have solved the eigenvalue problem in the range $E \in (-50, 0)$ with the boundary conditions

$$A_1 y(0) + B_1 y'(0) = 0, \tag{7.4.2}$$

$$A_2 y(x_f) + B_2 y'(x_f) = 0, \tag{7.4.3}$$

Table 7.2 Woods-Saxon potential: absolute errors ΔE_n at different steps

n	E_n	$h=1$	$h=1/2$	$h=1/4$
0	−49.45778872808258	−0.214*E*–06	−0.543*E*–10	−0.284*E*–13
1	−48.14843042000636	−0.192*E*–05	−0.498*E*–09	−0.142*E*–12
2	−46.29075395446608	−0.872*E*–05	−0.232*E*–08	−0.604*E*–12
3	−43.96831843181423	−0.257*E*–04	−0.727*E*–08	−0.195*E*–11
4	−41.23260777218022	−0.487*E*–04	−0.169*E*–07	−0.475*E*–11
5	−38.12278509672792	−0.435*E*–04	−0.303*E*–07	−0.908*E*–11
6	−34.67231320569966	0.371*E*–04	−0.415*E*–07	−0.138*E*–10
7	−30.91224748790885	0.152*E*–03	−0.373*E*–07	−0.161*E*–10
8	−26.87344891605987	0.155*E*–03	−0.351*E*–08	−0.121*E*–10
9	−22.58860225769321	0.618*E*–04	0.568*E*–07	0.112*E*–11
10	−18.09468828212442	0.182*E*–03	0.119*E*–06	0.216*E*–10
11	−13.43686904025008	0.507*E*–03	0.196*E*–06	0.399*E*–10
12	−8.67608167073655	0.390*E*–03	0.268*E*–06	0.399*E*–10
13	−3.90823248120623	−0.330*E*–03	0.957*E*–07	0.994*E*–11

where $A_1 = 1$, $B_1 = 0$, $A_2 = \sqrt{V(x_f) - E}$, $B_2 = 1$. The eigenvalue spectrum has 14 eigenenergies denoted $E_0, \dots, E_{13}$.

In Table 7.2 we list the exact eigenvalues, and the errors produced by the CP version of [16] at three values for the step width h. The examination of these data allows some interesting remarks. First, the error dependence with respect to the index n exhibits an oscillatory behaviour. This is in contrast with the classical methods (Numerov, for example) where the error is known to increase with n. Second, the maximal error along the spectrum decreases by three orders of magnitude at each halving of h, in agreement with the fact that the method is of order 12. Third, uncommonly big steps are sufficient to obtain very accurate eigenvalues. A step $h = 1$ is enough for obtaining all eigenvalues with at least three exact figures after the decimal point. Just for comparison, the Numerov method will require $h = 1/32$ to reach the same accuracy. As for the computational effort, all data in Table 7.2 required only a few seconds on a laptop with one processor at 1.73 GHz.

Acknowledgements This work was partially supported under contract IDEI-119 (Romania).

Appendix

Functions $\xi(Z), \eta_0(Z), \eta_1(Z), \dots$, originally introduced in [9] (they are denoted there as $\bar{\xi}(Z), \bar{\eta}_0(Z), \bar{\eta}_1(Z), \dots$), are defined as follows:

$$\xi(Z) = \begin{cases} \cos(|Z|^{1/2}) & \text{if } Z \le 0, \\ \cosh(Z^{1/2}) & \text{if } Z > 0, \end{cases} \tag{A.1}$$

$$\eta_0(Z) = \begin{cases} \sin(|Z|^{1/2})/|Z|^{1/2} & \text{if } Z < 0, \\ 1 & \text{if } Z = 0, \\ \sinh(Z^{1/2})/Z^{1/2} & \text{if } Z > 0, \end{cases} \tag{A.2}$$

$\eta_1(Z), \eta_2(Z), \ldots$, are constructed by recurrence:

$$\begin{aligned} \eta_1(Z) &= [\xi(Z) - \eta_0(Z)]/Z, \\ \eta_m &= [\eta_{m-2}(Z) - (2m-1)\eta_{m-1}(Z)]/Z, \quad m = 2, 3, \ldots. \end{aligned} \tag{A.3}$$

Some useful properties are as follows:

(i) Series expansion:

$$\eta_m(Z) = 2^m \sum_{q=0}^{\infty} \frac{g_{mq} Z^q}{(2q+2m+1)!}, \tag{A.4}$$

with

$$g_{mq} = \begin{cases} 1 & \text{if } m = 0, \\ (q+1)(q+2)\cdots(q+m) & \text{if } m > 0. \end{cases} \tag{A.5}$$

In particular,

$$\eta_m(0) = \frac{1}{(2m+1)!!}, \tag{A.6}$$

where $(2m+1)!! = 1 \times 3 \times 5 \times \cdots \times (2m+1)$.

(ii) Asymptotic behaviour at large $|Z|$:

$$\eta_m(Z) \approx \begin{cases} \xi(Z)/Z^{(m+1)/2} & \text{for odd } m, \\ \eta_0(Z)/Z^{m/2} & \text{for even } m. \end{cases} \tag{A.7}$$

(iii) Differentiation properties:

$$\xi'(Z) = \frac{1}{2}\eta_0(Z), \qquad \eta_m'(Z) = \frac{1}{2}\eta_{m+1}(Z), \quad m = 0, 1, 2, \ldots. \tag{A.8}$$

References

1. Abramowitz, M., Stegun, I.A.: Handbook of Mathematical Functions, 8th edn. Dover, New York (1972)
2. Calvo, M., Franco, J.M., Montijano, J.I., Ràndez, L.: Comput. Phys. Commun. **178**, 732–744 (2008)
3. Calvo, M., Franco, J.M., Montijano, J.I., Ràndez, L.: J. Comput. Appl. Math. **223**, 387–398 (2009)
4. Canosa, J., Gomes de Oliveira, R.: J. Comput. Phys. **5**, 188–207 (1970)
5. Gordon, R.G.: J. Chem. Phys. **51**, 14–25 (1969)

6. Ixaru, L.Gr.: The algebraic approach to the scattering problem. Internal Report IC/69/7, International Centre for Theoretical Physics, Trieste (1969)
7. Ixaru, L.Gr.: An algebraic solution of the Schrödinger equation. Internal Report IC/69/6, International Centre for Theoretical Physics, Trieste (1969)
8. Ixaru, L.Gr.: J. Comput. Phys. **9**, 159–163 (1972)
9. Ixaru, L.Gr.: Numerical Methods for Differential Equations and Applications. Reidel, Dordrecht/Boston/Lancaster (1984)
10. Ixaru, L.Gr.: J. Comput. Appl. Math. **125**, 347–357 (2000)
11. Ixaru, L.Gr.: Comput. Phys. Commun. **147**, 834–852 (2002)
12. Ixaru, L.Gr.: Comput. Phys. Commun. **177**, 897–907 (2007)
13. Ixaru, L.Gr., Rizea, M.: Comput. Phys. Commun. **19**, 23–27 (1980)
14. Ixaru, L.Gr., Rizea, M.: J. Comput. Phys. **73**, 306–324 (1987)
15. Ixaru, L.Gr., Rizea, M., Vertse, T.: Comput. Phys. Commun. **85**, 217–230 (1995)
16. Ixaru, L.Gr., De Meyer, H., Vanden Berghe, G.: J. Comput. Appl. Math. **88**, 289 (1998)
17. Ixaru, L.Gr., De Meyer, H., Vanden Berghe, G.: Comput. Phys. Commun. **118**, 259 (1999)
18. Kalogiratou, Z., Monovasilis, Th., Simos, T.E.: Comput. Phys. Commun. **180**, 167–176 (2009)
19. Ledoux, V., Van Daele, M., Vanden Berghe, G.: Comput. Phys. Commun. **162**, 151–165 (2004)
20. Ledoux, V., Ixaru, L.Gr., Rizea, M., Van Daele, M., Vanden Berghe, G.: Comput. Phys. Commun. **175**, 424–439 (2006)
21. Ledoux, V., Van Daele, M., Vanden Berghe, G.: Comput. Phys. Commun. **174**, 357–370 (2006)
22. Ledoux, V., Van Daele, M., Vanden Berghe, G.: Comput. Phys. Commun. **180**, 241–250 (2009)
23. NAG Fortran Library Manual: S17AGF, S17Astron. J.F, S17AHF, S17AKF, Mark 15, The Numerical Algorithms Group Limited, Oxford (1991)
24. Pruess, S.: SIAM J. Numer. Anal. **10**, 55–68 (1973)
25. Pryce, J.D.: Numerical Solution of Sturm-Liouville Problems. Oxford University Press, Oxford (1993)
26. Simos, T.E.: Comput. Phys. Commun. **178**, 199–207 (2008)
27. Vanden Berghe, G., Van Daele, M.: J. Comput. Appl. Math. **200**, 140–153 (2007)
28. Vanden Berghe, G., Van Daele, M.: Appl. Numer. Math. **59**, 815–829 (2009)

Chapter 8
Symplectic Partitioned Runge-Kutta Methods for the Numerical Integration of Periodic and Oscillatory Problems

Z. Kalogiratou, Th. Monovasilis, and T.E. Simos

Abstract In this work specially tuned Symplectic Partitioned Runge-Kutta (SPRK) methods have been considered for the numerical integration of problems with periodic or oscillatory solutions. The general framework for constructing exponentially/trigonometrically fitted SPRK methods is given and methods with corresponding order up to fifth have been constructed. The trigonometrically-fitted methods constructed are of two different types, fitting at each stage and Simos's approach. Also, SPRK methods with minimal phase-lag are derived as well as phase-fitted SPRK methods. The methods are tested on the numerical integration of Kepler's problem, Stiefel-Bettis problem and the computation of the eigenvalues of the Schrödinger equation.

T.E. Simos is active member of European Academy of Sciences and Arts. Please use the following address for all correspondence: Konitsis 10, Amfithea—Paleon Faliron, 175 64 Athens, Greece.

Z. Kalogiratou
Department of Informatics and Computer Technology, Technological Educational Institute of Western Macedonia at Kastoria, P.O. Box 30, 521 00, Kastoria, Greece

Th. Monovasilis
Department of International Trade, Technological Educational Institute of Western Macedonia at Kastoria, P.O. Box 30, 521 00, Kastoria, Greece

T.E. Simos (✉)
Laboratory of Computational Sciences, Department of Computer Science and Technology, Faculty of Science and Technology, University of Peloponnessos, 22100, Tripolis, Greece
e-mail: tsimos@mail.ariadne-t.gr

T.E. Simos
Department of Mathematics, College of Sciences, King Saud University, P.O. Box 2455, Riyadh 11451, Saudi Arabia

T.E. Simos (ed.), *Recent Advances in Computational and Applied Mathematics*,
DOI 10.1007/978-90-481-9981-5_8,

Keywords Partitioned-Runge-Kutta methods · Symplecticness · Exponentiall/trigonometricall fitting · Minimum phase-lag · Phase-fitting · Schrödinger equation · Hamiltonian problems

Mathematics Subject Classification (2000) 65L15 · 65L06 · 65L10 · 65P10

8.1 Introduction

In the last decades a lot of research has been performed in the area of numerical integration of Hamiltonian systems. Hamiltonian systems appear in many areas of mechanics, physics, chemistry, and elsewhere.

Symplecticity is a characteristic property of Hamiltonian systems and many authors developed and applied symplectic schemes for the numerical integration of such systems. Many authors constructed symplectic numerical methods based on the theory of Runge-Kutta methods these are symplectic Runge-Kutta (SRK) methods, symplectic Runge-Kutta-Nyström (SRKN) methods and symplectic Partitioned Runge-Kutta (SRRK) methods. The theory of these methods can be found in the books of Hairer et al. [5] and Sanz-Serna and Calvo [22].

Additionally the solution of Hamiltonian systems often has an oscillatory behavior and have been solved in the literature with methods which take into account the nature of the problem. There are two categories of such methods with coefficients depending on the problem and with constant coefficients. For the first category a good estimate of the period or of the dominant frequency is needed, such methods are exponentially and trigonometrically fitted methods, phase-fitted and amplification fitted methods. In the second category are methods with minimum phase-lag and P-stable methods and are suitable for every oscillatory problem with.

Originally exponentially/trigonometrically fitted multistep methods have been studied. In the last decade exponentially/trigonometrically fitted Runge-Kutta (EFRK, TFRK) and Runge-Kutta-Nyström (EFRKN, TFRKN) methods have been constructed by many authors. Simos [23], Vanden Berghe et al. [32] first constructed EFRK methods. The general theory of exponentially fitted methods can be found in the book of Ixaru and Vanden Berghe [6]. Also EFRKN methods have been studied by Simos [24], Franco [4], Kalogiratou and Simos [8]. There are two different approaches in exponentially fitted Runge-Kutta type methods. Fitting at each stages as suggested by Vanden Berghe et al. [32] and fitting the advance stage as suggested by Simos [23]. Both approaches perform better on different type of problems.

The idea of combining the exponentially fitting property and symplecticness arised some years later in the work of Simos and Vigo-Aguiar [25] where they constructed a symplectic modified RKN (SRKN) method with the trigonometrically fitted property. Also Van de Vyver [28] constructed another exponentially fitted SRKN method of second order. Aguiar and Tocino [27] gave order conditions for symplectic Runge-Kutta-Nyström methods. The authors constructed exponentially fitted symplectic partitioned Runge-Kutta (TFPRK) methods following both fitting approaches. Following Simos' approach they constructed methods up to fourth order (six stages) [12–17] and with the each stage approach methods up to fifth order [7, 10]. The authors also derived the order conditions for exponentially fitted SPRK methods [9]. Van de Vyver [29] constructed an exponentially fitted implicit symplectic RK (SRK) method based on the classical fourth order Gauss method. The phase-lag (or dispersion) property was introduced by Brusa and Nigro [2] and was extended to RK(N) methods by van der Houwen and Sommeijer [31]. Van de Vyver [30] constructed a symplectic Runge-Kutta-Nyström method with minimal phase-lag, the authors constructed SPRK methods with minimal phase-lag [19]. The idea of phase-fitting was introduced by Raptis and Simos [20]. Phase-fitted SPRK methods have been considered by the authors [18].

In this work (Sect. 8.2) we present the general framework for constructing exponentially/trigonometrically fitted symplectic PRK methods following the approach of Simos presented in [23] and the approach of Vanden Berghe et al. presented in [32]. We construct trigonometrically fitted SPRK methods of orders up to fifth (up to six stages). In Sect. 8.3 methods with minimum phase-lag and phase-fitted methods are presented. Numerical results are given in Sect. 8.4 and conclusions in Sect. 8.5. The Taylor expansions of the coefficients of the methods derived are given in the Appendix.

Let U be an open subset of $\Re^{2N}$, I an open subinterval of $\Re$ and $(p,q) \in U$, $x \in I$. Hamiltonian systems are of the general form

$$p_i' = -\frac{\partial H}{\partial q_i}(p,q,x), \qquad q_i' = \frac{\partial H}{\partial p_i}(p,q,x), \quad i = 1,\dots,N, \tag{8.1.1}$$

where the integer N is the number of degrees of freedom. The q variables are generalized coordinates, the p variables are the conjugated generalized momenta and $H(p,q,x)$ is the total mechanical energy.

The flow $\varphi_x : U \to \Re^{2N}$ of a Hamiltonian system is the mapping that advances the solution by time x $\varphi_t(p_0,q_0) = (p(p_0,q_0,x), q(p_0,q_0,x))$, where $p(p_0,q_0,x)$, $q(p_0,q_0,x)$ is the solution of the system coresponding to initial values $p(0) = p_0$, $q(0) = q_0$. The following result is due to Poincare and can be found in [22].

Let $H(p,q)$ be a twice continously differentiable function on $U \subset \Re^{2N}$. Then for each fixed x, the flow φ_x is a symplectic transformation wherever it is defined.

For each x, x_0 the solution operator $\Phi_H(x,x_0)$ of a Hamiltonian system is a symplectic transformation. A differentiable map $g : U \to \Re^{2N}$ is called symplectic if the Jacobian matrix $g'(p,q)$ satisfies

$$g'(p,q)^T J g'(p,q) = J, \quad \text{where } J = \begin{pmatrix} 0 & I \\ -I & 0 \end{pmatrix}.$$

We shall consider Hamiltonian systems with separable Hamiltonian

$$H(p,q,x) = T(p,x) + V(q,x),$$

where T is the kinetic energy and V is the potential energy. Then the Hamiltonian system can be written as:

$$p' = f(q,x), \qquad q' = g(p,x), \tag{8.1.2}$$

where

$$f(q,x) = -\frac{\partial H}{\partial q}(p,q,x) = -\frac{\partial V}{\partial q}(q,x),$$
$$g(p,x) = \frac{\partial H}{\partial p}(p,q,x) = \frac{\partial T}{\partial p}(p,x).$$

Partitioned Runge-Kutta methods are appropriate methods for the numerical integration of Hamiltonian systems with separable Hamiltonian.

A Partitioned Runge-Kutta (PRK) scheme is specified by two tableaux

$$\begin{array}{c|ccc} c_1 & a_{11} & \cdots & a_{1s} \\ \vdots & \vdots & \ddots & \vdots \\ c_s & a_{s1} & \cdots & a_{ss} \\ \hline & b_1 & \cdots & b_s \end{array} \qquad \begin{array}{c|ccc} C_1 & A_{11} & \cdots & A_{1s} \\ \vdots & \vdots & \ddots & \vdots \\ C_s & A_{s1} & \cdots & A_{ss} \\ \hline & B_1 & \cdots & B_s \end{array}$$

where

$$c_i = \sum_{j=1}^{s} a_{ij}, \quad \text{and} \quad C_i = \sum_{j=1}^{s} A_{ij}$$

or in matrix form

$$\begin{array}{c|c} c & a \\ \hline & b \end{array} \qquad \begin{array}{c|c} C & A \\ \hline & B \end{array}$$

where a, A are $s \times s$ matrices and c, C, b, B are s size vectors. Let $e = (1, 1, \ldots, 1)$ then $c = a.e$ and $C = A.e$. The first tableau is used for the integration of p components and the second tableau is used for the integration of the q components as follows:

$$P_i = p^n + h \sum_{j=1}^{s} a_{ij}\, f(Q_j, x + C_j h),$$
$$Q_i = q^n + h \sum_{j=1}^{s} A_{ij} g(P_j, x + c_j h), \tag{8.1.3}$$

$i = 1, 2, \ldots, s$, and

$$p^{n+1} = p^n + h \sum_{j=1}^{s} b_i\, f(Q_i, x + C_i h),$$
$$q^{n+1} = q^n + h \sum_{j=1}^{s} B_i g(P_i, x + c_i h).$$

Here P_i and Q_i are the stages for the p and q.

The order conditions for PRK methods are written in terms of bicoloured rooted trees $\beta\rho\tau$. These are trees with two kind of vertices coloured with white or black in such a way that adjacent vertices have different colours. The first tableau corresponds to the white vertices and the second to the black vertices. Then each order condition is defined in terms of a $\beta\rho\tau$

$$\Phi(\beta\rho\tau) = \frac{1}{\gamma(\beta\rho\tau)}, \tag{8.1.4}$$

where Φ is the elementary weight and γ is the density function.

The order conditions of first order are

$$\sum_{i=1}^{s} b_i = 1, \qquad \sum_{i=1}^{s} B_i = 1, \tag{8.1.5}$$

or

$$b.e = 1, \qquad B.e = 1$$

and correspond to the trees $\beta\rho\tau_{1,1,w}$ and $\beta\rho\tau_{1,1,b}$. The first subscript denotes the number of vertices, the last is w and b for the white or black root. The second is the numbering within the class of the trees with the same number of vertices. Following the numbering of Sanz-Serna ([22], Sect. 4.3) number one is given to the tree for which each father has only one son.

The second order conditions are

$$\sum_{i,j=1}^{s} b_i A_{ij} = \frac{1}{2}, \qquad \sum_{i,j=1}^{s} B_i a_{ij} = \frac{1}{2}, \tag{8.1.6}$$

or

$$b.A.e = 1/2, \qquad B.a.e = 1/2$$

and correspond to the trees $\beta\rho\tau_{2,1,w}$ and $\beta\rho\tau_{2,1,b}$.

Symplectic PRK methods have been considered Ruth [21], Forest and Ruth in [3] who derived the order conditions using Lie formalization. Also Abia and Sanz-Serna [1] considered symplectic PRK methods and gave the order conditions using graph theory according to the formalization of Butcher. The following theorem was found independently by Sanz-Serna and Suris and can be found in [22].

Theorem 8.1 *Assume that the coefficients of the PRK method (8.1.3) satisfy the relations*

$$b_i A_{ij} + B_j a_{ji} - b_i B_j = 0, \quad i, j = 1, 2, \ldots, s. \tag{8.1.7}$$

Then the method is symplectic when applied to Hamiltonian problems with separable Hamiltonian (8.1.2).

A RRK method that satisfies (8.1.7) is called symplectic PRK method (SPRK). We consider the order conditions of SPRK methods. To each bicolour tree $\beta\tau$ with $r > 1$ vertices correspond more than one bicolour rooted trees $\beta\rho\tau$. Abia and Sanz-Serna [1] proved that it is sufficient that for each $\beta\tau$ with r vertices, there is one $\beta\rho\tau$ associated with $\beta\tau$ for which (8.1.4) is satisfied. This result reduces the number of order conditions, for example for second order only one of the conditions (8.1.6) should be imposed. For a SPRK method there are two third order conditions (instead of four for a PRK method)

$$\sum_{i,j,k=1}^{s} b_i A_{ij} a_{ik} = \frac{1}{6}, \qquad \sum_{i,j=1}^{s} B_i a_{ij} A_{jk} = \frac{1}{6}, \tag{8.1.8}$$

or

$$b.A.a.e = 1/6, \qquad B.a.A.e = 1/6,$$

and correspond to the trees $\beta\rho\tau_{3,1,w}$ and $\beta\rho\tau_{3,1,b}$.

The advantage of using SPRK is that there exist explicit SPRK methods, while SRK methods can not be explicit. Assume the following explicit form $a_{ij} = 0$ for $i < j$ and $A_{ij} = 0$ for $i \leq j$. Then due to the symplecticness requirement (8.1.7) the coefficients a_{ij} and A_{ij} (and consequently c_i and C_i) are fully determined in terms of the coefficients b_i and B_i.

$$a_{ij} = b_j, \quad A_{ij} = B_j, \qquad c_i = \sum_{j=1}^{i} b_j, \quad C_i = \sum_{j=1}^{i-1} B_j, \quad i = 1, 2, \ldots, s. \tag{8.1.9}$$

The Butcher tableaux become

$$\begin{array}{c|ccccc} c_1 & b_1 & 0 & 0 & \cdots & 0 \\ c_2 & b_1 & b_2 & 0 & \cdots & 0 \\ c_3 & b_1 & b_2 & b_3 & \cdots & 0 \\ \vdots & \vdots & \vdots & \vdots & \ddots & \vdots \\ c_s & b_1 & b_2 & b_3 & \cdots & b_s \\ \hline & b_1 & b_2 & b_3 & \cdots & b_s \end{array} \qquad \begin{array}{c|ccccc} C_1 & 0 & 0 & 0 & \cdots & 0 \\ C_2 & B_1 & 0 & 0 & \cdots & 0 \\ C_3 & B_1 & B_2 & 0 & \cdots & 0 \\ \vdots & \vdots & \vdots & \vdots & \ddots & \vdots \\ C_s & B_1 & B_2 & B_3 & \cdots & 0 \\ \hline & B_1 & B_2 & B_3 & \cdots & B_s \end{array}$$

The SPRK method can be denoted by

$$[b_1, b_2, \ldots, b_s](B_1, B_2, \ldots, B_s).$$

The fact that the tableaux of the SPRK method (8.1.3) are constant along columns imply a favourable implementation of the method using only two d-dimensional vectors since P_{i+1} and Q_{i+1} can be overwritten on P_i and Q_i. The computation proceeds in the following form:

$$\begin{aligned} P_0 &= p^n, \\ Q_1 &= q^n, \quad \text{for } i = 1, \ldots, s \\ P_i &= P_{i-1} + hb_i f(Q_i, x_n + C_i h), \\ Q_{i+1} &= Q_i + hB_i g(P_i, x_n + c_i h), \\ p^{n+1} &= P_s, \\ q^{n+1} &= Q_{s+1}. \end{aligned} \tag{8.1.10}$$

We shall consider problems with Hamiltonians of the special form

$$H(p,q,x) = T(p) + V(q,x), \quad T(p) = \frac{1}{2}p^T p \tag{8.1.11}$$

in this case $g(P_i) = P_i$.

As it was mentioned above Ruth [21] suggested SPRK methods and derived order conditions up to third order. Also constructed a three stage third order method with the following coefficients

$$b_1 = \frac{7}{24}, \quad b_2 = \frac{3}{4}, \quad b_3 = -\frac{1}{24}, \qquad B_1 = \frac{2}{3}, \quad B_2 = -\frac{2}{3}, \quad B_3 = 1. \tag{8.1.12}$$

In an other work Forest and Ruth [3] derived conditions of fourth order and constructed a fourth order four stage SPRK. This method was also constructed independently by Yoshida [33] who suggested a different way of constructing SPRK methods.

$$\begin{aligned} &x_0 = 2^{\frac{1}{3}}, \quad x_1 = -\frac{x_0}{2-x_0}, \quad x_2 = \frac{1}{2-x0}, \\ &b_1 = b_4 = \frac{x_2}{2}, \quad b_2 = b_3 = \frac{x1+x2}{2}, \\ &B_1 = B_3 = x_2, \quad B_2 = x_1, \quad B_4 = 0. \end{aligned} \tag{8.1.13}$$

In the same work Yoshida suggested his well known two-stage second order method with coefficients

$$b_1 = 0, \quad b_2 = 1, \qquad B_1 = \frac{1}{2}, \quad B_2 = \frac{1}{2}. \tag{8.1.14}$$

McLachlan and Atela [11] also suggested a two-stage second order method with coefficients

$$b_1 = \frac{2-\sqrt{2}}{2}, \quad b_2 = \frac{\sqrt{2}}{2}, \qquad B_1 = \frac{\sqrt{2}}{2}, \quad B_2 = \frac{2-\sqrt{2}}{2}, \tag{8.1.15}$$

Monovasilis and Simos [13] constructed a third order method

$$\begin{aligned} &b_1 = \frac{9+3\,3^{\frac{1}{3}}+3^{\frac{2}{3}}}{24}, \quad b_2 = \frac{3-3\,3^{\frac{1}{3}}-3^{\frac{2}{3}}}{12}, \quad b_3 = b_1, \\ &B_1 = \frac{9-3\,3^{\frac{1}{3}}-4\,3^{\frac{2}{3}}}{30}, \quad B_2 = \frac{3+3^{\frac{2}{3}}}{6}, \quad B_3 = \frac{6+3\,3^{\frac{1}{3}}-3^{\frac{2}{3}}}{30}. \end{aligned} \tag{8.1.16}$$

McLachlan and Atela [11] also constructed a four stage fourth order method with the following coefficients

$$\begin{aligned} &b_1 = 0.1344962, \quad b_2 = -0.2248198, \quad b_3 = 0.75632, \quad b_4 = 0.3340036, \\ &B_1 = 0.515353, \quad B_2 = -0.085782, \quad B_3 = 0.441583, \quad B_4 = 0.128846. \end{aligned} \tag{8.1.17}$$

From any odd order (p) order method with s stages one can construct an even order method $(\geq p+1)$ with $2s$ stages with coefficients

$$\begin{array}{cccccccccc} \frac{b_1}{2}, & \frac{b_2}{2}, & \dots, & \frac{b_{s-1}}{2}, & \frac{b_s}{2}, & \frac{b_s}{2}, & \frac{b_{s-1}}{2}, & \dots, & \frac{b_2}{2}, & \frac{b_1}{2}, \\ \frac{B_1}{2}, & \frac{B_2}{2}, & \dots, & \frac{B_{s-1}}{2}, & B_s, & \frac{B_{s-1}}{2}, & \dots, & \frac{B_2}{2}, & \frac{B_1}{2}, & 0. \end{array} \tag{8.1.18}$$

In this way a method of fourth order is derived from any third order method above.

The six stage fifth order method of McLachlan and Atela [11]

$$\begin{aligned} B_1 &= 0.339839625839110000, & b_1 &= 0.1193900292875672758, \\ B_2 &= -0.088601336903027329, & b_2 &= 0.6989273703824752308, \\ B_3 &= 0.5858564768259621188, & b_3 &= -0.1713123582716007754, \\ B_4 &= -0.6030393565364911888, & b_4 &= 0.4012695022513534480, \\ B_5 &= 0.3235807965546976394, & b_5 &= 0.0107050818482359840, \\ B_6 &= 0.4423637942197494587, & b_6 &= -0.0589796254980311632. \end{aligned}$$

Phase-lag analysis of numerical methods for second order equations is based on the scalar test equation $q'' = -w^2 q$, where w is a real constant. For the numerical solution of this equation we can write

$$\begin{pmatrix} q_n \\ h\,p_n \end{pmatrix} = M_n \begin{pmatrix} q_0 \\ h\,p_0 \end{pmatrix}, \qquad M = \begin{pmatrix} A_s(v^2) & B_s(v^2) \\ C_s(v^2) & D_s(v^2) \end{pmatrix}, \qquad v = wh.$$

The eigenvalues of the M are called amplification factors of the method and are the roots of the characteristic equation

$$\xi^2 - \mathrm{tr}(M(v^2))\xi + \det(M(v^2)) = 0.$$

The phase-lag (dispersion) of the method is

$$\phi(v) = v - \arccos\left(\frac{\mathrm{tr}(M(v^2))}{2\sqrt{\det(M(v^2))}}\right),$$

and the dissipation (amplification error) is

$$\alpha(v) = 1 - \sqrt{\det(M(v^2))}.$$

For a symplectic PRK method the determinant of the amplification matrix is zero, so the methods we construct here are zero dissipative.

8.2 Construction of Trigonometrically Fitted SPRK Methods

8.2.1 Trigonometrically Fitted Symplectic PRK Methods

We want our method to integrate at each stage the $\exp(wx)$ and $\exp(-wx)$ (for the exponentially fitted case) and the functions $\sin(wx)$ and $\cos(wx)$ (for the trigonometrically fitted case). Following the idea of Vanden Berghe et al. [32] we consider the modified Runge-Kutta method with extra parameters γ_i and Γ_i for $i = 1, \ldots, s$

$$\begin{array}{c|c|ccc} c_1 & \gamma_1 & a_{11} & \cdots & a_{1s} \\ \vdots & \vdots & \vdots & \ddots & \vdots \\ c_s & \gamma_s & a_{s1} & \cdots & a_{ss} \\ \hline & & b_1 & \cdots & b_s \end{array} \qquad \begin{array}{c|c|ccc} C_1 & \Gamma_1 & A_{11} & \cdots & A_{1s} \\ \vdots & \vdots & \vdots & \ddots & \vdots \\ C_s & \Gamma_s & A_{s1} & \cdots & A_{ss} \\ \hline & & B_1 & \cdots & B_s \end{array}$$

then the method becomes

$$\begin{aligned} P_i &= \gamma_i p_n + h \sum_{j=1}^{s} a_{ij} f(Q_j, x_n + C_j h), \\ Q_i &= \Gamma_i q_n + h \sum_{j=1}^{s} A_{ij} g(P_j, x_n + c_j h), \qquad\qquad i = 1, 2, \ldots, s, \\ p_{n+1} &= p_n + h \sum_{i=1}^{s} b_i f(Q_i, x_n + C_i h), \\ q_{n+1} &= q_n + h \sum_{i=1}^{s} B_i g(P_i, x_n + c_i h). \end{aligned} \tag{8.2.1}$$

Theorem 8.2 *Assume that the coefficients of the PRK method (8.1.3) satisfy the relations*

$$\frac{b_i}{\Gamma_i} A_{ij} + \frac{B_j}{\gamma_j} a_{ji} - b_i B_j = 0, \quad i, j = 1, 2, \ldots, s. \tag{8.2.2}$$

Then the method is symplectic when applied to Hamiltonian problems with separable Hamiltonian (8.1.2).

It is easy to verify that (8.2.2) become (8.1.7) for $\Gamma_i = \gamma_i = 1$, for $i = 1, \ldots, s$. In the situation $a_{ij} = 0$ for $i < j$ and $A_{ij} = 0$ for $i \le j$ the coefficients a_{ij} and A_{ij} are fully determined by the coefficients b_i, B_i, γ_i and Γ_i in the following way

$$\begin{aligned} a_{i,j} &= \gamma_i b_j, \quad i = 1, \ldots, s, \quad j = 1, \ldots, i, \\ A_{i,j} &= \Gamma_i B_j, \quad i = 2, \ldots, s, \quad j = 1, \ldots, (i-1). \end{aligned}$$

The Butcher tableaux become

$$\begin{array}{c|cccccc}
c_1 & \gamma_1 b_1 & & & & \\
c_2 & \gamma_1 b_1 & \gamma_2 b_2 & & & \\
c_3 & \gamma_1 b_1 & \gamma_2 b_2 & \gamma_3 b_3 & & \\
\vdots & \vdots & \vdots & \vdots & \ddots & \\
c_s & \gamma_1 b_1 & \gamma_2 b_2 & \gamma_3 b_3 & \cdots & \gamma_s b_s \\
\hline
& b_1 & b_2 & b_3 & \cdots & b_s
\end{array}$$

$$\begin{array}{c|cccccc}
C_1 & & & & & & \\
C_2 & \Gamma_2 B_1 & & & & & \\
C_3 & \Gamma_3 B_1 & \Gamma_3 B_2 & & & & \\
\vdots & \vdots & \vdots & & & & \\
C_s & \Gamma_s B_1 & \Gamma_s B_2 & \Gamma_s B_3 & \cdots & \Gamma_s B_{s-1} & \\
\hline
& B_1 & B_2 & B_3 & \cdots & B_{s-1} & B_s
\end{array}$$

Rewritting the above method in the form of (8.1.11) we have

$$\begin{aligned}
P_0 &= p^n, \\
Q_1 &= \Gamma_1 q^n, \quad \text{for } i = 1, \dots, s \\
P_i &= \gamma_i \left(\frac{P_{i-1}}{\gamma_{i-1}} + h b_i f(Q_i, x_n + C_i h) \right), \\
Q_{i+1} &= \Gamma_{i+1} \left(\frac{Q_i}{\Gamma_i} + h B_i P_i \right), \\
p^{n+1} &= P_s, \\
q^{n+1} &= Q_{s+1}.
\end{aligned} \tag{8.2.3}$$

In other works the authors [15] considered the PRK method in the form (8.2.3) and inserted the parameters α_i, β_i, γ_i and δ_i as follows

$$\begin{aligned}
P_i &= \alpha_i P_{i-1} + h \gamma_i b_i f(Q_i, x_n + C_i h), \\
Q_{i+1} &= \beta_i Q_i + h \delta_i B_i P_i.
\end{aligned} \tag{8.2.4}$$

It was proved that the above method is symplectic if

$$\prod_{i=1}^{s} \alpha_i \beta_i = 1. \tag{8.2.5}$$

There is the following relation between these parameters

$$\alpha_i = \frac{\gamma_i}{\gamma_{i-1}}, \qquad \beta_i = \frac{\Gamma_{i+1}}{\Gamma_i}, \qquad \delta_i = \Gamma_{i+1}$$

and it is easy to verify that condition (8.2.5) holds.

In order to construct the trigonometrically fitted methods hereafter we shall consider the form

$$\begin{aligned} P_i &= \alpha_i P_{i-1} + h\widetilde{b}_i f(Q_i, x_n + C_i h), \\ Q_{i+1} &= \beta_i Q_i + h\widetilde{B}_i P_i. \end{aligned} \tag{8.2.6}$$

8.2.2 Final Stage Approach

Requiring the modified method to integrate exactly $\cos(wx)$ and $\sin(wx)$ we obtain a set of four equations. We can determine any four coefficients α_i, β_i, b_i, B_i. Here we shall determine the parameters α_1, β_1 we shall refer to them as α, β and take $\alpha_i = \beta_i = 1$ for $i = 2, \ldots, s$. Also we shall modify two of the coefficients b_i, B_i. The resulting equations are

$$\begin{aligned} &\alpha\left(1 - \sum_{k=1}^{s-1}(-1)^k \Phi(\beta\rho\tau_{2k,1,w})v^{2k}\right) = \cos(v), \\ &\sum_{k=1}^{s}(-1)^{k+1}\Phi^*(\beta\rho\tau_{2k-1,1,w})v^{2k-1} = \sin(v), \\ &\beta + \sum_{k=1}^{s-1}(-1)^k \Phi^*(\beta\rho\tau_{2k,1,b})v^{2k} = vB_s\sin(v) + \cos(v), \\ &\alpha\sum_{k=1}^{s-1}(-1)^{k+1}\Phi(\beta\rho\tau_{2k-1,1,b})v^{2k-1} = -vB_s\cos(v) + \sin(v), \end{aligned} \tag{8.2.7}$$

where with Φ^* we denote the elementary weight where instead of the vector e the vector

$$e^* = (1, \beta, \ldots, \beta)$$

is used.

For a two stage method the system (8.2.7) becomes

$$\begin{aligned} \alpha(1 - b_2B_1v^2) &= \cos(v), \\ (b_1 + \beta b_2)v - b_1b_2B_1v^3 &= \sin(v), \\ \beta + b_1B_1v^2 &= \cos(v) + B_2v\sin(v), \\ \alpha B_1 v &= -B_2v\cos(v) + \sin(v). \end{aligned} \tag{8.2.8}$$

This general form of the trigonometrically fitted symplectic method can generate several special methods if we use different sets of coefficients.

If we choose the Yoshida coefficients (8.1.14) ($b_2 = 1,\ B_1 = \frac{1}{2}$) we have the following method:

$$\beta = \frac{2 - v^2}{2\cos(v)}, \qquad \alpha = \frac{1}{\beta},$$
$$b_1 = \frac{\sin(2v)}{v(2 - v^2)} - \frac{1}{\cos(v)}, \tag{8.2.9}$$
$$B_2 = \frac{\tan(v)}{v} - \frac{1}{2 - v^2}.$$

If we take McLachlan's coefficients (8.1.15) ($b_2 = B_1 = \frac{\sqrt{2}}{2}$) the following method is derived:

$$\beta = \frac{2 - v^2}{2\cos(v)}, \qquad \alpha = \frac{1}{\beta},$$
$$b_1 = \frac{2\sin(v)}{v(2 - v^2)} - \frac{1}{\sqrt{2}\cos(v)}, \tag{8.2.10}$$
$$B_2 = \frac{\tan(v)}{v} - \frac{\sqrt{2}}{2 - v^2}.$$

For a three stage method the system (8.2.7) becomes

$$\alpha(-(b_2B_1 + b_3B_1 + b_3B_2)v^2 + b_2b_3B_1B_2v^4)$$
$$= \cos(v),$$
$$(b_1 + \beta(b_2 + b_3))v - (b_1(b_2B_1 + b_3(B_1 + B_2)) + \beta b_2b_3B_2)v^3 + b_1b_2b_3B_1B_2v^5$$
$$= \sin(v),$$
$$\beta - (b_1B_1 + b_2b_1B_2 + \beta b_2B_2)v^2 + b_1b_2B_1B_2v^4$$
$$= \cos(v) + B_3v\sin(v),$$
$$\alpha((B_1 + B_2)v - b_2B_1B_2v^3)$$
$$= -B_3v\cos(v) + \sin(v). \tag{8.2.11}$$

The modified third order method based on (8.1.12) is

$$\beta = \frac{v^4 - 36v^2 + 72}{72\cos(v)}, \qquad \alpha = \frac{1}{\beta},$$
$$b_1 = \frac{72\sin(v)}{v(v^4 - 36v^2 + 72)} + \frac{v^2 - 34}{48\cos(v)}, \tag{8.2.12}$$
$$B_3 = \frac{\tan(v)}{v} - \frac{24v^2}{v^4 - 36v^2 + 72}.$$

Similarly the modified third order method based on (8.1.16)

$$\begin{aligned}
k_1 &= 360 - 180v^2 + \left(9 - 3^{\frac{4}{3}} + 3^{\frac{2}{3}}\right)v^4,\\
k_2 &= -12(24 - 3^{\frac{4}{3}} + 3^{\frac{2}{3}}) + 6(4 - 3^{\frac{4}{3}} + 3^{\frac{2}{3}})v^2,\\
k_3 &= 2(-15 + 3^{\frac{4}{3}} + 3^{\frac{2}{3}}) - (1 + 3^{\frac{4}{3}} + 3^{\frac{2}{3}})v^2,\\
\beta &= \frac{k_1}{360\cos(v)}, \qquad \alpha = \frac{1}{\beta},\\
b_1 &= \frac{360\sin(v)}{vk_1} + \frac{k_3}{48\cos(v)},\\
B_1 &= \frac{\tan(v)}{v} + \frac{k_2}{k_1}.
\end{aligned} \tag{8.2.13}$$

For a method with four stage ($s = 4$) we obtain the following system:

$$\begin{aligned}
&\alpha(1 - (b_2B_1 + b_3(B_1 + B_2) + b_4(B_1 + B_2 + B_3))v^2\\
&\quad + (b_2(b_3 + b_4)B_1B_2 + (b_3b_4B_2 + b_2b_4B_1 + b_3b_4B_1)B_3)v^4\\
&\quad - b_2b_3b_4B_1B_2B_3v^6) = \cos(v),\\
&(b_1 + \beta(b_2 + b_3 + b_4))v - (b_1b_4(B_2 + B_3) + \beta(b_2b_4(B_2 + B_3) + b_3b_4B_3))v^3\\
&\quad + (b_1(b_2B_1B_2(b_3 + b_4) + b_3b_4B_3B_2 + (b_2 + b_3)b_4B_1B_3) + \beta b_2b_3b_4B_3B_2)v^5\\
&\quad + b_1b_2b_3b_4B_1B_2B_3v^7 = \sin(v),\\
&\beta - (b_1B_1 + b_1B_2 + b_1B_3 + \beta(b_2B_2 + b_2B_3 + b_3B_3))v^2\\
&\quad - (b_1((b_2B_1 + b_3B_3)B_2 + (b_2 + b_3)B_1B_3) + \beta b_2b_3B_3B_2)v^4\\
&\quad + b_1b_2b_3B_1B_2B_3v^6 = B_4v\sin(v) + \cos(v),\\
&\alpha((B_1 + B_2 + B_3)v - (b_2B_1(B_2 + B_3) + b_3B_3(B_1 + B_2))v^3\\
&\quad + b_2b_3B_1B_2B_3v^5) = \sin(v) - B_4v\cos(v).
\end{aligned} \tag{8.2.14}$$

The modified method based on the fourth order method (8.1.17) is

$$\begin{aligned}
k &= 1 - 0.5v^2 + 0.0416667v^4 - 0.00110868v^6,\\
\beta &= \frac{k}{\cos(v)}, \qquad \alpha = \frac{1}{\beta},\\
b_1 &= \frac{\sin(v)}{vk} - \frac{0.865504 - 0.0994186v^2 + 0.00215129v^4}{\cos(v)},\\
B_4 &= \frac{\tan(v)}{v} - \frac{0.871154 - 0.102244v^2 + 0.00331935v^4}{k}.
\end{aligned} \tag{8.2.15}$$

The modified method based on the fourth order method (8.1.13) is

$$\begin{aligned}
k_1 &= 8 - 4(x_1 + 2x_2)^2 v^2 + 2x_2(x_1 + x_2)^2(x_1 + 2x_2)v^4 \\
&\quad - x_1 x_2^3 (x_1 + x_2)^2 v^6, \\
k_2 &= 4(x_1 + 2x_2) - 4x_2(x_1 + x_2)^2 v^2 + x_1 x_2^2 (x_1 + x_2)^2 v^4, \\
\beta &= \frac{k_1}{8\cos(v)}, \qquad \alpha = \frac{1}{\beta}, \\
b_1 &= \frac{8\sin(v)}{v k_1} - \frac{(-2(-2 + v^2 x_1^2)(x_1 + x_2) + k_2)}{8\cos(v)}, \\
B_4 &= \frac{\tan(v)}{v} - \frac{2k_2}{k_1}.
\end{aligned} \tag{8.2.16}$$

In the case of six stages the system of equations is long and not presented here, two modified six stage methods are given below. The modified method based on the fourth order method based on the third order Ruth coefficients is the following

$$\begin{aligned}
k_1 &= 1 - \frac{v^2}{2} + \frac{v^4}{24} - \frac{79v^6}{55296} + \frac{31v^8}{1327104} - \frac{7v^{10}}{63700992}, \\
k_2 &= 1 - \frac{v^2}{6} + \frac{v^4}{144} - \frac{v^6}{6912} + \frac{v^8}{1327104}, \\
k_3 &= -\frac{41}{48} + \frac{3v^2}{32} - \frac{103v^4}{27648} + \frac{179v^6}{2654208} - \frac{7v^8}{21233664}, \\
\beta &= \frac{k_1}{\cos(v)}, \qquad \alpha = \frac{1}{\beta}, \\
b_1 &= \frac{\sin(v)}{v k_1} + \frac{k_3}{\cos(v)}, \\
B_6 &= \frac{\tan(v)}{v} - \frac{k_2}{k_1}.
\end{aligned} \tag{8.2.17}$$

Similarly, the modified method based on the fifth order method of McLachlan is

$$\begin{aligned}
k_1 &= 1. - 0.5v^2 + 0.0416667v^4 - 0.00120185v^6 + 0.0000131033v^8 \\
&\quad - 1.0442 \times 10^{-7} v^{10}, \\
k_2 &= 0.557636 + 0.0545152v^2 - 0.0100985v^4 + 0.000314933v^6 \\
&\quad - 1.77044 \times 10^{-6} v^8, \\
k_3 &= -1 - 0.00894v^2 + 0.0255521v^4 - 0.000904778v^6 \\
&\quad + 5.20963 \times 10^{-6} v^8, \\
k_4 &= \frac{1 + k_1 k_3}{v^2 k_2},
\end{aligned} \tag{8.2.18}$$

$$\beta = \frac{k_1}{\cos(v)}, \qquad \alpha = \frac{1}{\beta},$$
$$b_1 = \frac{\sin(v)}{v k_1} - \frac{k_4}{\cos(v)},$$
$$B_6 = \frac{\tan(v)}{v} - \frac{k_2}{k_1}.$$

8.2.3 Each Stage Approach

In this work we follow the idea presented in Vanden Berghe et al. [32] each internal stage of a RK method can be seen as a linear multistep method on a non-equidistant grid. We want our method to integrate exactly $\sin(w\,x)$ and $\cos(w\,x)$ at each stage. Then we have four equations for each stage and we determine the new parameters α_i, β_i as well as the modified coefficients $\widetilde{b}_i$ and $\widetilde{B}_i$. The resulting equations are

$$\begin{aligned}
\cos(c_i\, v) &= \alpha_i \cos(c_{i-1}\, v) - b_i\, v \sin(C_i\, v),\\
\sin(c_i\, v) &= \alpha_i \sin(c_{i-1}\, v) + b_i\, v \cos(C_i\, v),\\
\cos(C_{i+1}\, v) &= \beta_i \cos(C_i\, v) - B_i\, v \sin(c_i\, v),\\
\sin(c_{i+1}\, v) &= \beta_i \sin(C_i\, v) + B_i\, v \cos(c_i\, v),
\end{aligned} \tag{8.2.19}$$

where $v = wh$.

The above system of equations gives the following solution

$$\begin{aligned}
\alpha_i &= \frac{\cos((c_i - C_i)\, v)}{\cos((c_{i-1} - C_i)\, v)},\\
b_i &= \frac{\sin((c_i - c_{i-1}) v)}{v \cos((c_{i-1} - C_i)\, v)},\\
\beta_i &= \frac{\cos((c_i - C_{i+1})\, v)}{\cos((c_i - C_i)\, v)},\\
B_i &= \frac{\sin((C_{i+1} - C_i) v)}{v \cos((c_i - C_i)\, v)}.
\end{aligned} \tag{8.2.20}$$

The two stage methods modified here are the second order method of Yoshida (8.1.14) and McLachlan (8.1.15)

$$\begin{aligned}
&\alpha_1 = 1, \quad \beta_1 = \cos\left(\frac{v}{2}\right), \qquad b_1 = 0, \quad B_1 = \frac{1}{v}\sin\left(\frac{v}{2}\right),\\
&\alpha_2 = 1, \quad \beta_2 = \frac{1}{\cos(\frac{v}{2})}, \qquad b_2 = \frac{2}{v}\sin\left(\frac{v}{2}\right), \quad B_2 = \frac{1}{v}\tan\left(\frac{v}{2}\right)
\end{aligned} \tag{8.2.21}$$

and

$$
\begin{aligned}
&\alpha_1 = \cos\left(\frac{(2-\sqrt{2})}{2}v\right), \qquad \beta_1 = \frac{\cos((1-\sqrt{2})v)}{\cos(\frac{(2-\sqrt{2})}{2}v)},\\
&b_1 = \frac{\sin(\frac{(2-\sqrt{2})}{2}v)}{v}, \qquad B_1 = \frac{\sin(\frac{2}{\sqrt{2}}v)}{v\cos(\frac{(2-\sqrt{2})}{2}v)},\\
&\alpha_2 = \frac{1}{\beta_1}, \qquad \beta_2 = \frac{1}{\alpha_1},\\
&b_2 = \frac{\sin(\frac{2}{\sqrt{2}}v)}{v\cos((1-\sqrt{2})v)}, \qquad B_2 = \frac{1}{v}\tan\left(\frac{v(2-\sqrt{2})}{2}\right).
\end{aligned}
\tag{8.2.22}
$$

The three stage methods modified here are (8.1.12) and (8.1.16)

$$
\begin{aligned}
&\alpha_1 = \cos\left(\frac{7}{24}v\right), \quad \beta_1 = \frac{\cos(\frac{3}{8}v)}{\cos(\frac{7}{24}v)}, \quad b_1 = \frac{\sin(\frac{7}{24}v)}{v}, \quad B_1 = \frac{\sin(\frac{2}{3}v)}{v\cos(\frac{7}{24}v)},\\
&\alpha_2 = 1, \quad \beta_2 = \frac{\cos(\frac{25}{24}v)}{\cos(\frac{3}{8}v)}, \quad b_2 = \frac{2}{v}\sin\left(\frac{3}{8}v\right), \quad B_2 = -\frac{\sin(\frac{2}{3}v)}{v\cos(\frac{3}{8}v)},\\
&\alpha_3 = \frac{\cos(v)}{\cos(\frac{25}{24}v)}, \quad \beta_3 = \frac{1}{\cos(v)}, \quad b_3 = -\frac{\sin(\frac{1}{24}v)}{v\cos(\frac{25}{24}v)}, \quad B_3 = \frac{1}{v}\tan(v)
\end{aligned}
\tag{8.2.23}
$$

and

$$
\begin{aligned}
&\alpha_1 = \cos\left(\frac{1}{24}(9 + 3\,3^{\frac{1}{3}} + 3^{\frac{2}{3}})v\right), \qquad \beta_1 = \frac{\cos(\frac{1}{40}(3 + 9\,3^{\frac{1}{3}} + 7\,3^{\frac{2}{3}})v)}{\cos(\frac{1}{24}(9 + 3\,3^{\frac{1}{3}} + 3^2 3)v)},\\
&b_1 = \frac{1}{v}\sin\left(\frac{1}{24}(9 + 3\,3^{\frac{1}{3}} + 3^2 3)v\right), \qquad B_1 = \frac{\sin(\frac{1}{30}(9 - 3\,3^{\frac{1}{3}} - 4\,3^{\frac{2}{3}})v)}{v\cos(\frac{1}{24}(9 + 3\,3^{\frac{1}{3}} + 3^{\frac{2}{3}})v)},\\
&\alpha_2 = \frac{\cos(\frac{1}{120}(39 - 3\,3^{\frac{1}{3}} + 11\,3^{\frac{2}{3}})v)}{\cos(\frac{1}{40}(3 + 9\,3^{\frac{1}{3}} + 7\,3^{\frac{2}{3}})v)}, \qquad \beta_2 = \frac{\cos(\frac{1}{40}(7 + 3^{\frac{1}{3}} + 3\,3^{\frac{2}{3}})v)}{\cos(\frac{1}{120}(39 - 3\,3^{\frac{1}{3}} + 11\,3^{\frac{2}{3}})v)},\\
&b_2 = -\frac{\sin(\frac{1}{12}(-3 + 3\,3^{\frac{1}{3}} + 3^{\frac{2}{3}})v)}{v\cos(\frac{1}{40}(3 + 9\,3^{\frac{1}{3}} + 7\,3^{\frac{2}{3}})v)}, \qquad B_2 = \frac{\sin(\frac{1}{6}(3 + 3^{\frac{2}{3}})v)}{v\cos(\frac{1}{120}(39 - 3\,3^{\frac{1}{3}} + 11\,3^{\frac{2}{3}})v)},\\
&\alpha_3 = \frac{\cos(\frac{1}{30}(-6 - 3\,3^{\frac{1}{3}} + 3^{\frac{2}{3}})v)}{\cos(\frac{1}{40}(7 + 3^{\frac{1}{3}} + 3\,3^{\frac{2}{3}})v)}, \qquad \beta_3 = \sec\left(\frac{1}{30}(-6 - 3\,3^{\frac{1}{3}} + 3^{\frac{2}{3}})v\right),\\
&b_3 = \frac{\sin(\frac{1}{24}(9 + 3\,3^{\frac{1}{3}} + 3^{\frac{2}{3}})v)}{v\cos(\frac{1}{40}(7 + 3^{\frac{1}{3}} + 33^{\frac{2}{3}})v)}, \qquad B_3 = -\frac{1}{v}\tan\left(\frac{1}{30}(-6 - 3\,3^{\frac{1}{3}} + 3^{\frac{2}{3}})v\right).
\end{aligned}
\tag{8.2.24}
$$

The modified method based on (8.1.17) is

$$\alpha_1 = \cos(0.1344962v), \qquad \beta_1 = \frac{\cos(0.3808566v)}{\cos(0.1344962v)}$$

$$b_1 = \frac{\sin(0.1344962v)}{v}, \qquad B_1 = \frac{\sin(0.5153528v)}{v\cos(0.1344962v)},$$

$$\alpha_2 = \frac{\cos(0.6056764v)}{\cos(0.3808566v)}, \qquad \beta_2 = \frac{\cos(0.519894v)}{\cos(0.6056764v)},$$

$$b_2 = \frac{-\sin(0.2248198v)}{v\cos(0.3808566v)}, \qquad B_2 = \frac{-\sin(0.085782v)}{v\cos(0.6056764v)},$$

$$\alpha_3 = \frac{\cos(0.236426v)}{\cos(0.519894v)}, \qquad \beta_3 = \frac{\cos(0.205157v)}{\cos(0.236426v)},$$

$$b_3 = \frac{\sin(0.7563200v)}{v\cos(0.519894v)}, \qquad B_3 = \frac{\sin(0.4415830v)}{v\cos(0.236426v)},$$

$$\alpha_4 = \frac{\cos(0.1288462v)}{\cos(0.205157v)}, \qquad \beta_4 = \frac{1}{\cos(0.1288462v)},$$

$$b_4 = \frac{\sin(0.3340036v)}{v\cos(0.205157v)}, \qquad B_4 = \frac{\tan(0.1288462v)}{v}.$$

The trigonometrically fitted method based on the forth order method of Yoshida (8.1.13).

$$\alpha_1 = \cos\left(v\left(-x_1 - \frac{3x_2}{2} + 1\right)\right), \qquad \beta_1 = 1,$$

$$b_1 = \frac{1}{v}\sin\left(v\left(-x_1 - \frac{3x_2}{2} + 1\right)\right), \qquad B_1 = \frac{\sin(vx_2)}{v\sec(v(-x_1 - \frac{3x_2}{2} + 1))},$$

$$\alpha_2 = \frac{\cos(\frac{1}{2}v(x_1 + 4x_2 - 2))}{\cos(v(-x_1 - \frac{5x_2}{2} + 1))}, \qquad \beta_2 = 1,$$

$$b_2 = \frac{\sin(\frac{1}{2}v(x_1 + x_2))}{v\cos(v(-x_1 - \frac{5x_2}{2} + 1))}, \qquad B_2 = \frac{\sin(vx_1)}{v\sec(\frac{1}{2}v(x_1 + 4x_2 - 2))},$$

$$\alpha_3 = \frac{\cos(v(-x_1 - \frac{3x_2}{2} + 1))}{\cos(v(-\frac{3x_1}{2} - 2x_2 + 1))}, \qquad \beta_3 = 1,$$

$$b_3 = \frac{\sin(\frac{1}{2}v(x_1 + x_2))}{v\cos(v(-\frac{3x_1}{2} - 2x_2 + 1))}, \qquad B_3 = \frac{\sin(vx_2)}{v\sec(v(-x_1 - \frac{3x_2}{2} + 1))},$$

$$\alpha_4 = \frac{\cos(v(x_1 + 2x_2 - 1))}{\cos(v(-x_1 - \frac{5x_2}{2} + 1))}, \qquad \beta_4 = 1,$$

$$b_4 = \frac{\sin(\frac{vx_2}{2})}{v\sec(v(-x_1 - \frac{5x_2}{2} + 1))}, \qquad B_4 = 0.$$

The trigonometrically fitted method based on the fourth order (six stage) method derived by Ruth's method (8.1.12) with the technique (8.1.18).

$$\alpha_1 = \cos\left(\frac{7v}{48}\right), \qquad \beta_1 = \cos\left(\frac{3v}{16}\right)\sec\left(\frac{7v}{48}\right),$$

$$b_1 = \frac{1}{v}\sin\left(\frac{7v}{48}\right), \qquad B_1 = \frac{1}{v}\sec\left(\frac{7v}{48}\right)\sin\left(\frac{v}{3}\right),$$

$$\alpha_2 = 1, \qquad \beta_2 = \cos\left(\frac{25v}{48}\right)\sec\left(\frac{3v}{16}\right),$$

$$b_2 = \frac{2}{v}\sin\left(\frac{3v}{16}\right), \qquad B_2 = -\frac{1}{v}\sec\left(\frac{3v}{16}\right)\sin\left(\frac{v}{3}\right),$$

$$\alpha_3 = \cos\left(\frac{v}{2}\right)\sec\left(\frac{25v}{48}\right), \qquad \beta_3 = 1,$$

$$b_3 = -\frac{1}{v}\sec\left(\frac{25v}{48}\right)\sin\left(\frac{v}{48}\right), \qquad B_3 = \frac{2}{v}\sin\left(\frac{v}{2}\right),$$

$$\alpha_4 = \cos\left(\frac{25v}{48}\right)\sec\left(\frac{v}{2}\right), \qquad \beta_4 = \cos\left(\frac{3v}{16}\right)\sec\left(\frac{25v}{48}\right),$$

$$b_4 = -\frac{1}{v}\sec\left(\frac{v}{2}\right)\sin\left(\frac{v}{48}\right), \qquad B_4 = -\frac{1}{v}\sec\left(\frac{25v}{48}\right)\sin\left(\frac{v}{3}\right),$$

$$\alpha_5 = 1, \qquad \beta_5 = \frac{2\sin(\frac{3v}{16})}{v},$$

$$b_5 = \cos\left(\frac{7v}{48}\right)\sec\left(\frac{3v}{16}\right), \qquad B_5 = \frac{1}{v}\sec\left(\frac{3v}{16}\right)\sin\left(\frac{v}{3}\right),$$

$$\alpha_6 = \sec\left(\frac{7v}{48}\right), \qquad \beta_6 = 1,$$

$$b_6 = \frac{1}{v}\tan\left(\frac{7v}{48}\right), \qquad B_6 = 0.$$

The trigonometrically fitted method based on the fifth order (six stage) method of McLachlan and Atela.

$$\alpha_1 = \cos(b_1\, v), \qquad \beta_1 = \frac{\cos(0.22045\, v)}{\cos(b_1\, v)},$$

$$b_1 = \frac{\sin(b_1\, v)}{v}, \qquad B_1 = \frac{\sin(B_1\, v)}{v\cos(b_1\, v)},$$

$$\alpha_2 = \frac{\cos(0.478478\, v)}{\cos(0.22045\, v)}, \qquad \beta_2 = \frac{\cos(0.567079\, v)}{\cos(0.478478\, v)},$$

$$b_2 = \frac{\sin(b_2\, v)}{v\, \cos(0.22045\, v)}, \qquad B_2 = -\frac{\sin(B_2\, v)}{v\, \cos(0.478478\, v)},$$

$$\alpha_3 = \frac{\cos(0.395767\, v)}{\cos(0.567079\, v)}, \qquad \beta_3 = \frac{\cos(0.19009\, v)}{\cos(0.395767\, v)},$$

$$b_3 = \frac{-\sin(b_3\, v)}{v\, \cos(0.567079\, v)}, \qquad B_3 = \frac{\sin(B_3\, v)}{v\, \cos(0.395767\, v)},$$

$$\alpha_4 = \frac{\cos(0.21118\, v)}{\cos(0.19009\, v)}, \qquad \beta_4 = \frac{\cos(0.814219\, v)}{\cos(0.21118\, v)},$$

$$b_4 = \frac{\sin(b_4\, v)}{v\, \cos(0.19009\, v)}, \qquad B_4 = -\frac{\sin(B_4\, v)}{v\, \cos(0.21118\, v)},$$

$$\alpha_5 = \frac{\cos(0.824924\, v)}{\cos(0.814219\, v)}, \qquad \beta_5 = \frac{\cos(0.501343\, v)}{\cos(0.824924\, v)},$$

$$b_5 = \frac{\sin(b_5\, v)}{v\, \sec(0.814219\, v)}, \qquad B_5 = \frac{\sin(B_5\, v)}{v\, \sec(0.824924\, v)},$$

$$\alpha_6 = \frac{\cos(0.442364\, v)}{\cos(0.501343\, v)}, \qquad \beta_6 = \sec(0.442364\, v),$$

$$b_6 = \frac{-\sin(b_6\, v)}{v\, \sec(0.501343\, v)}, \qquad B_6 = \frac{\tan(B_6\, v)}{v}.$$

8.3 Construction of SPRK Methods with Minimum Phase-Lag and Methods with Infinite Phase-Lag Order

Three stage methods are considered here with algebraic order two and three and fifth phase-lag order. For the second order method we set $b_1 = b_3$ and solve the order conditions leaving b_1 and B_3 as free parameters. We substitute into

$$\phi(v) = v - \arccos\left(\frac{\mathrm{tr}(M(v^2))}{2\sqrt{\det(M(v^2))}}\right),$$

and take the Taylor expansion the constant term and the coefficients of v and v^2 are zero, the coefficient of v^3 is

$$pl_3 = \frac{24(B_3 - 1){b_1}^3 - 12({B_3}^2 + B_3 - 2){b_1}^2 + 4(3{B_3}^2 - 2)b_1 + 1}{48 b_1 - 24}$$

and the coefficient of v^5 is $pl_5 = t1/t2$, where

$$\begin{aligned} t1 = {} & 320(B_3 - 1)^2 b_1^6 + 320(B_3 - 2)(B_3 - 1)^2 b_1^5 \\ & + 80(B_3^4 - 6B_3^3 + 17B_3^2 - 16B_3 + 4)b_1^4 \end{aligned}$$

$$
\begin{aligned}
&- 80(2B_3^4 - 2B_3^3 + 8B_3^2 - 5B_3 - 1)b_1^3 \\
&+ 8(10B_3^4 + 25B_3^2 - 5B_3 - 14)b_1^2 - 8(5B_3^2 - 4)b_1 - 3, \\
t2 &= 640(1 - 2b_1)^2.
\end{aligned}
$$

We solve the equations $pl_3 = 0,\ pl_5 = 0$ and obtain the following coefficients

$$
\begin{aligned}
b_1 &= 0.5974665433347971, & b_2 &= -0.1949330866695942, \\
b_3 &= 0.5974665433347971, & B_1 &= -0.18554773759667065 \\
B_2 &= 0.9618767420574417, & B_3 &= 0.2236709955392291.
\end{aligned} \tag{8.3.1}
$$

For a third order method the first nonzero coefficient in the Taylor expansion of $\phi(v)$ is the coefficient of v^3. We solve the five order conditions with the additional condition that the coefficient of v^5 is zero and obtain the following coefficients

$$
\begin{aligned}
b_1 &= 0.2603116924199056, & b_2 &= 1.0941427983167422, \\
b_3 &= -0.35445449073664803, & B_1 &= 0.6308476929866689, \\
B_2 &= -0.0941427983167424, & B_3 &= 0.4632951053300734.
\end{aligned} \tag{8.3.2}
$$

In order to construct methods with phase-lag of order infinity $\phi(v)$ should be zero. This equation together with the five equations (of algebraic order 3) are solved and the second method is derived. For the derivation of the first method only the three equations of second order are solved together with the zero dispersion equation. In this case we use the coefficients c_1 and d_3 of method (8.3.2).

The coefficients found are long and omitted in this version of the work, alternatively we present the Taylor expansions of these coefficients.

Taylor expansions of the coefficients of the First Method.

$$
\begin{aligned}
b_2 &= -0.194933 + 0.0000445788v^4 + 2.27794 \times 10^{-6}v^6 \\
&\quad + 1.36987 \times 10^{-7}v^8 + O(v^{10}), \\
b_3 &= 0.597467 - 0.0000445788v^4 - 2.27794 \times 10^{-6}v^6 \\
&\quad - 1.36987 \times 10^{-7}v^8 + O(v^{10}), \\
B_1 &= -0.185548 - 0.000219969v^4 - 0.0000112403v^6 \\
&\quad - 7.26254 \times 10^{-7}v^8 + O(v^{10}), \\
B_2 &= 0.961877 + 0.000219969v^4 + 0.0000112403v^6 \\
&\quad + 7.26254 \times 10^{-7}v^8 + O(v^{10}).
\end{aligned} \tag{8.3.3}
$$

Taylor expansions of the coefficients of the Second Method.

$$
\begin{aligned}
b_1 &= 0.260311692419906 - 0.00141750768546037v^2 \\
&\quad + 0.0000171295629106582v^4 + 1.09317432962664 \times 10^{-6}v^6 \\
&\quad + 5.39795398526919 \times 10^{-8}v^8 + O(v^{10}), \\
b_2 &= 1.09414279831674 + 0.0482677717443141v^2 \\
&\quad + 0.00444453889658238v^4 + 0.000416871771038772v^6 \\
&\quad + 0.0000395369062137604v^8 + O(v^{10}), \\
b_3 &= -0.354454490736648 - 0.0468502640588538v^2 \\
&\quad - 0.00446166845949304v^4 - 0.000417964945368399v^6 \\
&\quad - 0.0000395908857536131v^8 + O(v^{10}), \\
B_1 &= 0.630847692986669 - 0.0000945321219474852v^2 \\
&\quad + 0.000159464702869448v^4 + 7.57889134172802 \times 10^{-6}v^6 \\
&\quad + 3.79501123670239 \times 10^{-7}v^8 + O(v^{10}), \\
B_2 &= -0.0941427983167423 + 0.0147689541113375v^2 \\
&\quad - 0.000465169491230152v^4 + 1.10473661738189 \times 10^{-6}v^6 \\
&\quad - 4.71876906682702 \times 10^{-7}v^8 + O(v^{10}), \\
B_3 &= 0.463295105330073 - 0.0146744219893900v^2 \\
&\quad + 0.000305704788360704v^4 - 8.68362795910990 \times 10^{-6}v^6 \\
&\quad + 9.23757830124625 \times 10^{-8}v^8 + O(v^{10}).
\end{aligned}
\tag{8.3.4}
$$

8.4 Numerical Results

8.4.1 The Two-Body Problem

The following system of equations is known as the two-body problem and is a standard symplectic testcase:

$$
p_1' = -\frac{q_1}{\sqrt{(q_1^2+q_2^2)^3}}, \quad q_1' = p_1, \quad p_2' = -\frac{q_2}{\sqrt{(q_1^2+q_2^2)^3}}, \quad q_2' = p_2
$$

Table 8.1 Two-body. The maximum absolute error of the Hamiltonian

h	(8.1.15)	(8.2.10)	(8.2.22)	(8.1.16)	(8.2.13)	(8.2.24)
1	1.05×10^{-2}	9.89×10^{-2}	1.84×10^{-12}	2.30×10^{-3}	2.05×10^{-3}	6.56×10^{-13}
1/2	8.99×10^{-4}	4.98×10^{-4}	1.19×10^{-12}	4.42×10^{-5}	2.53×10^{-5}	5.51×10^{-12}
1/4	6.23×10^{-5}	5.46×10^{-5}	5.05×10^{-12}	7.21×10^{-7}	1.66×10^{-7}	1.69×10^{-12}

Table 8.2 Two-body. The maximum absolute error of the Hamiltonian

h	(8.1.16)	(8.3.1)	(8.3.2)	(8.3.3)	(8.3.4)
1	2.30×10^{-3}	4.57×10^{-3}	9.07×10^{-4}	4.57×10^{-3}	8.74×10^{-4}
1/2	4.42×10^{-5}	9.61×10^{-5}	7.47×10^{-6}	9.61×10^{-5}	7.42×10^{-6}
1/4	7.21×10^{-7}	2.12×10^{-6}	5.47×10^{-8}	1.12×10^{-6}	5.48×10^{-8}

with initial conditions

$$p_1(0) = 0, \quad q_1(0) = 1 - e, \qquad p_2(0) = \sqrt{\frac{1+e}{1-e}}, \quad q_2(0) = 0.$$

The Hamiltonian of this problem is

$$H(p_1, p_2, q_1, q_2) = T(p_1, p_2) + V(q_1, q_2),$$

$$T(p_1, p_2) = \frac{1}{2}(p_1^2 + p_2^2), \quad \text{and} \quad V(q_1, q_2) = -\frac{1}{\sqrt{q_1^2 + q_2^2}}.$$

The exact solution is

$$q_1(x) = \cos(E) - e, \qquad q_2(x) = \sqrt{1 - e^2}\,\sin(E),$$

where e is the eccentricity of the orbit and the eccectricity anomaly E is expressed as an implicit function of x by Kepler's equation

$$x = E - e\,\sin(E).$$

For this problem we use $v = h$.

In Tables 8.1, 8.2 the maximum absolute error of the Hamiltonian is given with $e = 0$ and integration interval [0, 10000] for several stepsizes.

8.4.2 *An Orbit Problem Studied by Stiefel and Bettis*

We consider the following almost periodic orbit problem studied by Stiefel and Bettis [26]:

$$p_1' = -q_1 + 0.001\cos(x), \quad q_1' = p_1, \qquad p_2' = -q_2 + 0.001\sin(x), \quad q_2' = p_2$$

Table 8.3 Stiefel-Bettis. The maximum absolute error of the solution

h	(8.1.15)	(8.2.10)	(8.2.22)	(8.1.16)	(8.2.13)	(8.2.24)
1/2	–	3.85×10^{-4}	8.32×10^{-3}	9.92×10^{-3}	2.97×10^{-4}	8.26×10^{-4}
1/4	–	1.82×10^{-4}	1.27×10^{-3}	4.36×10^{-4}	1.76×10^{-5}	7.53×10^{-5}
1/8	3.21×10^{-1}	5.50×10^{-5}	2.72×10^{-4}	8.70×10^{-6}	1.43×10^{-6}	7.81×10^{-6}

Table 8.4 Stiefel-Bettis. The maximum absolute error of the solution

h	(8.1.16)	(8.3.1)	(8.3.2)	(8.3.3)	(8.3.4)
1/2	1.83×10^{-2}	4.87×10^{-3}	1.41×10^{-3}	4.73×10^{-3}	1.72×10^{-3}
1/4	11.09×10^{-3}	1.01×10^{-3}	2.05×10^{-4}	1.01×10^{-3}	2.10×10^{-4}
1/8	6.75×10^{-5}	2.36×10^{-4}	2.59×10^{-5}	2.36×10^{-4}	2.60×10^{-5}

with initial conditions

$$p_1(0) = 0, \quad q_1(0) = 1, \qquad p_2(0) = 0.9995, \quad q_2(0) = 0.$$

The analytical solution is given by

$$q(x) = \cos(x) + 0.0005x\sin(x), \qquad p(x) = \sin(x) - 0.0005x\cos(x).$$

In Tables 8.3, 8.4 the maximum absolute error of the solution is given with integration interval [0, 1000] for several stepsizes.

8.4.3 The Schrödinger Equation

We shall use our new methods for the computation of the eigenvalues of the one-dimensional time-independent Schrödinger equation. The Schrödinger equation may be written in the form

$$-\frac{1}{2}\psi'' + V(x)\psi = E\psi, \tag{8.4.1}$$

where E is the energy eigenvalue, $V(x)$ the potential, and $y(x)$ the wave function. The tested problems are the harmonic oscillator and the exponential potential.

8.4.3.1 Harmonic Oscillator

The potential of the one dimensional harmonic oscillator is

$$V(x) = \frac{1}{2}kx^2$$

Table 8.5 Absolute Error ($\times 10^{-6}$) of the eigenvalues of the harmonic oscillator

	(8.1.15)	(8.2.10)	(8.2.22)	(8.1.16)	(8.2.13)	(8.2.24)	h
E_0	152	102	102	0	0	2	0.1
E_{10}	–	104	109	81	3	31	
E_{30}	–	108	131	2178	11	105	
E_{50}	–	79	147	–	0	186	
E_{100}	–	28	32	–	3	22	0.05
E_{150}	–	27	35	–	3	34	
E_{200}	–	22	39	–	2	49	
E_{250}	–	10	42	–	2	63	
E_{300}	–	22	45	–	14	81	

Table 8.6 Absolute Error ($\times 10^{-6}$) of the eigenvalues of the harmonic oscillator with step size $h = 0.1$

	(8.1.16)	(8.3.1)	(8.3.2)	(8.3.3)	(8.3.4)
E_0	0	0	0	0	0
E_{10}	81	8	7	0	0
E_{15}	285	16	15	1	1
E_{20}	692	18	18	2	2
E_{30}	2178	121	120	4	4
E_{50}	–	842	841	6	6
E_{100}	–	1505	1505	22	10

with boundary conditions $\psi(-R) = \psi(R) = 0$. We consider $k = 1$.

The exact eigenvalues are given by

$$E_n = n + \frac{1}{2}, \quad n = 0, 1, 2, \ldots.$$

In Tables 8.5 and 8.6 we give the computed eigenvalues of the harmonic oscillator up to E_{300}.

8.4.3.2 Doubly Anharmonic Oscillator

The potential is $V(x) = \frac{1}{2}x^2 + \lambda_1 x^4 + \lambda_2 x^6$ we take $\lambda_1 = \lambda_2 = 1/2$. The integration interval is $[-R, R]$. In Tables 8.7 and 8.8 we give the computed eigenvalues up to E_{30} with step $h = 1/30$. The performance of all methods considered is similar with this the harmonic oscillator.

8.5 Conclusions

In this work symplectic partitioned Runge-Kutta methods have been considered specially tuned for the numerical integration of problems with periodic or oscillatory

Table 8.7 Absolute Error ($\times 10^{-6}$) of the eigenvalues of the doubly anharmonic oscillator with step size $h = 1/30$

	(8.1.15)	(8.2.10)	(8.2.22)	(8.1.16)	(8.2.13)	(8.2.24)
0.807447	70	55	55	0	0	0
5.553677	1814	229	244	1	1	1
12.534335	9014	281	432	4	3	5
21.118364	–	87	622	17	6	15
31.030942	–	1454	815	53	12	34
42.104446	–	4753	1013	130	21	62
54.222484	–	–	1212	273	29	99
67.29805	–	–	1418	521	42	151
81.262879	–	–	1629	920	57	215
96.061534	–	–	1845	1525	73	293
111.647831	–	–	2067	2410	93	390
127.982510	–	–	–	3658	114	502
145.031661	–	–	–	5365	133	627
162.765612	–	–	–	7643	147	772
181.158105	–	–	–	–	111	877
200.185694	–	–	–	–	146	–

Table 8.8 Absolute Error ($\times 10^{-6}$) of the eigenvalues of the doubly anharmonic oscillator with step size $h = 1/30$

	(8.1.16)	(8.3.1)	(8.3.2)	(8.3.3)	(8.3.4)
0.807447	0	0	0	0	0
5.553677	1	1	1	1	1
12.534335	4	2	2	2	2
21.118364	17	6	6	6	6
31.030942	53	12	11	12	11
42.104446	130	21	20	20	19
54.222484	273	32	31	30	28
67.298050	521	50	47	43	41
81.262879	920	74	71	60	57
96.061534	1525	106	102	79	75
111.647831	2410	152	147	102	97
127.982510	3658	216	210	128	122
145.031661	5365	301	294	145	149
162.765612	7643	412	402	179	171
181.158105	–	516	505	156	149
200.185694	–	457	444	85	89

solutions. The general framework for constructing trigonometrically fitted methods is given and methods with corresponding order up to fifth have been constructed following two different approaches. The methods that integrate the trigonometric functions at each stage are suitable for the integration of Kepler's problem as shown by the numerical results. Both type of methods show almost the same behavior on Stiefel-Bettis problem. Simos' approach is more favourable for the computation of the eigenvalues of the Schrödinger equation as shown with the two potentials used here, the phase-fitted methods also have superb behavior on this problem.

Appendix

Taylor expansion for method (8.2.9)

$$\beta = 1 - \frac{v^4}{24} - \frac{7v^6}{360} - \frac{323v^8}{40320} - \frac{2951v^{10}}{907200} + O(v^{11}),$$

$$\alpha = 1 + \frac{v^4}{24} + \frac{7v^6}{360} + \frac{131v^8}{13440} + \frac{4421v^{10}}{907200} + O(v^{11}),$$

$$b_1 = -\frac{v^2}{6} - \frac{v^4}{30} + \frac{13v^6}{5040} + \frac{211v^8}{22680} + \frac{315523v^{10}}{39916800} + O(v^{11}),$$

$$B_2 = \frac{1}{2} + \frac{v^2}{12} + \frac{v^4}{120} - \frac{43v^6}{5040} - \frac{851v^8}{90720} - \frac{67477v^{10}}{9979200} + O(v^{11}).$$

Taylor expansion for method (8.2.10)

$$\beta = 1 - \frac{v^4}{24} - \frac{7v^6}{360} - \frac{323v^8}{40320} - \frac{2951v^{10}}{907200} + O(v^{11}),$$

$$\alpha = 1 + \frac{v^4}{24} + \frac{7v^6}{360} + \frac{131v^8}{13440} + \frac{4421v^{10}}{907200} + O(v^{11}),$$

$$\begin{aligned} b_1 = {} & \left(1 - \frac{1}{\sqrt{2}}\right) + \left(\frac{1}{3} - \frac{1}{2\sqrt{2}}\right)v^2 + \left(\frac{7}{40} - \frac{5}{24\sqrt{2}}\right)v^4 \\ & + \left(\frac{11}{126} - \frac{61}{720\sqrt{2}}\right)v^6 + \left(\frac{2263}{51840} - \frac{277}{8064\sqrt{2}}\right)v^8 \\ & + \left(\frac{48403}{2217600} - \frac{50521}{3628800\sqrt{2}}\right)v^{10} + O(v^{11}), \end{aligned}$$

$$\begin{aligned} B_2 = {} & \left(1 - \frac{1}{\sqrt{2}}\right) + \left(\frac{1}{3} - \frac{1}{2\sqrt{2}}\right)v^2 + \left(\frac{2}{15} - \frac{1}{4\sqrt{2}}\right)v^4 \\ & + \left(\frac{17}{315} - \frac{1}{8\sqrt{2}}\right)v^6 + \left(\frac{62}{2835} - \frac{1}{16\sqrt{2}}\right)v^8 + \left(\frac{1382}{155925} - \frac{1}{32\sqrt{2}}\right)v^{10} \\ & + O(v^{11}). \end{aligned}$$

Taylor expansion for method (8.2.12)

$$\beta = 1 - \frac{v^4}{36} - \frac{v^6}{80} - \frac{619v^8}{120960} - \frac{3767v^{10}}{1814400} + O(v^{11}),$$
$$\alpha = 1 + \frac{v^4}{36} + \frac{v^6}{80} + \frac{2137v^8}{362880} + \frac{5027v^{10}}{1814400} + O(v^{11}),$$
$$b_1 = \frac{7}{24} + \frac{23v^4}{960} + \frac{1213v^6}{60480} + \frac{37921v^8}{2903040} + \frac{7507v^{10}}{985600} + O(v^{11}),$$
$$B_3 = 1 - \frac{v^4}{30} - \frac{187v^6}{7560} - \frac{43v^8}{2835} - \frac{256331v^{10}}{29937600} + O(v^{11}).$$

Taylor expansion for method (8.2.13)

$$\beta = 1 + \frac{1}{360}(-6 - 3\sqrt[3]{3} + 3^{2/3})v^4 + \frac{1}{720}(-5 - 3\sqrt[3]{3} + 3^{2/3})v^6$$
$$+ \frac{(-113 + \frac{70}{\sqrt[3]{3}} - 70\sqrt[3]{3})}{40320}v^8 + \frac{(-2059 - 1281\sqrt[3]{3} + 427\,3^{2/3})v^{10}}{1814400} + O(v^{12}),$$
$$\alpha = 1 + \frac{1}{360}(6 + 3\sqrt[3]{3} - 3^{2/3})v^4 + \frac{1}{720}(5 + 3\sqrt[3]{3} - 3^{2/3})v^6$$
$$+ \frac{(1779 + 1232\sqrt[3]{3} - 364\,3^{2/3})}{604800}v^8$$
$$+ \frac{(2227 + 1785\sqrt[3]{3} - 455\,3^{2/3})v^{10}}{1814400} + O(v^{12}),$$
$$b_1 = \frac{1}{24}(9 + 3\sqrt[3]{3} + 3^{2/3}) + \frac{1}{960}\left(9 - \frac{13}{\sqrt[3]{3}} + 3\sqrt[3]{3}\right)v^4$$
$$+ \frac{(185 + 91\sqrt[3]{3} - 63\,3^{2/3})v^6}{20160} + \frac{(88747 + 51321\sqrt[3]{3} - 25677\,3^{2/3})}{14515200}v^8$$
$$+ \frac{(833343 + 541629\sqrt[3]{3} - 220825\,3^{2/3})v^{10}}{239500800} + O(v^{12}),$$
$$B_3 = \frac{1}{10}\left(2 - \frac{1}{\sqrt[3]{3}} + \sqrt[3]{3}\right) + \frac{(-27 - 16\sqrt[3]{3} + 7\,3^{2/3})v^4}{1800}$$
$$+ \frac{(-897 - 546\sqrt[3]{3} + 217\,3^{2/3})v^6}{75600} + \frac{(-16256 - 10563\sqrt[3]{3} + 3801\,3^{2/3})}{2268000}v^8$$
$$+ \frac{(-193472 - 135751\sqrt[3]{3} + 44352\,3^{2/3})v^{10}}{49896000} + O(v^{12}).$$

Taylor expansion for method (8.2.15)

$$\beta = 1 - 1.52485 \times 10^{-7}v^2 - 6.2304 \times -8v^4 + 0.000280188v^6$$
$$+ 0.000115295v^8 + 0.0000462484v^{10} + O(v^{11}),$$

$$\alpha = 1 + 1.524845 \times 10^{-7} v^2 + 6.2304 \times 10^{-8} v^4 - 0.000280188 v^6 - 0.000115295 v^8 - 0.0000462485 v^{10} + O(v^{11}),$$

$$b_1 = 0.134496 + 1.43141 \times 10^{-7} v^2 + 0.000578139 v^4 - 2.74604 \times 10^{-6} v^6 - 0.0000946384 v^8 - 0.0000757574 v^{10} + O(v^{11}),$$

$$B_4 = 0.128846 - 1.70966 \times 10^{-7} v^2 - 0.000354727 v^4 + 0.0000472764 v^6 + 0.00010977 v^8 + 0.0000818646 v^{10} + O(v^{11}).$$

Taylor expansion for method (8.2.16)

$$\beta = 1 + \frac{(32 + 25\sqrt[3]{2} + 202^{2/3}) v^6}{1440} + \frac{(447 + 350\sqrt[3]{2} + 2802^{2/3}) v^8}{40320} + \frac{(16756 + 13125\sqrt[3]{2} + 105002^{2/3}) v^{10}}{3628800} + O(v^{11}),$$

$$\alpha = 1 + \frac{(-32 - 25\sqrt[3]{2} - 202^{2/3}) v^6}{1440} + \frac{(-447 - 350\sqrt[3]{2} - 2802^{2/3}) v^8}{40320} + \frac{(-8378 - \frac{13125}{2^{2/3}} - 52502^{2/3}) v^{10}}{1814400} + O(v^{11}),$$

$$b_1 = \frac{1}{12}(4 + 2\sqrt[3]{2} + 2^{2/3}) + \frac{1}{720}(6 + 5\sqrt[3]{2} + 52^{2/3}) v^4 + \frac{(-1076 - 826\sqrt[3]{2} - 6232^{2/3}) v^6}{60480} + \frac{(-12070 - \frac{14619}{\sqrt[3]{2}} - 9369\sqrt[3]{2}) v^8}{725760} + \frac{(-2518940 - 1961674\sqrt[3]{2} - 15444772^{2/3}) v^{10}}{239500800} + O(v^{11}),$$

$$B_4 = \frac{1}{720}(26 + 20\sqrt[3]{2} + 152^{2/3}) v^4 + \frac{(404 + 315\sqrt[3]{2} + 2452^{2/3}) v^6}{10080} + \frac{(9406 + 7350\sqrt[3]{2} + 57752^{2/3}) v^8}{362880} + \frac{(478917 + 373450\sqrt[3]{2} + 2945252^{2/3}) v^{10}}{39916800} + O(v^{11}).$$

Taylor expansion for method (8.2.17)

$$\beta = 1 - \frac{11 v^6}{276480} - \frac{991 v^8}{46448640} - \frac{98593 v^{10}}{11147673600} + O(v^{11}),$$

$$\alpha = 1 + \frac{11 v^6}{276480} + \frac{991 v^8}{46448640} + \frac{98593 v^{10}}{11147673600} + O(v^{11}),$$

$$c_1 = \frac{7}{48} - \frac{203 v^4}{138240} - \frac{57863 v^6}{92897280} - \frac{175937 v^8}{743178240} - \frac{7260959 v^{10}}{81749606400} + O(v^{11}),$$

$$d_6 = \frac{v^4}{720} + \frac{1163v^6}{1935360} + \frac{5339v^8}{23224320} + \frac{10575749v^{10}}{122624409600} + O(v^{11}).$$

Taylor expansion for method (8.2.18)

$$\beta = 1 + 1.7743047420193392 \times 10^{-8}v^2 + 6.204841085377666 \times 10^{-9}v^4$$
$$+ 0.000187041v^6 + 0.0000818221v^8 + 0.0000332888v^{10} + O(v^{11}),$$
$$\alpha = 1 - 1.7743047420193392 \times 10^{-8}v^2 - 6.204840780066334 \times 10^{-9}v^4$$
$$- 0.000187041v^6 - 0.0000818221v^8 - 0.0000332888v^{10} + O(v^{11}),$$
$$b_1 = 0.11939 - 1.2711962010802935 \times 10^{-8}v^2 - 8.788109379098685 \times 10^{-9}v^4$$
$$- 0.000182374v^6 - 0.000142348v^8 - 0.0000847776v^{10} + O(v^{11}),$$
$$B_6 = 0.442364 + 7.940184509891424 \times 10^{-9}v^2 + 9.158207150972153 \times 10^{-9}v^4$$
$$+ 0.000205354v^6 + 0.000152055v^8 + 0.0000887026v^{10} + O(v^{11}).$$

Taylor expansion for method (8.2.21)

$$\alpha_1 = 1,$$
$$\beta_1 = 1 - \frac{v^2}{8} + \frac{v^4}{384} - \frac{v^6}{46080} + \frac{v^8}{10321920} - \frac{v^{10}}{3715891200} + O(v^{11}),$$
$$B_1 = \frac{1}{2} - \frac{v^2}{48} + \frac{v^4}{3840} - \frac{v^6}{645120} + \frac{v^8}{185794560} - \frac{v^{10}}{81749606400} + O(v^{11}),$$
$$\alpha_2 = 1,$$
$$\beta_2 = 1 + \frac{v^2}{8} + \frac{5v^4}{384} + \frac{61v^6}{46080} + \frac{277v^8}{2064384} + \frac{50521v^{10}}{3715891200} + O(v^{11}),$$
$$b_2 = 1 - \frac{v^2}{24} + \frac{v^4}{1920} - \frac{v^6}{322560} + \frac{v^8}{92897280} - \frac{v^{10}}{40874803200} + O(v^{11}),$$
$$B_2 = \frac{1}{2} + \frac{v^2}{24} + \frac{v^4}{240} + \frac{17v^6}{40320} + \frac{31v^8}{725760} + \frac{691v^{10}}{159667200} + O(v^{11}).$$

Taylor expansion for method (8.2.22)

$$\alpha_1 = 1 + \left(-\frac{3}{4} + \frac{1}{\sqrt{2}}\right)v^2 + \frac{1}{96}(17 - 12\sqrt{2})v^4 + \frac{(-99 + 70\sqrt{2})}{5760}v^6$$
$$+ \frac{(577 - 408\sqrt{2})v^8}{645120} + \frac{(-3363 + 2378\sqrt{2})v^{10}}{116121600} + O(v^{11}),$$
$$\beta_1 = 1 + \left(-\frac{3}{4} + \frac{1}{\sqrt{2}}\right)v^2 + \frac{1}{32}(-17 + 12\sqrt{2})v^4 + \frac{37(-99 + 70\sqrt{2})}{5760}v^6$$
$$+ \frac{839(-577 + 408\sqrt{2})v^8}{645120} + \frac{30601(-3363 + 2378\sqrt{2})v^{10}}{116121600} + O(v^{11}),$$

$$b_1 = \left(1 - \frac{1}{\sqrt{2}}\right) + \frac{1}{24}(-10 + 7\sqrt{2})v^2 + \frac{1}{960}(58 - 41\sqrt{2})v^4 + \frac{(-338 + 239\sqrt{2})}{80640}v^6 + \frac{(1970 - 1393\sqrt{2})v^8}{11612160} + \frac{(-11482 + 8119\sqrt{2})v^{10}}{2554675200} + O(v^{11}),$$

$$B_1 = \frac{1}{\sqrt{2}} + \frac{1}{6}(-3 + 2\sqrt{2})v^2 + \left(-\frac{7}{12} + \frac{33}{40\sqrt{2}}\right)v^4 + \left(-\frac{497}{720} + \frac{41}{42\sqrt{2}}\right)v^6 + \frac{(-591816 + 418477\sqrt{2})v^8}{725760} + \frac{(-38445099 + 27184790\sqrt{2})v^{10}}{39916800} + O(v^{11}),$$

$$\alpha_2 = 1 + \left(\frac{3}{4} - \frac{1}{\sqrt{2}}\right)v^2 + \frac{1}{32}(51 - 36\sqrt{2})v^4 - \frac{217(-99 + 70\sqrt{2})}{5760}v^6 - \frac{9841(-577 + 408\sqrt{2})v^8}{645120} - \frac{717841(-3363 + 2378\sqrt{2})v^{10}}{116121600} + O(v^{11}),$$

$$\beta_2 = 1 + \left(\frac{3}{4} - \frac{1}{\sqrt{2}}\right)v^2 + \frac{1}{96}(85 - 60\sqrt{2})v^4 - \frac{61(-99 + 70\sqrt{2})}{5760}v^6 - \frac{277(-577 + 408\sqrt{2})v^8}{129024} - \frac{50521(-3363 + 2378\sqrt{2})v^{10}}{116121600} + O(v^{11}),$$

$$b_2 = \frac{1}{\sqrt{2}} + \left(-1 + \frac{17}{12\sqrt{2}}\right)v^2 + \left(-\frac{29}{12} + \frac{547}{160\sqrt{2}}\right)v^4 + \frac{(-461608 + 326409\sqrt{2})}{80640}v^6 + \frac{(-157064352 + 111061297\sqrt{2})v^8}{11612160} + \frac{(-81625595832 + 57718012769\sqrt{2})v^{10}}{2554675200} + O(v^{11}),$$

$$B_2 = \left(1 - \frac{1}{\sqrt{2}}\right) + \left(\frac{5}{6} - \frac{7}{6\sqrt{2}}\right)v^2 + \frac{1}{60}(58 - 41\sqrt{2})v^4 - \frac{17(-338 + 239\sqrt{2})}{5040}v^6 - \frac{31(-1970 + 1393\sqrt{2})v^8}{45360} - \frac{691(-11482 + 8119\sqrt{2})v^{10}}{4989600} + O(v^{11}).$$

Taylor expansion for method (8.2.23)

$$\alpha_1 = 1 - \frac{49v^2}{1152} + \frac{2401v^4}{7962624} - \frac{117649v^6}{137594142720} + \frac{823543v^8}{634033809653760} - \frac{40353607v^{10}}{3286831269245091840 0} + O(v^{11}),$$

$$\beta_1 = 1 - \frac{v^2}{36} - \frac{41v^4}{62208} - \frac{6091v^6}{268738560} - \frac{1691801v^8}{2167107747840} - \frac{755864131v^{10}}{28085716412006400} + O(v^{11}),$$

$$b_1 = \frac{7}{24} - \frac{343v^2}{82944} + \frac{16807v^4}{955514880} - \frac{117649v^6}{3302259425280} + \frac{5764801v^8}{136951302885212160} - \frac{282475249v^{10}}{8677234550807042457600} + O(v^{11}),$$

$$B_1 = \frac{2}{3} - \frac{109v^2}{5184} + \frac{121v^4}{59719680} - \frac{6669349v^6}{1444738498560} - \frac{8635447439v^8}{59916195012280320} - \frac{18974353194589v^{10}}{3796290115978081075200} + O(v^{11}),$$

$$\alpha_2 = 1,$$

$$\beta_2 = 1 - \frac{17v^2}{36} + \frac{935v^4}{62208} - \frac{87227v^6}{268738560} - \frac{1152821v^8}{433421549568} - \frac{7623177587v^{10}}{28085716412006400} + O(v^{11}),$$

$$b_2 = \frac{3}{4} - \frac{9v^2}{512} + \frac{81v^4}{655360} - \frac{243v^6}{587202560} + \frac{243v^8}{300647710720} - \frac{2187v^{10}}{2116559883468800} + O(v^{11}),$$

$$B_2 = -\frac{2}{3} + \frac{13v^2}{5184} - \frac{22201v^4}{59719680} - \frac{27692027v^6}{1444738498560} - \frac{65723908081v^8}{59916195012280320} - \frac{237325267297667v^{10}}{3796290115978081075200} + O(v^{11}),$$

$$\alpha_3 = 1 + \frac{49v^2}{1152} + \frac{124901v^4}{7962624} + \frac{936875149v^6}{137594142720} + \frac{1896071678543v^8}{634033809653760} + \frac{43219313660178607v^{10}}{32868312692450918400} + O(v^{11}),$$

$$\beta_3 = 1 + \frac{v^2}{2} + \frac{5v^4}{24} + \frac{61v^6}{720} + \frac{277v^8}{8064} + \frac{50521v^{10}}{3628800} + O(v^{11}),$$

$$b_3 = -\frac{1}{24} - \frac{937v^2}{41472} - \frac{609961v^4}{59719680} - \frac{6511231289v^6}{1444738498560} - \frac{7424833305271v^8}{3744762188267520} - \frac{103444872822456427v^{10}}{118634066124315033600} + O(v^{11}),$$

$$B_3 = 1 + \frac{v^2}{3} + \frac{2v^4}{15} + \frac{17v^6}{315} + \frac{62v^8}{2835} + \frac{1382v^{10}}{155925} + O(v^{11}).$$

Taylor expansion for method (8.2.24)

$$\alpha_1 = 1 + \frac{1}{384}(-33 - 19\sqrt[3]{3} - 9\,3^{2/3})v^2 + \frac{(2115 + 1497\sqrt[3]{3} + 955\,3^{2/3})v^4}{884736} + \frac{(-54883 - 38457\sqrt[3]{3} - 26331\,3^{2/3})v^6}{1698693120}$$

$$
\begin{aligned}
&+\frac{(4350345+3022795\sqrt[3]{3}+2093553\,3^{2/3})v^8}{18264348426240}\\
&+\frac{(-344509371-238934721\sqrt[3]{3}-165673459\,3^{2/3})v^{10}}{315607940805427200}+O(v^{11}),\\
\beta_1 &= 1+\frac{1}{600}(-21-8\sqrt[3]{3}-9\,3^{2/3})v^2+\frac{(-5343-\frac{7091}{\sqrt[3]{3}}-3389\sqrt[3]{3})v^4}{1440000}\\
&+\frac{(-30310189-20870847\sqrt[3]{3}-14343981\,3^{2/3})v^6}{51840000000}\\
&+\frac{(-337413172197-234059500031\sqrt[3]{3}-161907159213\,3^{2/3})v^8}{3483648000000000}\\
&+\frac{(-6082242830801181-4218418696110663\sqrt[3]{3}-2923800594515749\,3^{2/3})v^{10}}{37623398400000000000}\\
&+O(v^{11}),\\
b_1 &= \frac{1}{24}(9+3\sqrt[3]{3}+3^{2/3})+\frac{(-145-99\sqrt[3]{3}-57\,3^{2/3})v^2}{9216}\\
&+\frac{(10707+7561\sqrt[3]{3}+5067\,3^{2/3})v^4}{35389440}\\
&+\frac{(-846297-589755\sqrt[3]{3}-407233\,3^{2/3})v^6}{285380444160}\\
&+\frac{(22354489+15512283\sqrt[3]{3}+10753569\,3^{2/3})v^8}{1315033086689280}\\
&+\frac{(-589827737-\frac{1226986993}{3^{2/3}}-283597185\,3^{2/3})v^{10}}{9257832930292531200}+O(v^{11}),\\
B_1 &= \frac{1}{30}(9-3\sqrt[3]{3}-4\,3^{2/3})+\frac{(-971-333\sqrt[3]{3}-159\,3^{2/3})v^2}{72000}\\
&+\frac{(-9893067-6725041\sqrt[3]{3}-4669443\,3^{2/3})v^4}{6912000000}\\
&+\frac{(-110223290553-\frac{158020656611}{\sqrt[3]{3}}-76358768219\sqrt[3]{3})v^6}{464486400000000}\\
&+\frac{(-6350043743536681-4405010788277163\sqrt[3]{3}-3051258622095249\,3^{2/3})v^8}{160526499840000000000}\\
&+\frac{(-55997303261662691933 7-38831894009019017425 1\sqrt[3]{3}-26921176423093593627 3\,3^{2/3})v^{10}}{847579919155200000000000000}\\
&+O(v^{11}),\\
\alpha_2 &= 1+\frac{1}{120}(9+7\sqrt[3]{3}+3^{2/3})v^2+\frac{(22467+17241\sqrt[3]{3}+10043\,3^{2/3})v^4}{1728000}\\
&+\frac{(63724561+44978403\sqrt[3]{3}+30563769\,3^{2/3})}{20736000000}v^6
\end{aligned}
$$

$$+\frac{(1015271271657+705779397611\sqrt[3]{3}+488297833553\,3^{2/3})}{1393459200000000}v^8$$
$$+\frac{(5185245735246333+3596505417981159\sqrt[3]{3}+2493152211073157\,3^{2/3})v^{10}}{30098718720000000000}$$
$$+O(v^{11}),$$

$$\beta_2=1+\frac{1}{120}(3-\sqrt[3]{3}+2\,3^{2/3})v^2+\frac{(327+771\sqrt[3]{3}+983\,3^{2/3})v^4}{864000}$$
$$+\frac{(938671+1212933\sqrt[3]{3}+562959\,3^{2/3})}{10368000000}v^6$$
$$+\frac{(8718904887+6027766901\sqrt[3]{3}+3239842223\,3^{2/3})}{696729600000000}v^8$$
$$+\frac{(17304108560643+10653464795289\sqrt[3]{3}+7673078753147\,3^{2/3})v^{10}}{15049359360000000000}$$
$$+O(v^{11}),$$

$$b_2=\frac{1}{12}(3-3\sqrt[3]{3}-3^{2/3})+\frac{(-2747-1281\sqrt[3]{3}-2163\,3^{2/3})v^2}{115200}$$
$$+\frac{(-2847087-1868701\sqrt[3]{3}-1377223\,3^{2/3})}{442368000}v^4$$
$$-\frac{47(71319490701+49113289623\sqrt[3]{3}+34277991829\,3^{2/3})v^6}{2229534720000000}$$
$$+\frac{(-18215815214276273-12618020227459779\sqrt[3]{3}-8754867925251417\,3^{2/3})}{51368479948800000000}v^8$$
$$+\frac{(-113583473501900849022233-78744529286114007168859\sqrt[3]{3}-54601774342154239442573^{2/3})v^{10}}{135612787064832000000000000}$$
$$+O(v^{11}),$$

$$B_2=\frac{1}{6}(3+3^{2/3})+\frac{(-37-51\sqrt[3]{3}+27\,3^{2/3})v^2}{14400}$$
$$+\frac{(11703+22069\sqrt[3]{3}+14287\,3^{2/3})}{55296000}v^4$$
$$+\frac{(10155546543+8959120989\sqrt[3]{3}+4061194247\,3^{2/3})v^6}{278691840000000}$$
$$+\frac{(24922607251937+15440631924051\sqrt[3]{3}+9981610481673\,3^{2/3})}{642105999360000000}v^8$$
$$+\frac{(5720864832661462197+3685666714497670031\sqrt[3]{3}+2734652857842569213\,3^{2/3})v^{10}}{16951598383104000000000000}$$
$$+O(v^{11}),$$

$$\alpha_3=1+\frac{(21-17\sqrt[3]{3}+29\,3^{2/3})v^2}{1920}+\frac{(20787+50601\sqrt[3]{3}+33323\,3^{2/3})}{110592000}v^4$$

$$+ \frac{(144458419 + 104260137\sqrt[3]{3} + 61812651\, 3^{2/3})v^6}{5308416000000}$$
$$+ \frac{(2091346305117 + 1420085277191\sqrt[3]{3} + 994680593893\, 3^{2/3})}{14269022208000000000}v^8$$
$$+ \frac{(9752298693664947 + 6766245560458281\sqrt[3]{3} + 4700400283073963\, 3^{2/3})v^{10}}{123284351877120000000000}$$
$$+ O(v^{11}),$$

$$\beta_3 = 1 + \frac{1}{600}(6 + 13\sqrt[3]{3} - 3^{2/3})v^2 + \frac{(-42 + 159\sqrt[3]{3} + 157\, 3^{2/3})}{432000}v^4$$
$$+ \frac{61(1798 - 21\sqrt[3]{3} + 1017\, 3^{2/3})v^6}{6480000000} + \frac{277(50514 + 20197\sqrt[3]{3} + 4031\, 3^{2/3})}{21772800000000}v^8$$
$$+ \frac{50521(399702 + 765771\sqrt[3]{3} + 236233\, 3^{2/3})v^{10}}{29393280000000000000} + O(v^{11}),$$

$$b_3 = \frac{1}{24}(9 + 3\sqrt[3]{3} + 3^{2/3}) + \frac{(-143 - 189\sqrt[3]{3} + 153\, 3^{2/3})v^2}{115200}$$
$$+ \frac{(5619 + 5737\sqrt[3]{3} + 3051\, 3^{2/3})}{55296000}v^4$$
$$+ \frac{(1777353849 + 1192560027\sqrt[3]{3} + 805825121\, 3^{2/3})v^6}{278691840000000}$$
$$+ \frac{(136087408691 + 93914868393\sqrt[3]{3} + 65599498539\, 3^{2/3})}{401316249600000000}v^8$$
$$+ \frac{(9797155554558513 + 6798867873791299\sqrt[3]{3} + 4713547167196377\, 3^{2/3})v^{10}}{529737449472000000000000}$$
$$+ O(v^{11}),$$

$$B_3 = \frac{1}{10}\left(2 - \frac{1}{\sqrt[3]{3}} + \sqrt[3]{3}\right) + \frac{(-4 + 33\sqrt[3]{3} + 9\, 3^{2/3})v^2}{9000}$$
$$+ \frac{(228 + 119\sqrt[3]{3} + 487\, 3^{2/3})}{6750000}v^4 + \frac{17(20004 + 2217\sqrt[3]{3} + 4241\, 3^{2/3})v^6}{85050000000}$$
$$+ \frac{31(92924 + 86877\sqrt[3]{3} + 11421\, 3^{2/3})}{38272500000000}v^8$$
$$+ \frac{691(742332 + 1695011\sqrt[3]{3} + 1105003\, 3^{2/3})v^{10}}{631496250000000000} + O(v^{11}).$$

Taylor expansions for the trigonometrically fitted (each stage) method based on the fourth order method (8.1.13).

$$\alpha_1 = 1 - 0.22822v^2 + 0.00868074v^4 - 0.000132075v^6$$
$$+ 1.0765026331929995 \times 10^{-6}v^8$$
$$- 5.459545528244244 \times 10^{-9}v^{10} + O(v^{11}),$$

$$b_1 = 0.675604 - 0.0513954v^2 + 0.00117295v^4 - 0.0000127472v^6 + 8.08098944518828 \times 10^{-8}v^8 - 3.353171446634005 \times 10^{-10}v^{10} + O(v^{11}),$$

$$B_1 = 1.35121 - 0.102791v^2 + 0.00234589v^4 - 0.0000254943v^6 + 1.616197889037656 \times 10^{-7}v^8 - 6.70634289326801 \times 10^{-10}v^{10} + O(v^{11}),$$

$$\alpha_2 = 1 - 0.134057v^2 - 0.0174011v^4 - 0.00320379v^6 - 0.000592062v^8 - 0.000109513v^{10} + O(v^{11}),$$

$$b_2 = -0.175604 - 0.0391738v^2 - 0.00741726v^4 - 0.0013759v^6 - 0.000254606v^8 - 0.0000471009v^{10} + O(v^{11}),$$

$$B_2 = -1.70241 + 0.205582v^2 - 0.00744775v^4 + 0.000128483v^6 - 1.292958311230123 \times 10^{-6}v^8 + 8.516524610906462 \times 10^{-9}v^{10} + O(v^{11}),$$

$$\alpha_3 = 1 + 0.134057v^2 + 0.0353723v^4 + 0.0102784v^6 + 0.00301496v^8 + 0.00088524v^{10} + O(v^{11}),$$

$$b_3 = -0.175604 - 0.0627146v^2 - 0.0188803v^4 - 0.00556083v^6 - 0.0016335v^8 - 0.000479698v^{10} + O(v^{11}),$$

$$B_3 = 1.35121 - 0.102791v^2 + 0.00234589v^4 - 0.0000254943v^6 + 1.616197889037656 \times 10^{-7}v^8 - 6.70634289326801 \times 10^{-10}v^{10} + O(v^{11}),$$

$$\alpha_4 = 1 + 0.22822v^2 + 0.0434037v^4 + 0.00805655v^6 + 0.00149096v^8 + 0.000275822v^{10} + O(v^{11}),$$

$$b_4 = 0.675604 + 0.102791v^2 + 0.0187672v^4 + 0.00346723v^6 + 0.000641307v^8 + 0.000118633v^{10} + O(v^{11}).$$

Taylor expansions for the trigonometrically fitted (each stage) method based on the fourth order method (8.1.17).

$$\alpha_1 = 1 - 0.00904461v^2 + 0.0000136342v^4 - 8.221055656370937 \times 10^{-9}v^6 + 2.655581225772937 \times 10^{-12}v^8 - 5.337490419195165 \times 10^{-16}v^{10} + O(v^{12}),$$

$$\beta_1 = 1 - 0.0634813v^2 + 0.000288871v^4 - 7.522904550911909 \times 10^{-7}v^6$$

$$
\begin{aligned}
&- 2.879643476093273 \times 10^{-10} v^8 - 7.498877433437165 \times 10^{-12} v^{10} \\
&+ O(v^{12}), \\
b_1 &= 0.134496 - 0.000405489 v^2 + 3.6674890397071147 \times 10^{-7} v^4 \\
&- 1.5795724939577097 \times 10^{-10} v^6 \\
&+ 3.96850648508669 \times 10^{-14} v^8 - 6.52611071743779 \times 10^{-18} v^{10} \\
&+ O(v^{12}), \\
B_1 &= 0.515353 - 0.0181508 v^2 + 0.000131737 v^4 - 4.7237602431486547 \\
&\times 10^{-7} v^6 \\
&+ 8.469392903940952 \times 10^{-10} v^8 - 1.828526001573185 \times 10^{-12} v^{10} \\
&+ O(v^{12}), \\
\alpha_2 &= 1 - 0.110896 v^2 - 0.00331224 v^4 - 0.000207332 v^6 \\
&- 0.0000121651 v^8 - 7.151580951637622 \times 10^{-7} v^{10} + O(v^{12}), \\
\beta_2 &= 1 + 0.0482769 v^2 + 0.00629181 v^4 + 0.000924495 v^6 \\
&+ 0.000137286 v^8 + 0.0000204086 v^{10} + O(v^{12}), \\
b_2 &= -0.22482 - 0.0144114 v^2 - 0.000852893 v^4 - 0.0000501701 v^6 \\
&- 2.9495499139878988 \times 10^{-6} v^8 - 1.7339730988019144 \times 10^{-7} v^{10} \\
&+ O(v^{12}), \\
B_2 &= -0.085782 - 0.0156291 v^2 - 0.00238576 v^4 - 0.000355846 v^6 \\
&- 0.0000529255 v^8 - 7.869094771269424 \times 10^{-6} v^{10} + O(v^{12}), \\
\alpha_3 &= 1 + 0.107197 v^2 + 0.0115733 v^4 + 0.00126494 v^6 \\
&+ 0.00013853 v^8 + 0.0000151747 v^{10} + O(v^{12}), \\
\beta_3 &= 1 + 0.00690366 v^2 + 0.000136575 v^4 + 3.0573015286240117 \times 10^{-6} v^6 \\
&+ 6.917710386828629 \times 10^{-8} v^8 + 1.5669474854300608 \times 10^{-9} v^{10} \\
&+ O(v^{12}), \\
b_3 &= 0.75632 + 0.030108 v^2 + 0.00382896 v^4 + 0.000418471 v^6 \\
&+ 0.0000458476 v^8 + 5.02241305086368 \times 10^{-6} v^{10} \\
&+ O(v^{12}), \\
B_3 &= 0.441583 - 0.00200954 v^2 + 0.0000262685 v^4 \\
&+ 4.532774919129704 \times 10^{-7} v^6 \\
&+ 1.0413592845429705 \times 10^{-8} v^8 + 2.3583970107393747 \times 10^{-10} v^{10} \\
&+ O(v^{12}),
\end{aligned}
$$

$$\alpha_4 = 1 + 0.0127442v^2 + 0.000205868v^4 + 3.4889662628521454 \times 10^{-6}v^6 + 5.947253211683284 \times 10^{-8}v^8 + 1.0144177996172859 \times 10^{-9}v^{10} + O(v^{12}),$$

$$\beta_4 = 1 + 0.00830065v^2 + 0.0000574173v^4 + 3.876349986827062 \times 10^{-7}v^6 + 2.609136341938259 \times 10^{-9}v^8 + 1.755570645359568 \times 10^{-11}v^{10} + O(v^{12}),$$

$$b_4 = 0.334004 + 0.000818894v^2 + 0.000027219v^4 + 4.549550964670017 \times 10^{-7}v^6 + 7.766666845847533 \times 10^{-9}v^8 + 1.324886688646942 \times 10^{-10}v^{10} + O(v^{12}),$$

$$B_4 = 0.128846 + 0.000713003v^2 + 4.73471060426436 \times 10^{-6}v^4 + 3.181522149637934 \times 10^{-8}v^6 + 2.1403120215930684 \times 10^{-10}v^8 + 1.4400336343405398 \times 10^{-12}v^{10} + O(v^{12}).$$

Taylor expansions for the trigonometrically fitted (each stage) method based on the fourth order (six stages) method derived by Ruth's method (8.1.12) with the technique (8.1.18).

$$\alpha_1 = 1 - \frac{49v^2}{4608} + \frac{2401v^4}{127401984} - \frac{117649v^6}{8806025134080} + \frac{823543v^8}{1623126552713625600} - \frac{40353607v^{10}}{336571521970697404416000} + O(v^{12}),$$

$$\beta_1 = 1 - \frac{v^2}{144} - \frac{41v^4}{995328} - \frac{6091v^6}{17199267840} - \frac{1691801v^8}{554779583447040} - \frac{755864131v^{10}}{28759773605894553600} + O(v^{12}),$$

$$b_1 = \frac{7}{48} - \frac{343v^2}{663552} + \frac{16807v^4}{30576476160} - \frac{117649v^6}{422689206435840} + \frac{5764801v^8}{70119067077228625920} - \frac{282475249v^{10}}{17770976360052822953164800} + O(v^{12}),$$

$$B_1 = \frac{1}{3} - \frac{109v^2}{41472} + \frac{121v^4}{1911029760} - \frac{6669349v^6}{184926527815680} - \frac{8635447439v^8}{30677091846287523840} - \frac{18974353194589v^{10}}{7774802157523110042009600} + O(v^{12}),$$

$$\beta_2 = 1 - \frac{17v^2}{144} + \frac{935v^4}{995328} - \frac{87227v^6}{17199267840} - \frac{1152821v^8}{110955916689408} - \frac{7623177587v^{10}}{2875977360589453600} + O(v^{12}),$$

$$b_2 = \frac{3}{8} - \frac{9v^2}{4096} + \frac{81v^4}{20971520} - \frac{243v^6}{75161927680} + \frac{243v^8}{153931627888640} - \frac{2187v^{10}}{4334714641344102400} + O(v^{12}),$$

$$B_2 = -\frac{1}{3} + \frac{13v^2}{41472} - \frac{22201v^4}{1911029760} - \frac{27692027v^6}{184926527815680} - \frac{65723908081v^8}{3067709184628752384\,0} - \frac{237325267297667v^{10}}{777480215752311004200960\,0} + O(v^{12}),$$

$$\alpha_3 = 1 + \frac{49v^2}{4608} + \frac{124901v^4}{127401984} + \frac{936875149v^6}{8806025134080} + \frac{1896071678543v^8}{16231265527136256\,0} + \frac{43219313660178607v^{10}}{33657152197069740441600} + O(v^{12}),$$

$$b_3 = -\frac{1}{48} - \frac{937v^2}{331776} - \frac{609961v^4}{1911029760} - \frac{6511231289v^6}{184926527815680} - \frac{7424833305271v^8}{19173182403929702400} - \frac{103444872822456427v^{10}}{24296256742259718881280\,0} + O(v^{12}),$$

$$B_3 = 1 - \frac{v^2}{24} + \frac{v^4}{1920} - \frac{v^6}{322560} + \frac{v^8}{92897280} - \frac{v^{10}}{40874803200} + O(v^{12}),$$

$$\alpha_4 = 1 - \frac{49v^2}{4608} - \frac{110495v^4}{127401984} - \frac{763859089v^6}{8806025134080} - \frac{284943681589v^8}{32462531054272512} - \frac{29929168009554247v^{10}}{33657152197069740441600} + O(v^{12}),$$

$$\beta_4 = 1 + \frac{17v^2}{144} + \frac{12937v^4}{995328} + \frac{24571307v^6}{17199267840} + \frac{87131648377v^8}{554779583447040} + \frac{496587784106147v^{10}}{2875977360589453600} + O(v^{12}),$$

$$b_4 = -\frac{1}{48} - \frac{1727v^2}{663552} - \frac{8288641v^4}{30576476160} - \frac{81542922047v^6}{2958824445050880} - \frac{1371109788552961v^8}{490833469540600381440} - \frac{35210074731213970367v^{10}}{124396834520369760672153600} + O(v^{12}),$$

$$B_4 = -\frac{1}{3} - \frac{1619v^2}{41472} - \frac{8231161v^4}{1911029760} - \frac{87591429659v^6}{184926527815680} - \frac{1597512864247921v^8}{30677091846287523840} - \frac{44512125440565344099v^{10}}{7774802157523110042009600} + O(v^{12}),$$

$$\beta_5 = 1 + \frac{v^2}{144} + \frac{89v^4}{995328} + \frac{21691v^6}{17199267840} + \frac{9958409v^8}{554779583447040} + \frac{7354742131v^{10}}{2875977360589455360 0} + O(v^{12}),$$

$$b_5 = \frac{3}{8} - \frac{9v^2}{4096} + \frac{81v^4}{20971520} - \frac{243v^6}{75161927680} + \frac{243v^8}{153931627888640} - \frac{2187v^{10}}{4334714641344102400} + O(v^{12}),$$

$$B_5 = \frac{1}{3} - \frac{13v^2}{41472} + \frac{22201v^4}{1911029760} + \frac{27692027v^6}{184926527815680} + \frac{65723908081v^8}{306770918462875238 40} + \frac{237325267297667v^{10}}{7774802157523110042009600} + O(v^{12}),$$

$$\alpha_6 = 1 + \frac{49v^2}{4608} + \frac{12005v^4}{127401984} + \frac{7176589v^6}{8806025134080} + \frac{228121411v^8}{32462531054272512} + \frac{2038704579247v^{10}}{33657152197069740441600} + O(v^{12}),$$

$$b_6 = \frac{7}{48} + \frac{343v^2}{331776} + \frac{16807v^4}{1911029760} + \frac{2000033v^6}{26418075402240} + \frac{178708831v^8}{273902605770424320} + \frac{195190397059v^{10}}{34708938203228169830400} + O(v^{12}).$$

References

1. Abia, L., Sanz-Serna, J.M.: Partitioned Runge-Kutta methods for separable Hamiltonian problems. Math. Comput. **60**, 617–634 (1993)
2. Brusa, L., Nigro, L.: A one-step method for direct integration of structural dynamic equations. Int. J. Numer. Methods Eng. **14**, 685–699 (1980)
3. Forest, E., Ruth, R.D.: Fourth order symplectic integration. Physica D **43**, 105–117 (1990)
4. Franco, J.M.: Exponentially fitted explicit Runge-Kutta-Nyström methods. J. Comput. Appl. Math. **167**, 1–19 (2004)
5. Hairer, E., Lubich, Ch., Wanner, G.: Geometric Numerical Integration. Springer, Berlin (2002)
6. Ixaru, L.Gr., Vanden Berghe, G.: Exponential Fitting. Kluwer Academic, Amsterdam (2004)
7. Kalogiratou, Z.: Symplectic trigonometrically fitted partitioned Runge-Kutta methods. Phys. Lett. A **370**, 1–7 (2007)
8. Kalogiratou, Z., Simos, T.E.: Construction of trigonometrically and exponentially fitted Runge-Kutta-Nyström methods for the numerical solution of the Schrödinger equation and related problems—a method of 8th algebraic order. J. Math. Chem. **31**, 212–232 (2002)
9. Kalogiratou, Z., Monovasilis, Th., Simos, T.E.: A symplectic trigonometrically fitted modified partitioned Runge-Kutta method for the numerical integration of orbital problems. Appl. Numer. Anal. Comput. Math. **2**, 359–364 (2005)
10. Kalogiratou, Z., Monovasilis, Th., Simos, T.E.: A fifth-order symplectic trigonometrically fitted partitioned Runge-Kutta method. In: Proceedings of International Conference on Numer-

ical Analysis and Applied Mathematics, ICNAAM 2007, pp. 313–317. American Institute of Physics Corfu-Greece 16–20/09/2007
11. McLachlan, R., Atela, P.: The accuracy of symplectic integrators. Nonlinearity **5**, 541–562 (1992)
12. Monovasilis, Th., Simos, T.E.: Symplectic and trigonometrically fitted symplectic methods of second and third order. Phys. Lett. A **354**, 377–383 (2006)
13. Monovasilis, Th., Simos, T.E.: Symplectic methods for the numerical integration of the Schrödinger equation. Comput. Mater. Sci. **38**, 526–532 (2007)
14. Monovasilis, Th., Kalogiratou, Z., Simos, T.E.: Trigonometrically and exponentially fitted symplectic methods of third order for the numerical integration of the Schrödinger equation. Appl. Numer. Anal. Comput. Math. **2**(2), 238–244 (2005)
15. Monovasilis, Th., Kalogiratou, Z., Simos, T.E.: Trigonometrically fitted and exponentially fitted symplectic methods for the numerical integration of the Schrödinger equation. J. Math. Chem. **40**, 257–267 (2006)
16. Monovasilis, Th., Kalogiratou, Z., Simos, T.E.: Computation of the eigenvalues of the Schrödinger equation by symplectic and trigonometrically fitted symplectic partitioned Runge-Kutta methods. Phys. Lett. A **372**, 569–573 (2008)
17. Monovasilis, Th., Kalogiratou, Z., Simos, T.E.: A family of trigonometrically fitted partitioned Runge-Kutta symplectic methods. Appl. Math. Comput. **209**(1), 91–96 (2009)
18. Monovasilis, Th., Kalogiratou, Z., Simos, T.E.: A phase-fitted symplectic partitioned Runge-Kutta method for the numerical solution of the Schrödinger equation. AIP Conf. Proc. **1168**, 1595–1599 (2009)
19. Monovasilis, Th., Kalogiratou, Z., Simos, T.E.: Symplectic partitioned Runge-Kutta methods with minimal phase-lag. Comput. Phys. Commun. **181**, 1251–1254 (2010)
20. Rapris, A.D., Simos, T.E.: A four step phase-fitted method for the numerical integration of second order initial-value problems. BIT **31**, 160–168 (1991)
21. Ruth, R.D.: A canonical integration technique. IEEE Trans. Nuclear Sci. **30**, 2669–2671 (1983)
22. Sanz-Serna, J.M., Calvo, M.P.: Numerical Hamiltonian Problem. Chapman and Hall, London (1994)
23. Simos, T.E.: An exponentially-fitted Runge-Kutta method for the numerical integration of initial-value problems with periodic or oscillating solutions. Comput. Phys. Commun. **115**, 1–8 (1998)
24. Simos, T.E.: Exponentially fitted Runge-Kutta-Nyström method for the numerical solution of initial-value problems with oscillating solutions. Appl. Math. Lett. **15**, 217–225 (2002)
25. Simos, T.E., Vigo-Aguiar, J.: Exponentially fitted symplectic integrator. Phys. Rev. E **67**, 016701 (2003)
26. Stiefel, E., Bettis, D.G.: Stabilization of Cowell's method. Numer. Math. **13**, 154 (1969)
27. Tocino, A., Aguiar, J.V.: Symplectic conditions for exponential fitting Runge-Kutta-Nyström methods. Math. Comput. Model. **42**, 873–876 (2005)
28. Van de Vyver, H.: A symplectic exponentially fitted modified Runge-Kutta-Nyström method for the numerical integration of orbital problems. New Astron. **10**(4), 261–269 (2005)
29. Van de Vyver, H.: A fourth-order symplectic exponentially fitted integrator. Comput. Phys. Commun. **174**(4), 255–262 (2006)
30. Van de Vyver, H.: A symplectic Runge-Kutta-Nyström method with minimal phase-lag. Phys. Lett. A **367**(1–2), 16–24 (2007)
31. Van Der Houwen, P.J., Sommeijer, B.P.: Explicit Runge-Kutta(-Nyström) methods with reduced phase errors for computing oscillating solutions. SIAM J. Numer. Anal. **24**, 595–617 (1987)
32. Vanden Berghe, G., De Meyer, H., Van Daele, M., Van Hecke, T.: Exponentially fitted explicit Runge-Kutta methods. Comput. Phys. Commun. **123**, 7–15 (1999)
33. Yoshida, H.: Construction of higher order symplectic integrators. Phys. Lett. A **150**, 262–268 (1990)

Chapter 9
On the Klein-Gordon Equation on Some Examples of Conformally Flat Spin 3-Manifolds

Rolf Sören Kraußhar

Abstract In this paper we present an overview about our recent results on the analytic treatment of the Klein-Gordon equation on some conformally flat 3-tori and on 3-spheres.

In the first part of this paper we consider the time independent Klein-Gordon equation $(\Delta - \alpha^2)u = 0$ ($\alpha \in \mathbb{R}$) on some conformally flat 3-tori associated with a representative system of conformally inequivalent spinor bundles. We set up an explicit formula for the fundamental solution associated to each spinor bundle. We show that we can represent any solution to the homogeneous Klein-Gordon equation on such a torus as finite sum over generalized 3-fold periodic or resp. antiperiodic elliptic functions that are in the kernel of the Klein-Gordon operator. Furthermore, we prove Cauchy and Green type integral formulas and set up an appropriate Teodorescu and Cauchy transform for the toroidal Klein-Gordon operator on this spin tori. These in turn are used to set up explicit formulas for the solution to the inhomogeneous Klein-Gordon equation $(\Delta - \alpha^2)u = f$ on the 3-torus attached to the different choices of different spinor bundles. In the second part of the paper we present a unified approach to describe the solutions to the Klein-Gordon equation on 3-spheres. We give an explicit representation formula for the solutions in terms of hypergeometric functions and monogenic homogeneous polynomials.

Keywords Klein-Gordon equation · Hypercomplex integral operators · PDE on manifolds · Conformally flat spin 3-tori and spheres · Multiperiodic functions · Special functions

Mathematics Subject Classification (2000) 30G35 · 35Q40 · 35A08 · 35R01

R. Sören Kraußhar (✉)
Fachbereich Mathematik, Technische Universität Darmstadt, Schloßgartenstraße 7, 64289 Darmstadt, Germany
e-mail: krausshar@mathematik.tu-darmstadt.de

T.E. Simos (ed.), *Recent Advances in Computational and Applied Mathematics*,
DOI 10.1007/978-90-481-9981-5_9,

9.1 Introduction

The Klein-Gordon equation is a relativistic version of the Schrödinger equation. It describes the motion of a quantum scalar or pseudoscalar field, a field whose quanta are spinless particles. The Klein-Gordon equation describes the quantum amplitude for finding a point particle in various places, cf. for instance [6, 21]. It can be expressed in the form

$$\left(\Delta - \alpha^2 - \frac{1}{c^2}\frac{\partial^2}{\partial t^2}\right)u(\mathbf{x}; t) = 0,$$

where $\Delta := \sum_{i=1}^{3} \frac{\partial^2}{\partial x_i^2}$ is the usual Euclidean Laplacian in $\mathbb{R}^3$ and $\alpha = \frac{mc}{\hbar}$. Here, m represents the mass of the particle, c the speed of light and $\hbar$ is the Planck number. This equation correctly describes the spin-less pion which however appears in nature as a composite particle.

Since a long time it is well known that any solution to the Dirac equation, which describes the spinning electron, satisfies the Klein-Gordon equation. However, the converse is not true. In the time-independent case the homogeneous Klein-Gordon equation simplifies to a screened Poisson equation of the form

$$(\Delta - \alpha^2)u(\mathbf{x}) = 0.$$

The solution of this equation provides the first step to solve the more complicated time-dependent case. Therefore, the study of the time-independent solutions is very important. As explained extensively in the literature, see for example [10, 11, 15, 16] and elsewhere, the quaternionic calculus allows us to factorize the Klein-Gordon operator elegantly by

$$\Delta - \alpha^2 = -(\mathbf{D} - i\alpha)(\mathbf{D} + i\alpha),$$

where $\mathbf{D} := \sum_{i=1}^{3} \frac{\partial}{\partial x_i} e_i$ is the Euclidean Dirac operator associated to the spatial variable. Here the elements e_1, e_2, e_3 stand for the elementary quaternionic imaginary units. The study of the solutions to the original scalar second order equation is thus reduced to study vector valued eigensolutions to the first order Dirac operator associated to purely imaginary eigenvalues. For eigensolutions to the first order Euclidean Dirac operator it was possible to develop a powerful higher dimensional version of complex function theory, see for instance [10, 15, 20, 22, 24]. By means of these function theoretical methods it was possible to set up fully analytic representation formulas for the solutions to the homogeneous and inhomogeneous Klein-Gordon in the three dimensional Euclidean space in terms of quaternionic integral operators.

In the first part of this paper (Sects. 9.3–9.5) we present analogous methods for the Klein-Gordon equation on three-dimensional conformally flat tori associated to different conformally inequivalent spinor bundles. The torus model is one of the most important geometrical models in modern quantum gravity and cosmology.

We give an explicit formula for the fundamental solution in terms of an appropriately adapted three-fold periodic resp. anti-periodic generalization of the Weierstraß $\wp$-function associated to the operator $(\mathbf{D} - i\alpha)$. Then we show that we can represent any solution to the homogeneous Klein-Gordon equation on these classes of tori as a finite sum over generalized three-fold periodic resp. anti-periodic elliptic functions that are in the kernel of the Klein-Gordon operator. Furthermore we give a Green type integral formula and set up a Teodorescu and Cauchy transform for the toroidal Klein-Gordon operator. These in turn are used to set up explicit formulas for the solution to the inhomogeneous Klein-Gordon equation on this class of 3-tori. A non-zero right-hand side in the Klein-Gordon equation naturally arises in the context when including for instance quantum gravitational effects into the model.

In turn, the results of this paper refer to a very particular mathematical subcase that appears within the theory of generalized Helmholtz type equations with arbitrary complex parameters that we developed for the general framework of k dimensional cylinders in $\mathbb{R}^n$ with arbitrary spinor bundles in [5], jointly written with D. Constales. However, from the quantum mechanical view the case treated here in great detail has a very special meaning and in the three-dimensional case the Bessel functions simplify significantly to ordinary trigonometric functions.

Finally, in the remaining part of this paper (Sect. 9.6) we present a unified approach to describe the solutions to the Klein-Gordon equation on spheres. We give an explicit representation formula for the solutions in terms of hypergeometric functions and monogenic homogeneous polynomials. We also establish links to projective hyperbolic models and discuss some interesting limit cases.

This provides a counterpart contribution to the unified model that the author developed in joint work with I. Cação and D. Constales in [3] to treat the solutions to the time-harmonic Maxwell-equations in radially symmetric spaces.

9.2 Notations

Let $\{e_1, e_2, e_3\}$ be the standard basis of $\mathbb{R}^3$. We embed $\mathbb{R}^3$ into the quaternions $\mathbb{H}$ whose elements have the form $a = a_0 e_0 + \mathbf{a}$ with $\mathbf{a} = a_1 e_1 + a_2 e_2 + a_3 e_3$. In the quaternionic calculus one has the multiplication rules $e_1 e_2 = e_3 = -e_2 e_1$, $e_2 e_3 = e_1 = -e_3 e_2$, $e_3 e_1 = e_2 = -e_1 e_3$, and $e_j e_0 = e_0 e_j$ and $e_j^2 = -1$ for all $j = 1, 2, 3$. The quaternionic conjugate of a is defined by $\overline{a} = a_0 - \sum_{i=1}^{3} a_i e_i$, that means it switches the sign on the imaginary units $\overline{e}_j = -e_j$ for $j = 1, 2, 3$ and it leaves the scalar part invariant.

By $\mathbb{H} \otimes_{\mathbb{R}} \mathbb{C}$ we obtain the complexified quaternions. These will be denoted by $\mathbb{H}(\mathbb{C})$. Their elements have the form $\sum_{j=0}^{3} a_j e_j$ where a_j are complex numbers $a_j = a_{j_1} + i a_{j_2}$. The complex imaginary unit satisfies $i e_j = e_j i$ for all $j = 0, 1, 2, 3$. The scalar part $a_0 e_0$ of a (complex) quaternion will be denoted by $\mathrm{Sc}(a)$. On $\mathbb{H}(\mathbb{C})$ one considers a standard (pseudo)norm defined by $\|a\| = (\sum_{j=0}^{3} |a_j|^2)^{1/2}$ where $|\cdot|$ is the usual absolute value.

The complex imaginary unit i commutes with all basis elements e_j, i.e. we have $ie_j = e_j i$ for all $j = 1, 2, 3$. We denote the complex conjugate of a complex number $\lambda \in \mathbb{C}$ by $\lambda^\sharp$. For any elements $a \in \mathbb{H}(\mathbb{C})$ we have $(\overline{a})^\sharp = \overline{(a^\sharp)}$.

For simplicity we write in all that follows $\mathbf{D}$ for the Euclidean Dirac operator $\mathbf{D} := \sum_{i=1}^{3} \frac{\partial}{\partial x_i} e_i$ associated exclusively to the spatial variable.

9.3 Conformally Flat 3-Tori

In this section and the following two ones we treat conformally flat spin 3-tori with inequivalent spinor bundles. Let $\Omega := \mathbb{Z}e_1 + \mathbb{Z}e_2 + \mathbb{Z}e_3$ be the standard lattice in $\mathbb{R}^3$. Then, following e.g. [13] the topological quotient space $\mathbb{R}^3/\Omega$ realizes a 3-dimensional conformally flat torus denoted by T_3. This is contructed by gluing the equivalent vertices of the fundamental period cell together. However, as mentioned in [14] one can construct a number of conformally inequivalent spinor bundles over T_3.

We recall that in general different spin structures on a spin manifold M are detected by the number of distinct homomorphisms from the fundamental group $\Pi_1(M)$ to the group $\mathbb{Z}_2 = \{0, 1\}$. In this case we have that $\Pi_1(T_3) = \mathbb{Z}^3$. There are two homomorphisms of $\mathbb{Z}$ to $\mathbb{Z}_2$. The first one is $\theta_1 : \mathbb{Z} \to \mathbb{Z}_2 : \theta_1(n) \equiv 0 \mod 2$ while the second one is the homomorphism $\theta_2 : \mathbb{Z} \to \mathbb{Z}_2 : \theta_2(n) =\equiv 1 \mod 2$. Consequently there are 2^3 distinct spin structures on T_3. T_3 is a simple example of a Bieberbach manifold. Further details of spin structures on the n-torus and other Bieberbach manifolds can be found for instance in [9, 18, 19].

We shall now give an explicit construction for some of these spinor bundles over T_3. All the others are constructed similarly. First let l be an integer in the set $\{1, 2, 3\}$, and consider the sublattice $\mathbb{Z}^l = \mathbb{Z}e_1 + \cdots + \mathbb{Z}e_l$ where $(0 \leq l \leq 3)$. In the case $l = 0$ we have $\mathbb{Z}^0 := \emptyset$.

There is also the remainder lattice $\mathbb{Z}^{3-l} = \mathbb{Z}e_{l+1} + \cdots + \mathbb{Z}e_3$. In this case $\mathbb{Z}^3 = \{\underline{m} + \underline{n} : \underline{m} \in \mathbb{Z}^l \text{ and } \underline{n} \in \mathbb{Z}^{3-l}\}$. Suppose now that $\underline{m} = m_1 e_1 + \cdots + m_l e_l$. Let us now make the identification $(\mathbf{x}, X)$ with $(\mathbf{x} + \underline{m} + \underline{n}, (-1)^{m_1 + \cdots + m_l} X)$ where $\mathbf{x} \in \mathbb{R}^3$ and $X \in \mathbb{H}$. This identification gives rise to a quaternionic spinor bundle $E^{(l)}$ over T_3. For example in the case $l = 1$, we have the lattice decomposition $\mathbb{Z} \oplus \mathbb{Z}^2$ and we identify $(\mathbf{x}, X)$ with $(\mathbf{x} + m_1 e_1 + m_2 e_2 + m_3 e_3, (-1)^{m_1} X)$.

Notice that $\mathbb{R}^3$ is the universal covering space of T_3. Consequently, there exists a well-defined projection map $p : \mathbb{R}^3 \to T_3$. As explained for example in [13] every 3-fold periodic resp. anti-periodic open set $U \subset \mathbb{R}^3$ and every 3-fold periodic resp. anti-periodic section $f : U' \to E^{(l)}$, satisfying $f(\mathbf{x}) = (-1)^{m_1 + \cdots + m_l}(\mathbf{x} + \omega)$ for all $\omega \in \mathbb{Z}^l \oplus \mathbb{Z}^{3-l}$, descends to a well-defined open set $U' = p(U) \subset T_3$ (associated with the chosen spinor bundle) and a well-defined spinor section $f' := p(f) : U' \subset T_3 \to E^{(l)} \subset \mathbb{H}$, respectively.

9.4 The Klein-Gordon Equation on Conformally Flat 3-Tori

The study of the null-solutions to the first order operator $\mathbf{D} - i\alpha$ leads to a full understanding of the solutions to the Klein-Gordon equation. The null-solutions to this equation are also often called $i\alpha$-holomorphic, see for instance [15].

Following for instance [10, 22], in the three-dimensional case, the fundamental solution to $\mathbf{D} - i\alpha$ has the special form

$$e_{i\alpha}(\mathbf{x}) = \frac{1}{4\pi} e^{-\alpha\|\mathbf{x}\|_2} \left(\frac{i\alpha}{\|\mathbf{x}\|_2} - \frac{\mathbf{x}}{\|\mathbf{x}\|_2^3}(1 + \alpha\|\mathbf{x}\|_2) \right).$$

The projection map p induces a shifted Dirac operator and a Klein-Gordon operator on the torus T_3 (with the chosen spinor bundle) viz $\mathbf{D}'_{i\alpha;l} := p(\mathbf{D} - i\alpha)$ resp. $\Delta'_{i\alpha;l} := p(\Delta - \alpha^2)$. The projections of the 3-fold (anti-)periodization of the function $e_{i\alpha}(\mathbf{x})$ denoted by

$$\wp_{i\alpha;\mathbf{0};l}(\mathbf{x}) := \sum_{\omega \in \mathbb{Z}^l \oplus \mathbb{Z}^{3-l}} (-1)^{m_1 + \cdots + m_l} e_{i\alpha}(\mathbf{x} + \omega)$$

provides us with the fundamental section to the toroidal operator $\mathbf{D}'_{i\alpha;l}$ acting on the corresponding spinor bundle of the torus T_3. From the function theoretical point of view the function $\wp_{i\alpha;\mathbf{0};0}(\mathbf{x})$ can be regarded as the canonical generalization of the classical elliptic Weierstraß $\wp$ function to the context of the shifted Dirac operator $\mathbf{D} - i\alpha$ in three dimensions.

To prove the convergence of the series we use the following asymptotic estimate. We have

$$\|e_{i\alpha}(\mathbf{x})\|_2 \leq c \frac{e^{-\alpha\|\mathbf{x}\|_2}}{\|\mathbf{x}\|_2} \tag{9.4.1}$$

supposed that $\|\mathbf{x}\|_2 \geq r'$ where r' is a sufficiently large real. Now we decompose the total lattice $\mathbb{Z}^3$ into the following union of lattice points $\Omega = \bigcup_{m=0}^{+\infty} \Omega_m$ where

$$\Omega_m := \{\omega \in \mathbb{Z}^3 \mid \|\omega\|_{max} = m\}.$$

We further consider the following subsets of this lattice $L_m := \{\omega \in \mathbb{Z}^3 \mid \|\omega\|_{max} \leq m\}$. Obviously the set L_m contains exactly $(2m+1)^3$ points. Hence, the cardinality of Ω_m is $\sharp\Omega_m = (2m+1)^3 - (2m-1)^3$. The Euclidean distance between the set Ω_{m+1} and Ω_m has the value $d_m := \text{dist}_2(\Omega_{m+1}, \Omega_m) = 1$.

To show the normal convergence of the series, let us consider an arbitrary compact subset $\mathcal{K} \subset \mathbb{R}^3$. Then there exists a positive $r \in \mathbb{R}$ such that all $\mathbf{x} \in \mathcal{K}$ satisfy $\|\mathbf{x}\|_{max} \leq \|\mathbf{x}\|_2 < r$. Suppose now that $\mathbf{x}$ is a point of $\mathcal{K}$. To show the normal convergence of the series, we can leave out without loss of generality a finite set of lattice points. We consider without loss of generality only the summation over those lattice points that satisfy $\|\omega\|_{max} \geq [R] + 1$, where $R := \max\{r, r'\}$ In view of

$\|\mathbf{x}+\omega\|_2 \geq \|\omega\|_2 - \|\mathbf{x}\|_2 \geq \|\omega\|_{max} - \|\mathbf{x}\|_2 = m - \|\mathbf{x}\|_2 \geq m - r$ we obtain

$$\begin{aligned}
&\sum_{m=[R]+1}^{+\infty} \sum_{\omega \in \Omega_m} \|e_{i\alpha}(\mathbf{x}+\omega)\|_2 \\
&\leq c \sum_{m=[R]+1}^{+\infty} \sum_{\omega \in \Omega_m} \frac{e^{-\alpha\|\mathbf{x}+\omega\|_2}}{\|\mathbf{x}+\omega\|_2} \\
&\leq c \sum_{m=[R]+1}^{+\infty} [(2m+1)^3 - (2m-1)^3] \frac{e^{\alpha(R-m)}}{m-R},
\end{aligned}$$

where c is an appropriately chosen positive real constant, because $m - R \geq [R] + 1 - R > 0$. This sum clearly is absolutely uniformly convergent. Hence, the series

$$\wp_{i\alpha;\mathbf{0};l}(\mathbf{x}) := \sum_{\omega \in \mathbb{Z}^l \oplus \mathbb{Z}^{3-l}} (-1)^{m_1+\cdots+m_l} e_{i\alpha}(\mathbf{x}+\omega),$$

which can be written as

$$\wp_{i\alpha;\mathbf{0};l}(\mathbf{x}) := \sum_{m=0}^{+\infty} \sum_{\omega \in \Omega_m} (-1)^{m_1+\cdots+m_l} e_{i\alpha}(\mathbf{x}+\omega),$$

converges normally on $\mathbb{R}^3 \backslash \mathbb{Z}^3$. Since $e_{i\alpha}(\mathbf{x})$ belongs to Ker $(\mathbf{D} - i\alpha)$ in each $\mathbf{x} \in \mathbb{R}^3 \backslash \{0\}$ and has a pole of order 2 at the origin and exponential decrease for $\|\mathbf{x}\| \to +\infty$, the series $\wp_{i\alpha;\mathbf{0};l}(\mathbf{x})$ satisfies $(\mathbf{D} - i\alpha)\wp_{i\alpha;\mathbf{0};l}(\mathbf{x}) = 0$ in each $\mathbf{x} \in \mathbb{R}^3 \backslash \mathbb{Z}^3$ and has a pole of order 2 in each lattice point $\omega \in \mathbb{Z}^3$.

Obviously, by a direct rearrangement argument, one obtains that

$$\wp_{i\alpha;\mathbf{0};l}(\mathbf{x}) = (-1)^{m_1+\cdots+m_l} \wp_{i\alpha;\mathbf{0};l}(\mathbf{x}+\omega), \quad \forall \omega \in \Omega.$$

Remarks

- In the general n-dimensional case we will have estimates of the form

$$\begin{aligned}
&\sum_{m=[R]+1}^{+\infty} \sum_{\omega \in \Omega_m} \|e_{i\alpha}(\mathbf{x}+\omega)\|_2 \\
&\leq c \sum_{m=[R]+1}^{+\infty} \sum_{\omega \in \Omega_m} \frac{e^{-\alpha\|\mathbf{x}+\omega\|_2}}{\|\mathbf{x}+\omega\|_2^{(n-1)/2}} \\
&\leq c \sum_{m=[R]+1}^{+\infty} [(2m+1)^n - (2m-1)^n] \frac{e^{\alpha(R-m)}}{(m-R)^{n-1}}.
\end{aligned}$$

This provides a correction to the convergence proof given in [5] for the corresponding n-dimensional series for general complex λ. Notice that the majorant series

$$\sum_{m=[R]+1}^{+\infty} [(2m+1)^n - (2m-1)^n] \frac{e^{\alpha(R-m)}}{(m-R)}$$

is also still convergent. This is a consequence of the exponentially fast decreasing factor in the nominator. Therefore, all the claimed results in [5] remain valid in its full extent. In the convergence proof of [5] one simply has to replace the square root in the denominator by the $(n-1)$ power of that square root. This replacement however makes the convergence of the series even stronger, so that all constructions proposed in [5] remain well-defined and are true in its full extent.

- In the limit case $\alpha \to 0$ we will obtain a divergent series. Indeed, as shown for instance in [12], a non-constant n-fold periodic function satisfying $\mathbf{D} f = 0$ except at one pole of order $n-1$ in each period cell does not exist. One needs to have at least two singularities of order $n-1$ in a period cell or singularities of higher order.

Further elementary non-trivial examples of 3-fold (anti-) periodic $i\alpha$-holomorphic sections are the partial derivatives of $\wp_{i\alpha;\mathbf{0};l}$. These are denoted by $\wp_{i\alpha;\mathbf{m};l}(\mathbf{x}) := \frac{\partial^{|\mathbf{m}|}}{\partial \mathbf{x}^{\mathbf{m}}} \wp_{i\alpha;\mathbf{0};l}(\mathbf{x})$ where $\mathbf{m} \in \mathbb{N}_0^3$ is a multi-index. For each $\mathbf{x}, \mathbf{y} \in \mathbb{R}^3 \backslash \mathbb{Z}^3$ the function $\wp_{i\alpha;\mathbf{0};l}(\mathbf{y}-\mathbf{x})$ induces the Cauchy kernel for $\mathbf{D}'_{i\alpha;l}$ viz $G_{i\alpha;l}(\mathbf{y}'-\mathbf{x}')$ on T_3 where $\mathbf{x}' := p(\mathbf{x}), \mathbf{y}' := p(\mathbf{y})$. This attributes a key role to the functions $\wp_{i\alpha;\mathbf{0};l}(\mathbf{x})$. Notice that in the limit case $\alpha = 0$ we need to apply a modification of this construction involving two singularities of order $n-1$ to get a Cauchy kernel on these tori, as indicated in the previous remark above. This particular case is treated in detail in [14].

So, in what follows we always assume $\alpha \neq 0$.

From the 3-fold (anti-)periodic basic toroidal $i\alpha$-holomorphic function $\wp_{i\alpha;\mathbf{0};l}$ we can easily obtain 3-fold (anti-)periodic solutions to the Klein-Gordon operator $\Delta - \alpha^2 = -(\mathbf{D} - i\alpha)(\mathbf{D} + i\alpha)$. Let C_1, C_2 be arbitrary complex quaternions from $\mathbb{H}(\mathbb{C})$. Then the functions

$$\mathrm{Sc}(\wp_{i\alpha;\mathbf{0};l}(\mathbf{x}) C_1)$$

and

$$\mathrm{Sc}(\wp_{-i\alpha;\mathbf{0};l}(\mathbf{x}) C_2)$$

as well as all their partial derivatives are 3-fold (anti-)periodic and satisfy the homogeneous Klein-Gordon equation $(\Delta - \alpha^2)u = 0$ in the whole space $\mathbb{R}^3 \backslash \mathbb{Z}^3$.

As a consequence of the Borel-Pompeiu formula proved in [10, 22] for the Euclidean case we can readily prove a Green's integral formula for solutions to the homogeneous Klein-Gordon equation on this class of conformally flat 3-tori of the following form:

Theorem 9.1 *Suppose that a section $h : U' \to E^{(l)} \subset \mathbb{H}(\mathbb{C})$ is a solution to the toroidal Klein-Gordon operator $\Delta'_{i\alpha;l}$ in $U' \subset T_3$. Let V' be a relatively compact subdomain with $\mathrm{cl}(V') \subset U'$. Then provided the boundary of V' is sufficiently* *smooth we have*

$$h(\mathbf{y}) = \int_{\partial V'} (G_{-i\alpha;l}(\mathbf{x}' - \mathbf{y}')(d_{\mathbf{x}} p(n(\mathbf{x})))h(\mathbf{x})$$

$$+ [\mathrm{Sc}(G)_{-i\alpha;l}](\mathbf{y}' - \mathbf{x}')(d_{\mathbf{x}} p(n(\mathbf{x})))\mathbf{D}'_{+i\alpha;l} h(\mathbf{x}'))dS(\mathbf{x}') \quad (9.4.2)$$

for each $\mathbf{y}' \in V'$. *Here* $d_{\mathbf{x}}$ *stands for the derivative of* $p(n(\mathbf{x}))$ *with respect to* $\mathbf{x}$.

Notice that we only have one point singularity in each period cell. The reproduction of the function by the Green's integral hence follows by applying Cauchy's theorem and the Almansi-Fischer type decomposition. See also [20] for details.

One really striking property is that we can represent any solution to the homogeneous Klein-Gordon equation on T_3 as a finite sum over projections of a finite number of generalized three-fold (anti-)periodic elliptic functions that are in the kernel of the Klein-Gordon operator. We can prove

Theorem 9.2 *Let* $a'_1, a'_2, \ldots, a'_p \in T_3$ *be a finite set of points.*

Suppose that $f' : T_3 \backslash \{a'_1, \ldots, a'_p\} \to E^{(l)} \subset \mathbb{H}(\mathbb{C})$ *is a section in the kernel of the toroidal Klein-Gordon operator which has at most isolated poles at the points* a'_i *of the order* K_i. *Then there are constants* $b'_1, \ldots, b'_p \in \mathbb{H}(\mathbb{C})$ *such that this section can be represented by*

$$f'(\mathbf{x}') = \sum_{i=1}^{p} \sum_{m=0}^{K_i-1} \sum_{m=m_1+m_2+m_3} \left[\mathrm{Sc} \frac{\partial^{|\mathbf{m}|}}{\partial \mathbf{x}^{\mathbf{m}}} G_{i\alpha,\mathbf{0};l}(\mathbf{x}' - a'_i) \right] b'_i. \quad (9.4.3)$$

To establish this result we first need to prove the following lemmas:

Lemma 9.3 *Suppose that* f *is a* 3*-fold (anti-)periodic function that satisfies* $(\mathbf{D} - i\alpha)f = 0$ *in the whole space* $\mathbb{R}^3$. *Then* f *vanishes identically.*

Proof Suppose first that f is 3-fold periodic, that means, $f(\mathbf{x}) = f(\mathbf{x} + \omega)$ for all $\omega \in \Omega$. In this case it takes all its values in the fundamental period cell spanned with the edges $0, e_1, e_2, e_3, e_1 + e_2, e_1 + e_3, e_2 + e_3, e_1 + e_2 + e_3$. This is a compact set. Since f is continuous it must be bounded on the fundamental cell. As a consequence of the 3-fold periodicity, f must be a bounded function on the whole space $\mathbb{R}^3$. Since f is entire $i\alpha$-holomorphic, adapting from [22], it has a Taylor series representation

$$f(\mathbf{x}) = \sum_{q=0}^{+\infty} \|\mathbf{x}\|^{-q-1/2} \Big(e^{\pi i(q/2+1/4)} I_{q+1/2}(\alpha\|\mathbf{x}\|)$$

$$- \frac{\mathbf{x}}{\|\mathbf{x}\|} e^{\pi i(q/2+3/4)} I_{q+3/2}(\alpha\|\mathbf{x}\|) \Big) P_q(\mathbf{x}), \quad (9.4.4)$$

where P_q are the well-known inner spherical monogenics, as defined for instance in [7], is valid on the whole space $\mathbb{R}^n$. Since the Bessel functions I with real arguments are exponentially unbounded, the expression f can only be bounded if all spherical monogenics P_q vanish identically. Hence, $f \equiv 0$.

In the case where f is anti-periodic, i.e. $f(\mathbf{x}) = (-1)f(\mathbf{x}+\omega)$ for a primitive period ω, then f satisfies $f(\mathbf{x}) = f(\mathbf{x}+2\omega)$, so one has $f(\mathbf{x}) = f(\mathbf{x}+2\omega)$ for all $\omega \in \Omega$. It is hence periodic with respect to the double of any primitive period. Instead of the usual fundamental cell one considers the larger cell with the edges

$$0, 2e_1, 2e_2, 2e_3, 2(e_1+e_2), 2(e_1+e_3), 2(e_2+e_3), 2(e_1+e_2+e_3)$$

which then is a compact periodicity cell of f and the previously outlined arguments can be applied in the same way as in the periodic case. □

Lemma 9.4 *Let $a'_1, a'_2, \dots, a'_p \in T_3$ be a finite set of points.*

Suppose that the section $f' : T_3 \backslash \{a'_1, \dots, a'_p\} \to E^{(l)} \subset \mathbb{H}(\mathbb{C})$ is a solution to $\mathbf{D}'_{i\alpha} f' = 0$ which has at most isolated poles at the points a'_i of the order K_i. Then there are constants $b'_1, \dots, b'_p \in \mathbb{H}(\mathbb{C})$ such that

$$f'(\mathbf{x}') = \sum_{i=1}^{p} \sum_{m=0}^{K_i-2} \sum_{m=m_1+m_2+m_3} \left[\frac{\partial^{|\mathbf{m}|}}{\partial \mathbf{x}^{\mathbf{m}}} G_{i\alpha,\mathbf{0};l}(\mathbf{x}' - a'_i) \right] b'_i. \tag{9.4.5}$$

Proof Since $f = p^{-1}(f')$ is supposed to be $i\alpha$-holomorphic with isolated poles of order K_i at the points a_i, the singular parts of the local Laurent series expansions are of the form $e_{i\alpha,\mathbf{m}}(\mathbf{x} - a_i)b_i$ in each point $a_i + \Omega$, where $e_{i\alpha,\mathbf{m}}(\mathbf{y}) := \frac{\partial^{|\mathbf{m}|}}{\partial \mathbf{y}^{\mathbf{m}}} e_{i\alpha}(\mathbf{y})$. As a sum of 3-fold (anti-) periodic $i\alpha$-holomorphic functions, the expression

$$g(\mathbf{x}) = \sum_{i=1}^{p} \sum_{m=0}^{K_i-2} \sum_{m=m_1+m_2+m_3} \left[\wp_{i\alpha,\mathbf{m};l}(\mathbf{x} - a_i) b_i \right]$$

is also 3-fold (anti-) periodic and has also the same principal parts as f. Hence, the function $h := g - f$ is also a 3-fold periodic and $i\alpha$-holomorphic function, but without singular parts, since these are canceled out. So, the function h is an entire $i\alpha$-holomorphic 3-fold periodic function. Consequently, it vanishes identically as a consequence of the preceding lemma. □

The statement of Theorem 9.2 now follows as a direct consequence.

9.5 The Inhomogeneous Klein-Gordon Equation on Tori

We round off with discussing the inhomogeneous Klein-Gordon equation $(\Delta' - \alpha^2)u' = f'$ on a strongly Lipschitz domain on the surface of such a 3-torus $\Omega' \subset T_3$ with prescribed boundary data $u' = g'$ at $\partial\Omega'$. Non-zero terms on the right-hand side naturally appear for instance when we include gravitational effects in our consideration. To solve inhomogeneous boundary value problems of this type one can introduce a toroidal Teodorescu transform and an appropriate singular Cauchy transform for the operator $\mathbf{D}'_{i\alpha;l}$ by replacing the kernel $e_{i\alpha}$ by the projection of the 3-fold (anti-)periodic function $\wp_{i\alpha,\mathbf{0};l}$ in the corresponding integral formulas given in [10] for the Euclidean flat space. By means of these operators one can then also solve the inhomogeneous Klein-Gordon equation on these 3-tori with given boundary data explicitly using the same method as proposed in [1, 10, 11, 15] for analogous problems in the Euclidean flat space. Precisely, the proper analogies of the operators needed to meet these ends are defined as follows. The Teodorescu transform of toroidal $i\alpha$-holomorphic functions on the torus T_3 with values in the spinor bundle $E^{(l)}$ is defined by

$$T^{T_3}_{i\alpha;l} : W^p_{l,E^{(l)}}(\Omega') \to W^{p+1}_{l,E^{(l)}}(\Omega');$$

$$[T^{T_3}_{i\alpha;l} f'(\mathbf{x}')] = -\int_{\Omega'} G_{-i\alpha;l}(\mathbf{x}' - \mathbf{y}') f'(\mathbf{y}') dV'(\mathbf{y}'),$$

where $\mathbf{x}'$ and $\mathbf{y}'$ are distinct points on the 3-torus from Ω'. Here $W^p_{l,\mathbb{H}(\mathbb{C})}(\Omega')$ denotes as usual the Sobolev space of $E^{(l)}$-valued L^p functions defined in Ω' which are l-times differentiable in the sense of Sobolev.

Next, the toroidal $i\alpha$-holomorphic Cauchy transform has the mapping properties

$$F^{T_3}_{i\alpha;l} : W^{p-1}_{l-\frac{1}{p},E^{(l)}}(\partial\Omega') \to W^p_{l,E^{(l)}}(\Omega') \cap \operatorname{Ker} \mathbf{D}'_{i\alpha;l};$$

$$[F^{T_3}_{i\alpha;l} f'(\mathbf{y}')] = \int_{\partial V'} G_{-i\alpha;l}(\mathbf{x}' - \mathbf{y}') n(\mathbf{x}') d_{\mathbf{x}} p(n(\mathbf{x})) f'(\mathbf{x}') dS'(\mathbf{x}'),$$

where dS' is the projected scalar surface Lebesgue measure on the surface of the torus. Using the toroidal Teodorescu transform, a direct analogy of the Borel-Pompeiu formula for the shifted Dirac operator $\mathbf{D}'_{i\alpha;l}$ on the 3-torus can now be formulated in the classical form

$$f' = F^{T_3}_{i\alpha;l} f' + T^{C_k}_{i\alpha;l} \mathbf{D}'_{i\alpha;l} f',$$

as formulated for the Euclidean case in [10, 11]. Adapting the arguments from [10, p. 80] that were explicitly worked out for the Euclidean case, one can show that the space of square integrable functions over a domain Ω' of the 3-torus, admits the orthogonal decomposition

$$L^2(\Omega', E^{(l)}) = \operatorname{Ker} \mathbf{D}'_{i\alpha;l} \cap L^2(\Omega', E^{(l)}) \oplus \mathbf{D}'_{i\alpha;l} \mathring{W}^1_{2,E^{(l)}}(\Omega').$$

The space $\operatorname{Ker} \mathbf{D}'_{i\alpha;l} \cap L^2(\Omega', E^{(l)})$ is a Banach space endowed with the L^2 inner product

$$\langle f', g' \rangle := \int_{\Omega'} \overline{f'(\mathbf{x}')}^{\sharp} g(\mathbf{x}') dV(\mathbf{x}'),$$

as used in [4].

As a consequence of the Cauchy integral formula for the toroidal $i\alpha$-holomorphic sections and of the Cauchy-Schwarz inequality we can show that this space has a continuous point evaluation and does hence possess a reproducing kernel, say $B(\mathbf{x}', \mathbf{y}')$. If f' is any arbitrary section from $L^2(\Omega', E^{(l)})$, then the operator

$$[P^{T_3}_{i\alpha;l} f'(\mathbf{y}')] = \int_{V'} B(\mathbf{x}', \mathbf{y}') f(\mathbf{x}') dV(\mathbf{x}')$$

produces the ortho-projection from $L^2(\Omega', E^{(l)})$ into $\operatorname{Ker} \mathbf{D}'_{i\alpha;l} \cap L^2(\Omega', E^{(l)})$. It will be called the toroidal $i\alpha$-holomorphic-Bergman projector. With these operators we can represent in complete analogy to the Euclidean case treated in [10] the solutions to the inhomogeneous Klein-Gordon equation on these 3-tori:

Theorem 9.5 *Let $\alpha > 0$. Let Ω' be a domain on the flat 3-torus T_3 (associated with one of the above described spinor bundles) with a strongly Lipschitz boundary. Let $f' \in W^p_{2,E^{(l)}}(\Omega')$ and $g' \in W^{p+3/2}_{2,E^{(l)}}(\partial\Omega')$ be two sections. Let $\Delta'_{i\alpha;l}$ stand for the toroidal Klein-Gordon operator. Then the system*

$$\Delta'_{i\alpha;l} u' = f' \quad \text{in } V', \tag{9.5.1}$$

$$u' = g' \quad \text{at } \partial V', \tag{9.5.2}$$

always has a unique solution $u \in W^{p+2,loc}_{2,E^{(l)}}(V')$ of the form

$$u' = F^{T_3}_{i\alpha} g' + T^{T_3}_{-i\alpha} P^{T_3}_{i\alpha;l} \mathbf{D}'_{i\alpha;l} h' - T^{T_3}_{-i\alpha}(I - P^{T_3}_{i\alpha;l}) T^{T_3}_{i\alpha;l} f', \tag{9.5.3}$$

where h' is the unique $W^{p+2}_{E^{(l)}}$ extension of g'.

To the proof one can apply the same calculation steps as in [10, p. 81] involving now the properly adapted version of the Borel-Pompeiu formula for the toroidal shifted Dirac operator $\mathbf{D}'_{i\alpha;l}$ and the adapted integral transform. Notice that we have for all values $\alpha > 0$ always a unique solution, because the Laplacian has only negative eigenvalues. Notice further that we can represent any solution to the toroidal Klein-Gordon equation by the scalar parts of a finite number of the basic $i\alpha$-holomorphic generalized elliptic functions $\frac{\partial^{|\mathbf{m}|}}{\partial \mathbf{x}^{\mathbf{m}}} \wp_{i\alpha,\mathbf{0};l}(\mathbf{x} - \mathbf{a}) b_{\mathbf{m}}$, such as indicated in Theorem 9.2. The Bergman kernel can be hence approximated by applying for instance the Gram-Schmidt algorithm to a sufficiently large set of those $i\alpha$-holomorphic generalized elliptic functions series that have no singularities inside the domain.

9.6 The Homogeneous Klein-Gordon Equation on Spheres

Aim In this section we present a unified approach to treat the solutions to the Klein-Gordon equation on spheres of ray $R \in]0, +\infty]$ in terms of a Dirac type operator.

Model To this end we add the radial symmetric Euler operator $E := \sum_{i=1}^{3} x_i \frac{\partial}{\partial x_i}$ in the shifted Dirac equation. We recover the solutions to the time-independent Klein-Gordon operator on a sphere of ray $R \in]0, +\infty]$ as the solutions to the system

$$\left[\mathbf{D} - i\alpha - \frac{1}{R}E\right] f = 0. \tag{9.6.1}$$

The cases $R \in]0, +\infty[$ deal with a radially symmetric universe model with ray R (at time $t = t_0$).

In the case $R \to +\infty$ we deal with an Euclidean flat universe of infinite extension.

9.6.1 Representation of the Regular Solutions

The following theorem provides us with an explicit representation of the solutions to the system (9.6.1) in terms of hypergeometric functions and the basic homogeneous monogenic polynomials that we introduced previously. More precisely, adapting from [3] in the particular three dimensional case we obtain for the regular part of the solutions the following representation:

Theorem 9.6 *Let f be a $\mathbb{H}(\mathbb{C})$-valued function that satisfies in the 3-dimensional open ball $\|\mathbf{x}\| < r_2$ ($r_2 > 0$) the system of differential equations $(\mathbf{D} - i\alpha - \beta E)f = 0$ for $\beta, \alpha \in \mathbb{R}\backslash\{0\}$. Then there exists a sequence of monogenic homogeneous polynomials of total degree $m = 0, 1, 2, \ldots$, say $P_m(\mathbf{x})$, such that in each open ball $\|\mathbf{x}\| < r$ with $0 < r < r_2$*

$$f(\mathbf{x}) = \sum_{m=0}^{+\infty} \left(a_m(\|\mathbf{x}\|) + b_m(\|\mathbf{x}\|) \frac{\mathbf{x}}{\|\mathbf{x}\|} \right) P_m(\mathbf{x}),$$

where

$$a_m(\|\mathbf{x}\|) = {}_2F_1\left(\frac{i\alpha}{2\beta} + \frac{m}{2}, \frac{i\alpha}{2\beta} + \frac{m+1}{2}; m + \frac{3}{2}; -\beta^2 \|\mathbf{x}\|^2 \right), \tag{9.6.2}$$

$$b_m(\|\mathbf{x}\|) = -\frac{i\alpha + \beta m}{2m+3} |\mathbf{x}| \, {}_2F_1\left(1 + \frac{i\alpha}{2\beta} + \frac{m}{2}, \frac{i\alpha}{2\beta} + \frac{m+1}{2}; \right.$$
$$\left. m + \frac{5}{2}; -\beta^2 \|\mathbf{x}\|^2 \right). \tag{9.6.3}$$

Here, ${}_2F_1$ denotes the standard hypergeometric function reading explicitly

$${}_2F_1(a, b; c; z) = \sum_{n=0}^{+\infty} \frac{(a)_n (b)_n}{(c)_n} \frac{z^n}{n!},$$

$(a)_n$, etc. stand for the Pochhammer symbol.

Proof Suppose that f is a $\mathbb{H}(\mathbb{C})$-valued function defined in the 3-dimensional open ball $\|\mathbf{x}\| < r_2$ $(r_2 > 0)$ that satisfies the differential equation $(\mathbf{D} - i\alpha - \beta E)f = 0$ with some arbitrarily chosen $\beta, \alpha \in \mathbb{R}\backslash\{0\}$.

Since the operator $\mathbf{D} - i\alpha - \beta E$ is elliptic, its null-solutions are real-analytic in the open ball $\|\mathbf{x}\| < r_2$ $(r_2 > 0)$. Therefore f admits the series expansion

$$f(\mathbf{x}) = \sum_{m=0}^{+\infty} R_m(\mathbf{x}), \tag{9.6.4}$$

converging normally on compact subsets of the open ball $\|\mathbf{x}\| < r_2$, and where R_m are $\mathbb{H}(\mathbb{C})$-valued homogeneous polynomials of degree m. When applying the Fischer decomposition one obtains

$$R_m(\mathbf{x}) = \sum_{j=0}^{m} \mathbf{x}^{m-j} P_{m,j}(\mathbf{x}),$$

where the expressions $P_{m,j}$ are homogeneous monogenic polynomials of total degree j (for each $m = 0, 1, 2, \ldots$).

Since f is a solution to $(\mathbf{D} - i\alpha - \beta E)f = 0$, each single term from (9.6.4) of the same degree of homogeneity has to solve this equation, i.e.

$$\mathbf{D}R_{m+1}(\mathbf{x}) - (i\alpha + \beta E)R_m(\mathbf{x}) = 0. \tag{9.6.5}$$

As a consequence of (9.6.5) there exists a sequence of homogeneous monogenic polynomials of total degree j, say $(P_j)_{j=0,1,\ldots,+\infty}$, such that

$$P_{m,j}(\mathbf{x}) = \beta_{m,j} P_j(\mathbf{x})$$

for all $m = 0, 1, \ldots$ with uniquely defined scalars $\beta_{m,j}$ where in particular $\beta_{j,j} = 1$. So one obtains that

$$R_m(\mathbf{x}) = \sum_{j=0}^{m} \mathbf{x}^{m-j} \beta_{m,j} P_j(\mathbf{x}).$$

This results into

$$f(\mathbf{x}) = \sum_{m=0}^{+\infty} \left(\sum_{j=0}^{m} \mathbf{x}^{j} \beta_{m,m-j} P_{m-j}(\mathbf{x}) \right). \tag{9.6.6}$$

Now one introduces spherical coordinates (r, ω), where $r = \|\mathbf{x}\|$ and $\omega = \frac{\mathbf{x}}{\|\mathbf{x}\|} \in S^2 := \{\mathbf{x} \in \mathbb{R}^3 \mid \|\mathbf{x}\| = 1\}$. Next one can rewrite the representation (9.6.6) in the form

$$f(\mathbf{x}) = \sum_{m=0}^{+\infty} a_m(r) P_m(r\,\omega) + b_m(r)\,\omega P_m(r\,\omega), \tag{9.6.7}$$

where

$$a_m(r) = \sum_{k=0}^{+\infty} \beta_{2k+m,m}(-r^2)^k$$

and

$$b_m(r) = r \sum_{k=0}^{+\infty} \beta_{2k+1+m,m}(-r^2)^k.$$

Notice that these sums actually only encompass finitely many terms. Again as a consequence of (9.6.5), for each degree $m = 0, 1, 2, \ldots$ each associated homogeneity term of the series representation (9.6.7) satisfies

$$[\mathbf{D} - i\alpha - \beta E](a_m(r)P_m(r\,\omega) + b_m(r)\,\omega P_m(r\,\omega)) = 0. \tag{9.6.8}$$

Notice, this is not an ansatz. This equality needs to hold necessarily. The expressions P_m, multiplied from the left with $a_m(r)$ or $b_m(r)\omega$, are all mapped under the three operators $\mathbf{D}$, $i\alpha$ and E to an expression of the same form. One never obtains a polynomial of an other degree being different from P_m. If we fix r and vary ω, after applying $\mathbf{D} - i\alpha - \beta E$ we still end up with the same $P_m(r\,\omega)$, but with new coefficients $a_m(r)$ and $b_m(r)$.

Next one needs to compute the action of the Euler operator and the Dirac operator on each single term. For simplicity we use the notation

$$a'_m(r) := \frac{\partial}{\partial r} a_m(r) \quad \text{and} \quad b'_m(r) := \frac{\partial}{\partial r} b_m(r).$$

As in [2] one determines:

$$E[a_m(r)\,P_m(r\,\omega)] = r\,a'_m(r)\,P_m(r\,\omega) + m\,a_m(r)\,P_m(r\,\omega), \tag{9.6.9}$$

$$E[b_m(r)\,\omega\,P_m(r\,\omega)] = r\,b'_m(r)\,\omega\,P_m(r\,\omega) + m\,b_m(r)\,\omega\,P_m(r\,\omega), \tag{9.6.10}$$

$$\mathbf{D}[a_m(r)\,P_m(r\,\omega)] = a'_m(r)\omega\,P_m(r\,\omega), \tag{9.6.11}$$

$$\mathbf{D}[b_m(r)\,\omega\,P_m(r\,\omega)] = \frac{-2-2m}{r}\,b_m(r)\,P_m(r\,\omega) - b'_m(r)P_m(r\,\omega). \tag{9.6.12}$$

Next one applies the calculus rules (9.6.9), (9.6.10), (9.6.11) and (9.6.12) to (9.6.8). We obtain

$$\begin{aligned}
&[(\mathbf{D} - i\alpha) - \beta E](a_m(r)P_m(r\omega) + b_m(r)\omega P_m(r\omega))\\
&\quad = \mathbf{D}[a_m(r)P_m(r\omega)] + \mathbf{D}[b_m(r)\omega P_m(r\omega)] - i\alpha a_m(r)P_m(r\omega)\\
&\qquad - i\alpha b_m(r)\omega P_m(r\omega)\\
&\qquad - \beta E[a_m(r)P_m(r\omega)] - \beta E[b_m(r)\omega P_m(r\omega)]\\
&\quad = a'_m(r)\omega P_m(r\omega) + \frac{-2-2m}{r} b_m(r)P_m(r\omega) - b'_m(r)P_m(r\omega)
\end{aligned}$$

$$
\begin{aligned}
&- i\alpha a_m(r)P_m(r\omega) - i\alpha b_m(r)\omega P_m(r\omega)\\
&- \beta r a'_m(r)P_m(r\omega) - \beta m a_m(r)P_m(r\omega)\\
&- \beta r b'_m(r)\omega P_m(r\omega) - \beta m b_m(r)\omega P_m(r\omega).
\end{aligned}
$$

Now we collect all the terms that belong to $P_m(r\omega)$ and $\omega P_m(r\omega)$, respectively:

$$
\begin{aligned}
&[(\mathbf{D} - i\alpha) - \beta E](a_m(r)P_m(r\omega) + b_m(r)\omega P_m(r\omega))\\
&= \left[-b'_m(r) + \frac{-2-2m}{r}b_m(r) - (i\alpha + \beta m)a_m(r) - \beta r a'_m(r)\right]P_m(r\omega)\\
&\quad + [a'_m(r) - \beta r b'_m(r) - (\beta m + i\alpha)b_m(r)]\omega P_m(r\omega).
\end{aligned}
$$

Due to the linear independence of the functions $P_m(r\,\omega)$ and $\omega\,P_m(r\,\omega)$ we obtain the following coupled first order linear system of ODE

$$
-b'_m(r) - \beta\, r\, a'_m(r) - (\beta\, m + i\alpha)\, a_m(r) - \frac{2+2m}{r}\, b_m(r) = 0, \quad (9.6.13)
$$

$$
a'_m(r) - \beta\, r\, b'_m(r) - (\beta\, m + i\alpha)\, b_m(r) = 0. \quad (9.6.14)
$$

To solve this system we first solve equation (9.6.14) to $a'_m(r)$. We have

$$
a'_m(r) = \beta r b'_m(r) + (\beta m + i\alpha)b_m(r). \quad (9.6.15)
$$

Now we insert this expression for $a'_m(r)$ into (9.6.13):

$$
\begin{aligned}
&-b'_m(r) - \beta r(\beta r b'_m(r) + (\beta m + i\alpha)b_m(r))\\
&-(\beta m + i\alpha)a_m(r) - \frac{2+2m}{r}b_m(r) = 0.
\end{aligned}
$$

Thus

$$
\begin{aligned}
(\beta m + i\alpha)a_m(r) &= (-\beta^2 r^2 - 1)b'_m(r)\\
&\quad - \beta r(\beta m + i\alpha)b_m(r) - \frac{2+2m}{r}b_m(r). \quad (9.6.16)
\end{aligned}
$$

In view of $\alpha, \beta \in \mathbb{R}\backslash\{0\}$ and $m \in \mathbb{N}_0$ we always have that $(\beta m + i\alpha) \neq 0$. Then (9.6.16) is equivalent to

$$
a_m(r) = -\frac{\beta^2 r^2 + 1}{\beta m + i\alpha}b'_m(r) + \left(-\beta r - \frac{2+2m}{r(\beta m + i\alpha)}\right)b_m(r). \quad (9.6.17)
$$

Deriving this expression to the variable r yields

$$
\begin{aligned}
a'_m(r) &= -\frac{\beta^2 r^2 + 1}{\beta m + i\alpha}b''_m(r) - \left[\frac{2\beta^2 r}{\beta m + i\alpha} + \beta r + \frac{2+2m}{r(\beta m + i\alpha)}\right]b'_m(r)\\
&\quad + \left[\frac{2+2m}{r^2(\beta m + i\alpha)} - \beta\right]b_m(r). \quad (9.6.18)
\end{aligned}
$$

Inserting (9.6.18) into equation (9.6.14) leads to the following second order linear ODE for $b_m(r)$:

$$-\frac{\beta^2 r^2+1}{\beta m+i\alpha} b''_m(r) - \left[\frac{2\beta^2 r}{\beta m+i\alpha} + 2\beta r + \frac{2+2m}{r(\beta m+i\alpha)}\right] b'_m(r)$$
$$+\left[\frac{2+2m}{r^2(\beta m+i\alpha)} - \beta - (\beta m+i\alpha)\right] b_m(r) = 0. \qquad (9.6.19)$$

The part of the solution that is regular around the origin turns out to be

$$b_m(r) = -\frac{i\alpha+\beta m}{2m+3} r\, {}_2F_1\left(1+\frac{i\alpha}{2\beta}+\frac{m}{2}, \frac{i\alpha}{2\beta}+\frac{m+1}{2}; m+\frac{5}{2}; -\beta^2 r^2\right). \quad (9.6.20)$$

Inserting this expression into (9.6.17) gives the stated formula for $a_m(r)$. □

9.6.2 Limit and Special Cases

- Unit sphere: In the particular case $\beta = 1$ in which the expressions (9.6.2) and (9.6.3) simplify to

$$a_m(\|\mathbf{x}\|) = {}_2F_1\left(\frac{i\alpha}{2}+\frac{m}{2}, \frac{i\alpha}{2}+\frac{m+1}{2}; m+\frac{3}{2}; -\|\mathbf{x}\|^2\right),$$

$$b_m(\|\mathbf{x}\|) = -\frac{i\alpha+m}{2m+3}\|\mathbf{x}\|\, {}_2F_1\left(1+\frac{i\alpha}{2}+\frac{m}{2}, \frac{i\alpha}{2}+\frac{m+1}{2}; m+\frac{5}{2}; -\|\mathbf{x}\|^2\right).$$

Here we recognize the regular solutions of the Dirac equation on the unit sphere and on the projective space $\mathbb{R}^{1,2}$ discussed by P. Van Lancker [23], D. Eelbode, F. Sommen [8], H. Liu, J. Ryan [17], and others.

Physical meaning: These solutions exactly represent the solutions to time dependent Klein-Gordon equations on the unit sphere or on the hyperbolic projective space, respectively. Applying a re-scaling argument, in the cases where $\beta = \frac{1}{R} > 0$ is an arbitrary positive number, the solutions to (9.6.1) may be physically interpreted as solutions to the time-independent Klein-Gordon equation on the sphere of arbitrary radius $R > 0$.

Notice that α may be any arbitrary non-zero real number. Thus, indeed all vector-valued solutions to (9.6.1) are solutions to the time-independent Klein-Gordon equations on the sphere of radius R when one puts $\beta = \frac{1}{R}$ and vice versa all solutions to the time-independent Klein-Gordon equation appear as vector-valued solutions of (9.6.1) because (9.6.1) is a first order equation.

- In the other limit case $\beta \to 0$ we obtain by asymptotic analysis that the expression $a_m(r)$ tends (up to a multiple) to the expression

$$\Gamma\left(m+\frac{3}{2}\right)\left(\frac{i\alpha\|\mathbf{x}\|}{2}\right)^{-m-1/2} \exp(\pi i(m/2+1/4)) I_{m+1/2}(\alpha\|\mathbf{x}\|).$$

Similarly the expression of $b_m(r)$ tends asymptotically for $\beta \to 0$ to a multiple of the expression

$$|\mathbf{x}|^{-\frac{1}{2}-m} \exp(\pi(m/2+3/4)) I_{m+3/2}(\alpha\|\mathbf{x}\|).$$

We hence recognize the representation formulas for regular solutions to $(\mathbf{D}-i\alpha)$ from [22] around the origin described for example by Sommen and Xu. In this case we are dealing with the regular part of the solutions to the time-independent Klein-Gordon equation inside a ball of an infinite Euclidean flat universe.

- In the case where $\alpha = 0$ in which the equation simplifies to the system $(\mathbf{D} - \beta E)f = 0$ the functions $a_m(r)$ and $b_m(r)$ read as follows:

$$a_m(r) = {}_2F_1\left(\frac{m}{2}, \frac{m+1}{2}; m+\frac{3}{2}; -\beta^2 r^2\right),$$

$$b_m(r) = -\frac{\beta m}{2m+3} r\, {}_2F_1\left(\frac{2+m}{2}, \frac{m+1}{2}; m+\frac{5}{2}; -\beta^2 r^2\right).$$

In this case only hypergeometric functions ${}_2F_1(a,b;c;z)$ with integer resp. half-integer parameters a, b, c do appear.

In the limit case where $\alpha = 0$ and $\beta \to 0$ (which deals with the equation $\mathbf{D}f = 0$) the expression $a_m(r)$ reduces to the constant value 1 and $b_m(r) \equiv 0$. Hence, in this case we obtain that f must be of form

$$f(\mathbf{x}) = \sum_{m=0}^{+\infty} P_m(\mathbf{x}).$$

This is the well-known representation of a monogenic function as a Taylor series, cf. [7].

- In the case $\beta = 1$ and $\alpha = 0$ we deal with the equation $\mathbf{D}f = Ef$. Its regular solutions around the origin have the form

$$f(\mathbf{x}) = \sum_{m=0}^{+\infty}\Bigg({}_2F_1\left(\frac{m}{2}, \frac{m+1}{2}; m+\frac{3}{2}; -\|\mathbf{x}\|^2\right) P_m(\mathbf{x})$$
$$- \frac{m}{2m+3}\|\mathbf{x}\|\, {}_2F_1\left(\frac{m+2}{2}, \frac{m+1}{2}; m+\frac{5}{2}; -\|\mathbf{x}\|^2\right) P_m(\mathbf{x})\Bigg).$$

In turn these solutions again appear as special regular solutions of the Dirac equation on the hyperbolic projective space $\mathbb{R}^{1,2}$ treated by D. Eelbode and F. Sommen [8] and of the Dirac equation on the unit sphere treated by P. Van Lancker [23] and H. Liu and J. Ryan [17].

Acknowledgements This work reflects the current state of the art in this research field. It is based on previous results which were obtained in collaborations with Dr. D. Constales (Ghent University), Prof. I. Cação (University of Aveiro) and Prof. Ryan (University of Arkansas) to whom the author wishes to express his profound gratitude. The author id also very thankful to the very helpful suggestions by the anonymous referees.

References

1. Bahmann, H., Gürlebeck, K., Shapiro, M., Sprößig, W.: On a modified teodorescu transform. Integral Transforms Spec. Funct. **12**(3), 213–226 (2001)
2. Cação, I., Constales, D., Kraußhar, R.S.: On the role of arbitrary order Bessel functions in higher dimensional Dirac type equations. Arch. Math. **87**(5), 468–477 (2006)
3. Cação, I., Constales, D., Kraußhar, R.S.: Explicit representations of the regular solutions of the time-harmonic Maxwell equations combined with the radial symmetric Euler operator. Math. Methods Appl. Sci. **32**(1), 1–11 (2009)
4. Constales, D., Kraußhar, R.S.: Hilbert spaces of solutions to polynomial Dirac equations, Fourier transforms and reproducing kernel functions for cylindrical domains. Z. Anal. Ihre Anwend. **24**(3), 611–636 (2005)
5. Constales, D., Kraußhar, R.S.: Multiperiodic eigensolutions to the Dirac operator and applications to the generalized Helmholtz equation on flat cylinders and on the n-torus. Math. Methods Appl. Sci. **32**, 2050–2070 (2009)
6. Davydov, A.S.: Quantum Mechanics, 2nd edn. Pergamon, Elmsford (1976)
7. Delanghe, R., Sommen, F., Souček, V.: Clifford Algebra and Spinor Valued Functions. Kluwer, Dordrecht/Boston/London (1992)
8. Eelbode, D., Sommen, F.: The fundamental solution of the hyperbolic Dirac operator on $\mathbb{R}^{1,m}$: a new approach. Bull. Belg. Math. Soc.—Simon Stevin **12**(1), 23–37 (2005)
9. Friedrich, T.: Zur Abhängigheit des Dirac-operarors von der Spin-Struktur. Colloq. Math. **48**, 57–62 (1984)
10. Gürlebeck, K., Sprößig, W.: Quaternionic Analysis and Elliptic Boundary Value Problems. Birkhäuser, Basel (1990)
11. Gürlebeck, K., Sprößig, W.: Quaternionic and Clifford Calculus for Physicists and Engineers. Wiley, Chichester/New York (1997)
12. Kraußhar, R.S.: Generalized Analytic Automorphic Forms in Hypercomplex Spaces. Frontiers in Mathematics. Birkhäuser, Basel (2004)
13. Kraußhar, R.S., Ryan, J.: Clifford and harmonic analysis on cylinders and tori. Rev. Mat. Iberoam. **21**, 87–110 (2005)
14. Kraußhar, R.S., Ryan, J.: Some conformally flat spin manifolds, Dirac operators and automorphic forms. J. Math. Anal. Appl. **325**(1), 359–376 (2007)
15. Kravchenko, V.V., Shapiro, M.: Integral Representations for Spatial Models of Mathematical Physics. Addison-Wesley/Longman, Harlow (1996)
16. Kravchenko, V.V., Castillo, P.R.: On the kernel of the Klein-Gordon operator. Z. Anal. Anwend. **17**(2), 261–265 (1998)
17. Liu, H., Ryan, J.: Clifford analysis techniques for spherical pde. J. Fourier Anal. Appl. **8**, 535–564 (2002)
18. Miatello, R., Podesta, R.: Spin structures and spectra of $\mathbb{Z}_2$ manifolds. Math. Z. **247**, 319–335 (2004)
19. Pfäffle, F.: The Dirac spectrum of Bieberbach manifolds. J. Geom. Phys. **35**, 367–385 (2000)
20. Ryan, J.: Cauchy-Green type formulae in Clifford analysis. Trans. Am. Math. Soc. **347**(4), 1331–1341 (1995)
21. Sakurai, J.J.: Advanced Quantum Mechanics. Addison-Wesley, Reading (1967)
22. Xu, Z.: A function theory for the operator $(D - \lambda)$. Complex Var. **16**(1), 27–42 (1991)
23. Van Lancker, P.: Clifford analysis on the sphere. In: Dietrich, V., et al. (eds.) Clifford Algebras and Their Applications in Mathematical Physics, pp. 201–215. Kluwer, Dordrecht (1998)
24. Xu, Z.: Helmholtz equations and boundary value problems. In: Partial Differential Equations with Complex Analysis. Pitman Res. Notes Math. Ser., vol. 262, pp. 204–214. Longman Sci. Tech., Harlow (1992)

Chapter 10
A Survey of *hp*-Adaptive Strategies for Elliptic Partial Differential Equations

William F. Mitchell and Marjorie A. McClain

Abstract The *hp* version of the finite element method (*hp*-FEM) combined with adaptive mesh refinement is a particularly efficient method for solving partial differential equations because it can achieve a convergence rate that is exponential in the number of degrees of freedom. *hp*-FEM allows for refinement in both the element size, h, and the polynomial degree, p. Like adaptive refinement for the h version of the finite element method, *a posteriori* error estimates can be used to determine where the mesh needs to be refined, but a single error estimate can not simultaneously determine whether it is better to do the refinement by h or by p. Several strategies for making this determination have been proposed over the years. In this paper we summarize these strategies and demonstrate the exponential convergence rates with two classic test problems.

Keywords Elliptic partial differential equations · Finite elements · *hp*-adaptive strategy · *hp*-FEM

Mathematics Subject Classification (2000) 65N30 · 65N50

10.1 Introduction

The numerical solution of partial differential equations (PDEs) is the most compute-intensive part of a wide range of scientific and engineering applications. Consequently the development and application of faster and more accurate methods for

---

W.F. Mitchell (✉) · M.A. McClain
Mathematical and Computational Sciences Division, National Institute of Standards and Technology, Gaithersburg, MD 20899-8910, USA
e-mail: william.mitchell@nist.gov

T.E. Simos (ed.), *Recent Advances in Computational and Applied Mathematics*,
DOI 10.1007/978-90-481-9981-5_10,
Published by Springer Netherlands

solving partial differential equations has received much attention in the past fifty years. Self-adaptive methods to determine a quasi-optimal grid are a critical component of the improvements, and have been studied for nearly 30 years now. They are often cast in the context of finite element methods, where the domain of the PDE is partitioned into a mesh consisting of a number of elements (in two dimensions, usually triangles or rectangles), and the approximate solution is a polynomial over each element. Most of the work has focused on h-adaptive methods. In these methods, the mesh size, h, is adapted locally by means of a local error estimator with the goal of placing the smallest elements in the areas where they will do the most good. In particular, elements that have a large error estimate get refined so that ultimately the error estimates are approximately equal over all elements.

Recently, the research community has begun to focus more attention on *hp*-adaptive methods. In these methods, one not only locally adapts the size of the mesh, but also the degree of the polynomials, p. The attraction of *hp*-adaptivity is that the error converges at an exponential rate in the number of degrees of freedom, as opposed to a polynomial rate for fixed p. Much of the theoretical work showing the advantages of *hp*-adaptive methods was done in the 1980's, but it wasn't until the 1990's that practical implementation began to be studied. The new complication is that the local error estimator is no longer sufficient to guide the adaptivity. It indicates which elements should be refined, but it does not indicate whether it is better to refine the element by h or by p. A method for making that determination is called an *hp*-adaptive strategy. A number of strategies have been proposed. In this paper we summarize 15 such *hp*-adaptive strategies.

The remainder of the paper is organized as follows. In Sect. 10.2 we define the equation to be solved, present the finite element method, and give some *a priori* error estimates. In Sect. 10.3 we give the details of an *hp*-adaptive finite element algorithm. Section 10.4 defines the *hp*-adaptive strategies. Section 10.5 contains numerical results to demonstrate the convergence achieved by the different strategies. Finally, we draw our conclusions in Sect. 10.6.

10.2 The Finite Element Method

For simplicity, consider the Poisson boundary value problem

$$-\frac{\partial^2 u}{\partial x^2} - \frac{\partial^2 u}{\partial y^2} = f(x, y) \quad \text{in } \Omega \tag{10.2.1}$$

$$u = g(x, y) \quad \text{on } \partial\Omega \tag{10.2.2}$$

where Ω is a bounded, connected, open region in $\mathbb{R}^2$. Note, however, that everything in this paper applies equally well to a general second order elliptic PDE with mixed boundary conditions. The data in (10.2.1–10.2.2) are assumed to satisfy the usual ellipticity and regularity assumptions.

Denote by $L^2(\Omega)$ the space of square integrable functions over Ω with inner product

$$\langle u, v\rangle_2 = \iint_\Omega uv$$

and norm

$$\|v\|_2^2 = \langle v, v\rangle_2.$$

$H^m(\Omega)$ denotes the usual Sobolev spaces of functions whose derivatives up to order m are in $L^2(\Omega)$. The Sobolev spaces have inner products

$$\langle u, v\rangle_{H^m(\Omega)} = \iint_\Omega \sum_{|\alpha|\le m} D^\alpha u D^\alpha v$$

and norms

$$\|v\|^2_{H^m(\Omega)} = \langle v, v\rangle_{H^m(\Omega)}$$

where

$$D^\alpha v = \frac{\partial^{|\alpha|} v}{\partial^{\alpha_1} x \partial^{\alpha_2} y} \qquad \alpha = (\alpha_1, \alpha_2),\ \alpha_i \in \mathbb{N},\ |\alpha| = \alpha_1 + \alpha_2.$$

Let $H_0^m(\Omega) = \{v \in H^m(\Omega) : v = 0 \text{ on } \partial\Omega\}$. Let $\tilde{u}_D$ be a lift function satisfying the Dirichlet boundary conditions in (10.2.2) and define the affine space $\tilde{u}_D + H_0^1(\Omega) = \{\tilde{u}_D + v : v \in H_0^1(\Omega)\}$. Define the bilinear form

$$B(u, v) = \iint_\Omega \frac{\partial u}{\partial x}\frac{\partial v}{\partial x} + \frac{\partial u}{\partial y}\frac{\partial v}{\partial y}$$

and the linear form

$$L(v) = \iint_\Omega fv.$$

Then the variational form of the problem is to find the unique $u \in \tilde{u}_D + H_0^1(\Omega)$ that satisfies

$$B(u, v) = L(v) \quad \forall v \in H_0^1(\Omega).$$

The energy norm of $v \in H_0^1$ is defined by $\|v\|^2_{E(\Omega)} = B(v, v)$.

The finite element space is defined by partitioning Ω into a grid (or mesh), G_{hp}, consisting of a set of N_T triangular elements, $\{T_i\}_{i=1}^{N_T}$ with $\bar{\Omega} = \bigcup_{i=1}^{N_T} \bar{T}_i$. If a vertex of a triangle is contained in the interior of an edge of another triangle, it is called a hanging node. We only consider compatible grids with no hanging nodes, i.e. $\bar{T}_i \cap \bar{T}_j$, $i \neq j$, is either empty, a common edge, or a common vertex. The diameter of the element is denoted h_i. With each element we associate an integer degree $p_i \ge 1$. The finite element space V_{hp} is the space of continuous piecewise polynomial functions on Ω such that over element T_i it is a polynomial of degree p_i. The degree of an edge is determined by applying the minimum rule, i.e. the edge is assigned the minimum of the degrees of the adjacent elements.

The finite element solution is the unique function $u_{hp} \in \tilde{u}_D + V_{hp}$ that satisfies

$$B(u_{hp}, v_{hp}) = L(v_{hp}) \quad \forall v_{hp} \in V_{hp}.$$

The error is defined by $e_{hp} = u - u_{hp}$.

The finite element solution is expressed as a linear combination of basis functions $\{\phi_i\}_{i=1}^{N_{\mathrm{dof}}}$ that span $\tilde{u}_D + V_{hp}$,

$$u_{hp}(x, y) = \sum_{i=1}^{N_{\mathrm{dof}}} \alpha_i \phi_i(x, y)$$

N_{dof} is the number of degrees of freedom in the solution. The p-hierarchical basis of Szabo and Babuška [31], which is based on Legendre polynomials, is used in the program used for most of the results in Sect. 10.5. The basis functions are hierarchical in the sense that the basis functions for a space of degree p are a subset of the basis functions for a space of degree $p+1$. For an element of degree p_i with edge degrees $p_{i,j}$, $j = 1, 2, 3$ there is one linear basis function associated with each vertex, $p_{i,j} - 1$ basis functions, of degree $2, 3, \ldots, p_{i,j}$, associated with edge j, and $q - 2$ basis functions of degree q for $q = 3, 4, \ldots, p_i$ (for a total of $(p_i - 1)(p_i - 2)/2$) whose support is the interior of the triangle.

The discrete form of the problem is a linear system of algebraic equations

$$Ax = b \tag{10.2.3}$$

where the matrix A is given by $A_{ij} = B(\phi_i, \phi_j)$ and the right hand side is given by $b_i = L(\phi_i)$. The solution x consists of the α_i's.

If h and p are uniform over the grid, $u \in H^m(\Omega)$, and the other usual assumptions are met, then the *a priori* error bound is [6, 7]

$$\|e_{hp}\|_{H^1(\Omega)} \le C \frac{h^\mu}{p^{m-1}} \|u\|_{H^m(\Omega)} \tag{10.2.4}$$

where $\mu = \min(p, m-1)$ and C is a constant that is independent of h, p and u, but depends on m.

With a suitably chosen hp mesh, and other typical assumptions, the error can be shown [13] to converge exponentially in the number of degrees of freedom,

$$\|e_{hp}\|_{H^1(\Omega)} \le C_1 e^{-C_2 N_{\mathrm{dof}}^{1/3}} \tag{10.2.5}$$

for some $C_1, C_2 > 0$ independent of N_{dof}.

begin with a very coarse grid in h with small p
form and solve the linear system
repeat
  determine which elements to coarsen and whether to coarsen by h or p
  coarsen elements
  determine which elements to refine and whether to refine by h or p
  refine elements
  form and solve the linear system
until some termination criterion is met

Fig. 10.1 Basic form of an hp-adaptive algorithm

10.3 *hp*-Adaptive Refinement Algorithm

One basic form of an hp-adaptive algorithm is given in Fig. 10.1. There are a number of approaches to each of the steps of the algorithm. In this paper, the following approaches are used.

Triangles are h-refined by the newest node bisection method [18]. Briefly, a parent triangle is h-refined by connecting one of the vertices to the midpoint of the opposite side to form two new child triangles. The most recently created vertex is chosen as the vertex to use in this bisection. Triangles are always refined in pairs (except when the edge to be refined is on the boundary) to maintain compatibility of the grid. This may require first refining a neighbor triangle to create the second triangle of the pair. The h-refinement level, l_i, of a triangle T_i is one more than the h-refinement level of the parent, with level 1 assigned to the triangles of the initial coarse grid. p-refinement is performed by increasing the degree of the element by one, followed by enforcing the minimum rule for the edges. Coarsening of elements means reversing the refinement.

Adaptive refinement is guided by a local *a posteriori* error indicator computed for each element. There are several choices of error indicators; see for example [2, 32]. For this paper, the error indicator for element T_i is given by solving a local Neumann residual problem:

$$-\frac{\partial^2 e_i}{\partial x^2} - \frac{\partial^2 e_i}{\partial y^2} = f - \frac{\partial^2 u_{hp}}{\partial x^2} - \frac{\partial^2 u_{hp}}{\partial y^2} \quad \text{in } T_i \tag{10.3.1}$$

$$e_i = 0 \quad \text{on } \partial T_i \cap \partial\Omega \tag{10.3.2}$$

$$\frac{\partial e_i}{\partial n} = -\frac{1}{2}\left[\frac{\partial u_{hp}}{\partial n}\right] \quad \text{on } \partial T_i \setminus \partial\Omega \tag{10.3.3}$$

where $\frac{\partial}{\partial n}$ is the unit outward normal derivative and $[\frac{\partial u_{hp}}{\partial n}]$ is the jump in the outward normal derivative of u_{hp} across the element boundary. The approximate solution, $e_{i,hp}$ of (10.3.1–10.3.3) is computed using the hierarchical bases of exact degree $p_i + 1$, where p_i is the degree of T_i. The error indicator for element T_i is then given by

$$\eta_i = \|e_{i,hp}\|_{E(T_i)}.$$

A global energy norm error estimate is given by

$$\eta = \left(\sum_{i=1}^{N_T} \eta_i^2 \right)^{1/2}.$$

The criterion for program termination is that the relative error estimate be smaller than a prescribed error tolerance τ, i.e. $\eta / \|u_{hp}\|_{E(\Omega)} < \tau$. Elements are selected for coarsening if $\eta_i < \max_i \eta_i / 100$ and for refinement if $\eta_i > \tau \|u_{hp}\|_{E(\Omega)} / \sqrt{N_T}$. Note that if every element had $\eta_i = \tau \|u_{hp}\|_{E(\Omega)} / \sqrt{N_T}$ then $\eta / \|u_{hp}\|_{E(\Omega)} = \tau$, hence the $\sqrt{N_T}$ factor.

10.4 The *hp*-Adaptive Strategies

In this section, the *hp*-adaptive strategies that have been proposed in the literature are presented. In some cases, these strategies were developed in the context of 1D problems, rectangular elements, or other settings that are not fully compatible with the context of this paper. In those cases, the strategy is appropriately modified for 2D elliptic PDEs and newest node bisection of triangles.

10.4.1 Use of a priori Knowledge of Solution Regularity

It is well known that for smooth solutions p-refinement will produce an exponential rate of convergence, but near singularities p-refinement is less effective than h-refinement. This is a consequence of the *a priori* error bounds in (10.2.4) and (10.2.5). For this reason, many of the *hp* strategies use h-refinement in areas where the solution is irregular (i.e., locally fails to be in H^m for some finite m, also called nonsmooth) or nearly irregular, and p-refinement elsewhere. The simplest strategy is to use any *a priori* knowledge about irregularities. For example, it is known that linear elliptic PDEs with smooth coefficients and piecewise analytic boundary data will have point singularities only near reentrant corners of the boundary and where boundary conditions change [4]. Another example would be a situation where one knows the approximate location of a shock in the interior of the domain.

An *hp*-adaptive strategy of this type was presented by Ainsworth and Senior [4]. In this approach they simply flag vertices in the initial mesh as being possible trouble spots. During refinement an element is refined by h if it contains a vertex that is so flagged, and by p otherwise. We will refer to this strategy by the name APRIORI.

We extend this strategy to allow more general regions of irregularity, and to provide the strength of the irregularity. The user provides a function that, given an element T_i as input, returns a regularity value for that element. For true singularities, it would ideally return the maximum value of m such that $u \in H^m(T_i)$. But it can also indicate that a triangle intersects an area that is considered to be nearly irregular, like a boundary layer or sharp wave front. Based on the definition of μ in (10.2.4), if the current degree of the triangle is p_i and the returned regularity value is m_i, we do p-refinement if $p_i \le m_i - 1$ and h-refinement otherwise. The same approach is used in all the strategies that estimate the local regularity m_i.

10.4.2 Estimate Regularity Using Smaller p Estimates

Süli, Houston and Schwab [30] presented a strategy based on (10.2.4) and an estimate of the convergence rate in p using error estimates based on $p_i - 2$ and $p_i - 1$. We will refer to this strategy as PRIOR2P. This requires $p_i \ge 3$, so we always use p-refinement in elements of degree 1 and 2.

Suppose the error estimate in (10.2.4) holds on individual elements and that the inequality is an approximate equality. Let η_{i,p_i-2} and η_{i,p_i-1} be *a posteriori* error estimates for partial approximate solutions over triangle T_i using the bases up to degree $p_i - 2$ and $p_i - 1$, respectively. Then

$$\frac{\eta_{i,p_i-1}}{\eta_{i,p_i-2}} \approx \left(\frac{p_i - 1}{p_i - 2}\right)^{-(m_i-1)}$$

and thus the regularity is estimated by

$$m_i \approx 1 - \frac{\log(\eta_{i,p_i-1}/\eta_{i,p_i-2})}{\log((p_i - 1)/(p_i - 2))}.$$

Use p-refinement if $p_i \le m_i - 1$ and h-refinement otherwise.

Thanks to the p-hierarchical basis, the computation of the error estimates is very inexpensive. For $1 \le j < p_i$,

$$u_{hp}|_{T_i} = \sum_{\text{supp}(\phi_k)\cap T_i \ne \emptyset} \alpha_k \phi_k = \sum_{\substack{\text{supp}(\phi_k)\cap T_i \ne \emptyset \\ \deg(\phi_k) \le p_i - j}} \alpha_k \phi_k + \sum_{\substack{\text{supp}(\phi_k)\cap T_i \ne \emptyset \\ \deg(\phi_k) > p_i - j}} \alpha_k \phi_k$$

where $\text{supp}(\phi_k)$ is the support of ϕ_k and $\deg(\phi_k)$ is the degree of ϕ_k. The last sum is the amount by which the solution changed when the degree of the element was increased from $p_i - j$ to p_i, and provides an estimate of the error in the partial approximate solution of degree $p_i - j$ given in the next to last sum. (Indeed, the local Neumann error estimator of (10.3.1–10.3.3) approximates this quantity for the increase from degree p_i to p_{i+1}.) Thus the error estimates are

$$\eta_{i,p_i-j} = \left\| \sum_{\substack{\text{supp}(\phi_k)\cap T_i \ne \emptyset \\ \deg(\phi_k) > p_i - j}} \alpha_k \phi_k \right\|_{H^1(T_i)}$$

which only involves computing the norm of known quantities.

10.4.3 Type Parameter

Gui and Babuška [12] presented an *hp*-adaptive strategy using what they call a type parameter, γ. This strategy is also used by Adjerid, Aiffa and Flaherty [1]. We will refer to this strategy as TYPEPARAM.

Given the error estimates η_{i,p_i} and η_{i,p_i-1}, define

$$R(T_i) = \begin{cases} \frac{\eta_{i,p_i}}{\eta_{i,p_i-1}} & \eta_{i,p_i-1} \neq 0 \\ 0 & \eta_{i,p_i-1} = 0. \end{cases}$$

By convention, $\eta_{i,0} = 0$, which forces p-refinement if $p_i = 1$.

R is used to assess the perceived solution smoothness. Given the type parameter, $0 \leq \gamma < \infty$, element T_i is h-refined if $R(T_i) > \gamma$, and p-refined if $R(T_i) \leq \gamma$. Note that $\gamma = 0$ gives pure h-refinement and $\gamma = \infty$ gives pure p-refinement.

For the error estimates, we use the local Neumann error estimate of (10.3.1–10.3.3) for η_{i,p_i}, and the η_{i,p_i-1} of Sect. 10.4.2. We use $\gamma = 0.3$ in the results of Sect. 10.5.

10.4.4 Estimate Regularity Using Larger p Estimates

Another approach that estimates the regularity is given by Ainsworth and Senior [3]. This strategy uses three error estimates based on spaces of degree $p_i + 1$, $p_i + 2$ and $p_i + 3$, so we refer to it as NEXT3P.

The error estimate used to approximate the regularity is a variation on the local Neumann residual error estimate given by (10.3.1–10.3.3) in which (10.3.3) is replaced by

$$\frac{\partial e_i}{\partial n} = g_i \quad \text{on } \partial T_i \setminus \partial\Omega$$

where g_i is an approximation of $\frac{\partial u}{\partial n}$ that satisfies an equilibrium condition. This is the equilibrated residual error estimator in [2].

The local problem is solved on element T_i three times using the spaces of degree $p_i + q$, $q = 1, 2, 3$, to obtain error estimates $e_{i,q}$. In contrast to the local Neumann residual error estimate, the whole space over T_i is used, not just the p-hierarchical bases of degree greater than p_i. These approximations to the error converge to the true solution of the residual problem at the same rate the approximate solution converges to the true solution of (10.2.1–10.2.2), i.e.

$$\|e_i - e_{i,q}\|_{E(T_i)} \approx C(p_i + q)^{-\alpha}$$

where C and α are positive constants that are independent of q but depend on T_i. Using the Galerkin orthogonality

$$\|e_i\|^2_{E(T_i)} = \|e_i - e_{i,q}\|^2_{E(T_i)} + \|e_{i,q}\|^2_{E(T_i)}$$

this can be rewritten

$$\|e_i\|^2_{E(T_i)} - \|e_{i,q}\|^2_{E(T_i)} \approx C^2(p_i+q)^{-2\alpha}.$$

We can compute $\|e_{i,q}\|^2_{E(T_i)}$ and p_i+q for $q=1,2,3$ from the approximate solutions, so the three constants $\|e_i\|_{E(T_i)}$, C and α can be approximated by fitting the data. Then, using the *a priori* error estimate in (10.2.4), the approximation of the local regularity is $m_i = 1+\alpha$. Use p-refinement if $p_i \le m_i - 1$ and h-refinement otherwise.

10.4.5 Texas 3 Step

The Texas 3 Step strategy [8, 20, 21] first performs h-refinement to get an intermediate grid, and follows that with p-refinement to reduce the error to some given error tolerance, τ. We will refer to this strategy as T3S. Note that for this strategy the basic form of the *hp*-adaptive algorithm is different than that in Fig. 10.1.

The first step is to create an initial mesh with uniform p and nearly uniform h such that the solution is in the asymptotic range of convergence in h. This may be accomplished by performing uniform h-refinements of some very coarse initial mesh until the asymptotic range is reached. The resulting grid has N_0 elements with sizes h_i, degrees p_i and *a posteriori* error estimates η_i, and approximate solution u_0. The results in Sect. 10.5 begin with $p=1$ and assume the initial grid is sufficiently fine in h.

The second step is to perform adaptive h-refinement to reach an intermediate error tolerance $\gamma\tau$ where γ is a given parameter. In the references, γ is in the range 5–10, usually 6 in the numerical results. This intermediate grid is created by computing a desired number of children for each element T_i by the formula

$$n_i = \left[\frac{\Lambda_i^2 N_I h_i^{2\mu_i}}{p_i^{2(m_i-1)}\eta_I^2}\right]^{\frac{1}{\beta\mu_i+1}} \tag{10.4.1}$$

where $N_I = \sum n_i$ is the number of elements in the intermediate grid, m_i is the local regularity of the solution, $\mu_i = \min(p_i, m_i - 1)$, $\eta_I = \gamma\tau\|u_0\|_{E(\Omega)}$, $\beta = 1$ for 2D problems, $\eta_0^2 = \sum \eta_i^2$ and

$$\Lambda_i = \frac{\eta_i \Lambda}{\eta_0}$$

where

$$\Lambda = \frac{\eta_0 p_i^{m_i-1}}{h_i^{\mu_i}}.$$

See any of the above references for the derivation of this formula. It is based on the *a priori* error estimate in (10.2.4). Inserting the expression for Λ_i into (10.4.1) and using $\beta = 1$ we arrive at

$$n_i = \left[\frac{\eta_i^2 N_I}{\eta_I^2} \right]^{\frac{1}{\mu_i + 1}}$$

N_I is not known at this point, since it is the sum of the n_i. Successive iterations are used to solve for n_i and N_I simultaneously. We use 5 iterations, which preliminary experiments showed to be sufficient (convergence was usually achieved in 3 or 4 iterations). Once the n_i have been determined, we perform $\lfloor 0.5 + \log_2 n_i \rfloor$ uniform h-refinements (bisections) of each element T_i to generate approximately n_i children, and solve the discrete problem on the intermediate grid.

The third step is to perform adaptive p-refinement to reduce the error to the desired tolerance τ. The new degree for each element is given by

$$\hat{p}_i = p_i \left[\frac{\eta_{I,i} \sqrt{N_I}}{\eta_T} \right]^{\frac{1}{m_i - 1}}$$

where $\eta_{I,i}$ is the *a posteriori* error estimate for element T_i of the intermediate grid and $\eta_T = \tau \|u_0\|_{E(\Omega)}$. Again, the formula is a simple reduction of the equations derived in the references. p-refinement is performed to increase the degree of each element T_i to $\hat{p}_i$, and the discrete problem is solved on the final grid.

In the results of Sect. 10.5, if $n_i < 2$ or $\hat{p}_i < p_i$ then refinement is not performed. Also, to avoid excessive refinement, the number of h-refinements done to any element in Step 2 and number of p-refinements in Step 3 is limited to 3.

The strategy of performing all the h-refinement in one step and all the p-refinement in one step is adequate for low accuracy solutions (e.g. 1%), but is not likely to work well with high accuracy solution (e.g. 10^{-8} relative error) [22]. We extend the Texas 3 Step strategy to high accuracy by cycling through Steps 2 and 3 until the final tolerance τ_{final} is met. τ in the algorithm above is now the factor by which one cycle of Steps 2 and 3 should reduce the error. Toward this end, before Step 2 the error estimate η_0 is computed for the current grid. The final (for this cycle) and intermediate targets are now given by $\eta_T = \tau \eta_0$ and $\eta_I = \gamma \eta_T$. In the results of Sect. 10.5 we use $\tau = 0.1$ and $\gamma = 6$. For the local regularity m_i we use the same routine as the APRIORI strategy (Sect. 10.4.1).

10.4.6 Alternate h and p

This strategy, which will be referred to as ALTERNATE, is a variation on T3S that is more like the algorithm of Fig. 10.1. The difference is that instead of predicting the number of refinements needed to reduce the error to the next target, the usual adaptive refinement is performed until the target is reached. Thus in Step 2 all elements with an error indicator larger than $\eta_I / \sqrt{N_0}$ are h-refined. The discrete problem is solved and the new error estimate compared to η_I. This is repeated until the error estimate is smaller than η_I. Step 3 is similar except adaptive p-refinement is performed and the target is η_T. Steps 2 and 3 are repeated until the final error tolerance is achieved.

10.4.7 Nonlinear Programming

Patra and Gupta [23] proposed a strategy for *hp* mesh design using nonlinear programming. We refer to this strategy as NLP. They presented it in the context of quadrilaterals with one level of hanging nodes, i.e., an element edge is allowed to have at most one hanging node. Here it is modified slightly for newest node bisection of triangles with no hanging nodes. This is another approach that does not strictly follow the algorithm in Fig. 10.1.

Given a current grid with triangles $\{T_i\}$, degrees $\{p_i\}$, h-refinement levels $\{l_i\}$, error estimates $\{\eta_i\}$, and element diameters

$$h_i = \left(\frac{1}{\sqrt{2}}\right)^{l_i} H_{0,i}$$

where $H_{0,i}$ is the diameter of the element in the initial grid that contains T_i, the object is to determine new mesh parameters $\{\hat{p}_i\}$ and $\{\hat{l}_i\}$, $i = 1..N_T$, by solving an optimization problem. The new grid is obtained by refining T_i $\hat{l}_i - l_i$ times (or coarsening if $\hat{l}_i < l_i$) and assigning degree $\hat{p}_i$ to the $2^{\hat{l}_i - l_i}$ children. The size of the children of T_i is

$$\hat{h}_i = \left(\frac{1}{\sqrt{2}}\right)^{\hat{l}_i} H_{0,i}.$$

There are two forms of the optimization problem, which can be informally stated as (1) minimize the number of degrees of freedom (or some other measure of grid size) subject to the error being less than a given tolerance and other constraints, and (2) minimize the error subject to the number of degrees of freedom being less than a given limit and other constraints. We will only consider the first form here; the second form simply reverses the objective function and constraint.

Computationally, the square of the error is approximated by $\sum_{i=0}^{N_T} \hat{\eta}_i^2$ where $\hat{\eta}_i$, to be defined later, is an estimate of the error in the refined grid over the region covered by T_i. The number of degrees of freedom associated with a triangle of degree p is taken to be $3/6$ (one for each vertex with an average of six triangles sharing a vertex) plus $3(p-1)/2$ ($p-1$ for each edge with two triangles sharing an edge) plus $(p-1)(p-2)/2$ (for the interior), which is $p^2/2$. Thus the number of degrees of freedom over the children of T_i is $2^{\hat{l}_i - l_i}\hat{p}_i^2/2$. We can now formally state the optimization problem as

$$\underset{\{\hat{l}_i\},\{\hat{p}_i\}}{\text{minimize}} \quad \sum_{i=1}^{N_T} 2^{\hat{l}_i - l_i} \frac{\hat{p}_i^2}{2} \tag{10.4.2}$$

$$\text{s.t.} \quad \sum_{i=1}^{N_T} \hat{\eta}_i^2 \le \hat{\tau}^2 \tag{10.4.3}$$

$$\hat{l}_j - 1 \le \hat{l}_i \le \hat{l}_j + 1$$

$$\forall j \text{ such that } T_j \text{ shares an edge with } T_i \tag{10.4.4}$$

$$1 \le \hat{l}_i \le l_{\max} \tag{10.4.5}$$

$$1 \le \hat{p}_i \le p_{\max} \tag{10.4.6}$$

$$l_i - \delta l_{\text{dec}} \le \hat{l}_i \le l_i + \delta l_{\text{inc}} \tag{10.4.7}$$

$$p_i - \delta p_{\text{dec}} \le \hat{p}_i \le p_i + \delta p_{\text{inc}} \tag{10.4.8}$$

where $\hat{\tau}$ is the error tolerance for this refinement phase. We use $\hat{\tau} = \eta/4$ where η is the global error estimate on the current grid. The divisor 4 is arbitrary and could be replaced by some other value. In practice, (10.4.3) is divided through by τ^2 so that the numbers are $O(1)$. Equation (10.4.4) is a necessary condition for compatibility of the grid (in [23] it enforces one level of hanging nodes). It is not a sufficient condition, however any violations of compatibility while this condition is met are cases where only one triangle of a compatibly divisible pair was refined, and it is a slight adjustment to the optimal solution to also refine the other one to maintain compatibility. Equation (10.4.5) insures that coarsening does not go beyond the initial grid, and that the refinement level of an element does not exceed a prescribed limit $l_{\max}$. Similarly, (10.4.6) insures that element degrees do not go below one or exceed a prescribed limit $p_{\max}$. Also, because many quantities are only approximate, it is wise to limit the amount of change that occurs to any element during one phase of refinement. Equations (10.4.7) and (10.4.8) restrict the amount of decrease in l and p to prescribed limits δl_{dec} and δp_{dec}, and the amount of increase to δl_{inc} and δp_{inc}. In the results in Sect. 10.5 we used $\delta l_{\text{dec}} = \delta p_{\text{dec}} = 1$, $\delta l_{\text{inc}} = 5$, and $\delta p_{\text{inc}} = 2$.

Since the $\hat{l}_i$ and $\hat{p}_i$ are naturally integers, the optimization problem is a mixed integer nonlinear program, which is known to be NP-hard. To make the problem tractable, the integer requirement is dropped to give a nonlinear program which can be solved by one of several software packages. For the results in Sect. 10.5, we used the program ALGENCAN[1] Version 2.2.1 [5, 9]. Following solution of the nonlinear program, the $\hat{l}_i$ and $\hat{p}_i$ are rounded to the nearest integer.

It remains to define $\hat{\eta}_i$, the estimate of the error in the refined grid over the region covered by T_i. Assuming approximate equality in the *a priori* error estimate of

[1]The mention of specific products, trademarks, or brand names is for purposes of identification only. Such mention is not to be interpreted in any way as an endorsement or certification of such products or brands by the National Institute of Standards and Technology. All trademarks mentioned herein belong to their respective owners.

(10.2.4), we have

$$\eta_i \approx C \frac{h_i^{\mu_i}}{p_i^{m_i-1}} \|u\|_{H^m(T_i)}$$

and

$$\hat{\eta}_i \approx C \frac{\hat{h}_i^{\mu_i}}{\hat{p}_i^{m_i-1}} \|u\|_{H^m(T_i)}$$

where m_i is the local regularity over T_i and $\mu_i = \min(p_i, m_i - 1)$. Combining these leads to

$$\hat{\eta}_i \approx \frac{\hat{h}_i^{\mu_i}}{\hat{p}_i^{m_i-1}} \frac{p_i^{m_i-1}}{h_i^{\mu_i}} \eta_i = \left(\frac{1}{\sqrt{2}}\right)^{\mu_i(\hat{l}_i - l_i)} \left(\frac{p_i}{\hat{p}_i}\right)^{m_i-1} \eta_i$$

and thus the constraint in (10.4.3) is

$$\sum_{i=1}^{N_T} \left(\frac{1}{2}\right)^{\min(p_i, m_i-1)(\hat{l}_i - l_i)} \left(\frac{p_i}{\hat{p}_i}\right)^{2(m_i-1)} \eta_i^2 < \hat{\tau}^2$$

in which the only remaining quantity to be determined is m_i. Patra and Gupta suggest estimating m_i by using the observed convergence rate from two grids, with a formula very similar to that used in the PRIOR2P strategy of Sect. 10.4.2. However, this requires that p_i be at least three in every element, so instead we use the estimate of m_i from the NEXT3P strategy of Sect. 10.4.4 which allows $p_i = 1$.

10.4.8 Another Optimization Strategy

Another strategy based on the formulation and solution of an optimization problem is given in Novotny et al. [19]. However, it turns out that (1) the optimization does not work near singularities, so *a priori* knowledge of singularities must be used to force h-refinement near singularities, and (2) for the finite element method and class of problems considered in this paper, the strategy always chooses p-refinement except for extremely large elements. Thus, this strategy is (nearly) identical to the APRIORI strategy, and will not be considered further in this paper.

10.4.9 Predict Error Estimate on Assumption of Smoothness

Melenk and Wohlmuth [16] proposed a strategy based on a prediction of what the error should be if the solution is smooth. We call this strategy SMOOTH_PRED.

When refining element T_i, assume the solution is locally smooth and that the optimal convergence rate is obtained. If h-refinement is performed and the degree of T_i is p_i, then the expected error on the two children of T_i is reduced by a factor of $\sqrt{2}^{\,p_i}$ as indicated by the *a priori* error estimate of (10.2.4). Thus if η_i is the error estimate for T_i, predict the error estimate of the children to be $\gamma_h \eta_i / \sqrt{2}^{\,p_i}$ where γ_h is a user specified parameter. If p-refinement is performed on T_i, exponential convergence is expected, so the error should be reduced by some constant factor $\gamma_p \in (0,1)$, i.e., the predicted error estimate is $\gamma_p \eta_i$. When the actual error estimate of a child becomes available, it is compared to the predicted error estimate. If the error estimate is less than or equal to the predicted error estimate, then p-refinement is indicated for the child. Otherwise, h-refinement is indicated since presumably the assumption of smoothness was wrong. For the results in Sect. 10.5 we use $\gamma_h = 4$ and $\gamma_p = \sqrt{0.4}$.

10.4.10 Larger of h-Based and p-Based Error Estimates

In 1D, Schmidt and Siebert [25] proposed a strategy that uses two *a posteriori* error estimates to predict whether h-refinement or p-refinement will reduce the error more. We extend this strategy to bisected triangles and refer to it as H&P_ERREST.

The local Neumann residual error estimate given by (10.3.1–10.3.3) is actually an estimate of how much the error will be reduced if T_i is p-refined. This is because the solution of the local problem is estimated using the p-hierarchical bases that would be added if T_i is p-refined, so it is an estimate of the actual change that would occur. Using the fact that the current space is a subspace of the refined space and Galerkin orthogonality, it can be shown that

$$\|u - \hat{u}_{hp}\|^2 = \|u - u_{hp}\|^2 - \|\hat{u}_{hp} - u_{hp}\|^2$$

where $\hat{u}_{hp}$ is the solution in the refined space. Thus the change in the solution indicates how much the error will be reduced.

A second error estimate for T_i can be computed by solving a local Dirichlet residual problem

$$-\frac{\partial^2 e_i}{\partial x^2} - \frac{\partial^2 e_i}{\partial y^2} = f - \frac{\partial^2 u_{hp}}{\partial x^2} - \frac{\partial^2 u_{hp}}{\partial y^2} \quad \text{in } T_i \cup T_i^{\text{mate}} \qquad (10.4.9)$$

$$e_i = g - u_{hp} \qquad \text{on } \partial(T_i \cup T_i^{\text{mate}}) \cap \partial\Omega \qquad (10.4.10)$$

$$e_i = 0 \qquad \text{on } \partial(T_i \cup T_i^{\text{mate}}) \setminus \partial\Omega \qquad (10.4.11)$$

where T_i^{mate} is the element that is refined along with T_i in the newest node bisection method [18]. The solution to this problem is approximated by an h-refinement of the two elements using only the new basis functions. The error estimate obtained by taking the norm of this approximate solution is actually an estimate of how much the solution will change, or the error will be reduced, if h-refinement is performed.

The two error estimates can be divided by the associated increase in the number of degrees of freedom to obtain an approximate error reduction per degree of freedom, and/or be multiplied by a user specified constant to bias the refinement toward h- or p-refinement. In the results of Sect. 10.5 the p-based error estimate is multiplied by 2, which seemed to work best on the largest number of test problems.

The type of refinement that is used is the one that corresponds to the larger of the two modified error estimates.

10.4.11 Legendre Coefficient Strategies

There are three *hp*-adaptive strategies that are based on the coefficients in an expansion of the solution in Legendre polynomials. In 1D, the approximate solution in element T_i with degree p_i can be written

$$u_i(x) = \sum_{j=0}^{p_i} a_j P_j(x)$$

where P_j is the jth degree Legendre polynomial scaled to the interval of element T_i.

Mavriplis [15] estimates the decay rate of the coefficients by a least squares fit of the the last four coefficients a_j to $Ce^{-\sigma j}$. Elements are refined by p-refinement where $\sigma > 1$ and by h-refinement where $\sigma \leq 1$. We refer to this strategy as COEF_DECAY. When four coefficients are not available, we fit to whatever is available. If only one coefficient is available, we use p-refinement.

Houston et al. [14] present the other two approaches which use the Legendre coefficients to estimate the regularity of the solution. One approach estimates the regularity using the root test yielding

$$m_i = \frac{\log(\frac{2p_i+1}{2a_{p_i}^2})}{2\log p_i}.$$

If $p_i = 1$, use p-refinement. Otherwise, use p-refinement if $p_i \leq m_i - 1$ and h-refinement if $p_i > m_i - 1$. We refer to this strategy as COEF_ROOT.

They also present a second way of estimating the regularity from the Legendre coefficients using the ratio test. However, they determined the ratio test is inferior to the root test, so it will not be considered further in this paper.

Both Mavriplis and Houston et al. presented the strategies in the context of one dimension and used the Legendre polynomials as the local basis so the coefficients are readily available. In [14] it is extended to 2D for rectangular elements with a tensor product of Legendre polynomials, and the regularity is estimated in each dimension separately, so the coefficients are still readily available. In this paper we are using triangular elements which have a basis that is based on Legendre polynomials [31]. In this basis there are $3 + \max(j - 2, 0)$ basis functions of exact degree j over an element, so we don't have a single Legendre polynomial coefficient to use. Instead, for the coefficients a_j we use the ℓ_1 norm of the coefficients of the degree

j basis functions, i.e.

$$a_j = \sum_{\substack{k \text{ s.t. } \deg(\phi_k)=j \\ \text{supp}(\phi_k)\cap T_i \neq \emptyset}} |\alpha_k|.$$

10.4.12 Reference Solution Strategies

Demkowicz and his collaborators developed an *hp*-adaptive strategy over a number of years, presented in several papers and books, e.g. [10, 11, 24, 28]. In its full glory, the method is quite complicated. Here we present only the basic ideas of the algorithm and how we have adapted it for bisected triangles (it is usually presented in the context of rectangular elements with some reference to quadrisected triangles), and refer to the references for further details. We refer to this strategy as REFSOLN_EDGE because it relies on computing a reference solution and bases the refinement decisions on edge refinements. Note that for this strategy the basic form of the *hp*-adaptive algorithm is different than that in Fig. 10.1.

The algorithm is first presented for 1D elliptic problems. Given the current existing (coarse) mesh, $G_{h,p} := G_{hp}$, and current solution, $u_{h,p} := u_{hp}$, a uniform refinement in both h and p is performed to obtain a fine mesh $G_{h/2,p+1}$. The equation is solved on the fine mesh to obtain a reference solution $u_{h/2,p+1}$. The norm of the difference between the current solution and reference solution is used as the global error estimate, i.e.,

$$\eta = \|u_{h/2,p+1} - u_{h,p}\|_{H^1(\Omega)}.$$

The next step is to determine the optimal refinement of each element. This is done by considering a p-refinement and all possible (bisection) h-refinements that give the same increase in the number of degrees of freedom as the p-refinement. In 1D, this means that the sum of the degrees of the two children must be $p+1$, resulting in a total of p h-refinements and one p-refinement to be examined. For each possibility, the error decrease rate is computed as

$$\frac{|u_{h/2,p+1} - \Pi_{hp,i}u_{h/2,p+1}|^2_{H^1(T_i)} - |u_{h/2,p+1} - \Pi_{\text{new},i}u_{h/2,p+1}|^2_{H^1(T_i)}}{N_{\text{new}} - N_{hp}}$$

where $\Pi_{hp,i}u_{h/2,p+1}$ is the projection-based interpolant of the reference solution in element T_i, computed by solving a local Dirichlet problem, and likewise $\Pi_{\text{new},i}$ is the projection onto the resulting elements from any one of the candidate refinements. $|\cdot|_{H^1(T_i)}$ is the H^1 seminorm over T_i. The refinement with the largest error decrease rate is selected as the optimal refinement. In the case of h-refinement, the degrees may be increased further by a process known as following the biggest subelement error refinement path, which is also used to determine the guaranteed element rate; see [10] for details.

Elements that have a guaranteed rate larger than $1/3$ the maximum guaranteed rate are selected for refinement, although the factor $1/3$ is somewhat arbitrary.

The 2D algorithm also begins by computing a reference solution on the globally *hp*-refined grid $G_{h/2,p+1}$. (For bisected triangles, we should use the subscript $h/\sqrt{2}$, $p+1$ for the fine grid and solution, but for simplicity we will use the original notation.) Then for each edge in the grid, the choice between p- and h-refinement, the determination of the guaranteed edge rate, and the selection of edges to refine are done exactly as in 1D, except that a weighted H^1 seminorm is used instead of the more natural $H^{1/2}$ seminorm which is difficult to work with. In the case of bisected triangles, we only consider edges that would be refined by the bisection of an existing triangle.

The h-refinement of edges determines the h-refinement of elements. It remains to determine the degree of each element. As a starting point, element degrees are assigned to satisfy the minimum rule for edge degrees, using the edge degrees determined in the previous step. Then the biggest subelement error refinement path is followed to determine the guaranteed element rate and assignment of element degrees. We again refer to [10] for details. Finally, the minimum rule for edge degrees is enforced by increasing edge degrees as necessary.

A related, but simpler, approach was developed by Šolín et al. [29]. We refer to this strategy as REFSOLN_ELEM since it also begins by computing a reference solution, $u_{h/2,p+1}$, on $G_{h/2,p+1}$, but bases the refinement on elements. The basic form of the *hp*-adaptive algorithm is different than that in Fig. 10.1 for this strategy, also.

The local error estimate is given by the norm of the difference between the reference solution and the current solution,

$$\eta_i = \|u_{h/2,p+1} - u_{h,p}\|_{H^1(T_i)}$$

and the elements with the largest error estimates are refined. If T_i is selected for refinement, let $p_0 = \lfloor (p_i+1)/2 \rfloor$ and consider the following options:

- p-refine T_i to degree $p_i + 1$,
- p-refine T_i to degree $p_i + 2$,
- h-refine T_i and consider all combinations of degrees p_0, p_0+1 and p_0+2 in the children.

In all cases the minimum rule is used to determine edge degrees. In [29], quadrisection of triangles is used leading to 83 options to consider. With bisection of triangles, there are only 11 options. Also, since the object of dividing by two to get p_0 is to make the increase in degrees of freedom from h-refinement comparable to that of p-refinement, we use $p_0 = \lfloor (p_i+1)/\sqrt{2} \rfloor$ since there are only two children instead of four. Šolín et al. allow an unlimited number of hanging nodes, so they have no issue of how to assign the degrees of children that were created to maintain compatibility or one level of hanging nodes. For the newest node bisection of triangles algorithm, we assign $\lfloor (p+1)/\sqrt{2} \rfloor$ to both children of a triangle of degree p that is refined only for the sake of compatibility.

For each candidate, the standard H^1 projection $\Pi^{H^1(T_i)}_{\text{candidate}}$ of $u_{h/2,p+1}$ onto the corresponding space is performed, and the projection error in the H^1 norm, $\zeta_{\text{candidate}}$, is computed,

$$\zeta_{\text{candidate}} = \|u_{h/2,p+1} - \Pi^{H^1(T_i)}_{\text{candidate}} u_{h/2,p+1}\|_{H^1(T_i)}$$

as well as the projection error of the projection onto T_i, ζ_i.

The selection of which candidate to use is not simply the candidate with the smallest projection error [27]. Let N_i be the number of degrees of freedom in the space corresponding to T_i, and $N_{\text{candidate}}$ be the number of degrees of freedom in the space corresponding to a candidate. For simplicity, when computing N_i and $N_{\text{candidate}}$ we apply the minimum rule for edge degree ignoring the degrees of the neighbors of T_i, e.g. $N_i = (p_i + 1)(p_i + 2)/2$ regardless of what the actual edge degrees of T_i are.

Candidates with $\zeta_{\text{candidate}} > \zeta_i$ are discarded. We also discard any of the h-refined candidates for which the degrees are both greater than p_i since the reference solution is (locally) in that space. Let n be the number of remaining candidates. Compute the average and standard deviation of the base 10 logarithms of the ζ's

$$\bar{\zeta} = \frac{1}{n} \sum_{\text{candidates}} \log \zeta_{\text{candidate}}$$

$$\sigma = \sqrt{\frac{1}{n} \sum_{\text{candidates}} (\log \zeta_{\text{candidate}})^2 - \bar{\zeta}^2}.$$

Finally, to determine which candidate to use, select an above-average candidate with the steepest error decrease, i.e., from among the candidates with $\log \zeta_{\text{candidate}} < \bar{\zeta} + \sigma$ and $N_{\text{candidate}} > N_i$, select the candidate that maximizes

$$\frac{\log \zeta_i - \log \zeta_{\text{candidate}}}{N_{\text{candidate}} - N_i}.$$

Following the refinement that is indicated by the selected candidate, the minimum rule for edge degrees is applied.

10.5 Numerical Results

In this section we demonstrate the performance of the hp-adaptive strategies using two problems that are commonly used in the adaptive refinement literature. It is not the intention of this paper to compare the strategies against each other. In this paper, we merely demonstrate the ability of the strategies to achieve exponential convergence rates on a problem with a regular, but nearly irregular, solution and a problem with a point singularity.

Fig. 10.2 The solution of the wave front problem. Colors represent the function value, with blue the minimum value and red the maximum

The computations were performed on an Intel Core 2 based PC operating under the 32 bit CentOS 4 distribution of Linux with kernel 2.6.18-128.1.10.el5. Programs were compiled with Intel Fortran Version 10.1 and gcc Version 4.1.2.

Results for REFSOLN_EDGE were computed using Demkowicz's code hp2d, which was obtained from the CD in Demkowicz's book [10]. For h-refinement of triangles this code uses quadrisection with one level of hanging nodes. The maximum degree for the polynomials is 7. Results for REFSOLN_ELEM were computed using Šolín's code Hermes Version 0.99 [26]. For h-refinement of triangles this code uses quadrisection with unlimited levels of hanging nodes. The maximum degree for the polynomials is 9. Results for all other strategies were computed using PHAML Version 1.6 [17]. This code uses newest node bisection of triangles. The maximum h-refinement level was set to 53 and the maximum degree was set to 21.

To observe the convergence rates, we apply the algorithm in Fig. 10.1 with a series of tolerances, $\tau = 0.1, 0.05, 0.025, 0.01, 0.005, \ldots, 10^{-8}$. For each run we record N_{dof} and $\|e_{hp}\|_{E(\Omega)}$ for the final grid and solution. A least squares fit to the exponential form

$$\|e_{hp}\|_{E(\Omega)} = Ae^{-BN_{\text{dof}}^C}$$

is computed to determine the rate of convergence. According to (10.2.5), C is optimally $1/3$. Slightly smaller values of C still indicate exponential convergence, although not quite optimal, but very small values of C indicate that exponential convergence was not obtained.

The first test problem is Poisson's equation given in (10.2.1–10.2.2) on the unit square with the right hand sides f and g chosen so the solution is

$$u(x, y) = \tan^{-1}\left(\alpha(\sqrt{(x - x_c)^2 + (y - y_c)^2} - r_0)\right).$$

The solution has a sharp circular wave front of radius r_0 centered at (x_c, y_c) as shown in Fig. 10.2. α determines the sharpness of the wave front. For this paper we use $\alpha = 200$, $(x_c, y_c) = (-0.05, -0.05)$ and $r_0 = 0.7$. The center of the circle is taken to be slightly outside the domain because the solution has a mild singularity at the center of the circle and we want to see how the strategies handle the wave front, not the singularity. For the regularity function for the APRIORI strategy we return 3.0 if the element touches the circle on which the wave front is centered, and a very large number otherwise. This causes h-refinement with cubic elements along the wave front and p-refinement elsewhere. The choice of cubics was arbitrary.

The convergence results are shown in Figs. 10.3–10.7 where the norm of the error is plotted against the number of degrees of freedom on a log-log scale. The circles show the actual results for the sequence of values of τ, and the lines are the exponential least squares fit to that data. The curvature of the lines is indicative of the exponential rate of convergence. Higher curvature indicates a larger exponent on N_{dof}, and a straight line would indicate a polynomial rate of convergence.

Table 10.1 contains the exponents C from the exponential least squares fit. All strategies exhibit exponential rates of convergence, as none of the exponents are far from the theoretical $1/3$. The differences from $1/3$, both smaller and larger, may be due to effects such as suboptimal performance of the strategy, data points that are not in the asymptotic range of convergence, etc. Note that if the early (coarse grid, low accuracy) data points are suboptimal, this causes an increased curvature as the accuracy "catches up" in the finer grids, which can create exponents larger than $1/3$.

The second test problem is Laplace's equation, i.e. (10.2.1) with the right hand side $f = 0$, on the L-shaped domain of Fig. 10.8. The reentrant corner induces a singularity such that the exact solution, which is also shown in Fig. 10.8, in polar coordinates is

$$u(r,\theta) = r^{2/3}\sin(2\theta/3).$$

Dirichlet boundary conditions are set accordingly. The solution is known to be in $H^{1+2/3}$ in any neighborhood of the reentrant corner, so the regularity function for APRIORI returns $1+2/3$ if the element touches the reentrant corner and a very large number otherwise. This results in h-refinement with linear elements at the reentrant corner and p-refinement elsewhere.

The convergence results are shown in Figs. 10.9–10.13, and the exponents from the least squares fit are given in Table 10.2. Again, all strategies achieved exponential rates of convergence with a few of them achieving an exponent of about $1/3$ or more.

10.6 Conclusion and Future Work

Several hp-adaptive strategies have been presented in this paper. Although they were presented in the context of Poisson's equation in 2D, the strategies either apply directly to other classes of PDEs or are easily modified for other classes. Numerical results with two classic test problems demonstrate that all of the strategies can achieve exponential rates of convergence, although the $1/3$ in the theoretical $N^{1/3}$ is not always achieved.

The purpose of this paper is to summarize the proposed strategies in one source and demonstrate that exponential rates of convergence can be achieved. It would be of interest to know which, if any, of the strategies consistently outperform the

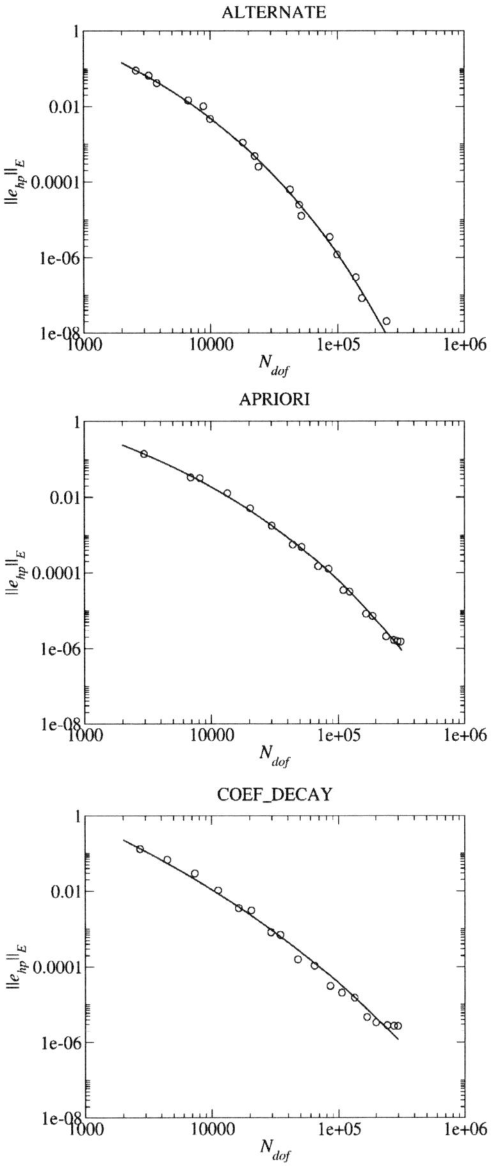

Fig. 10.3 Convergence plots for the wave front problem

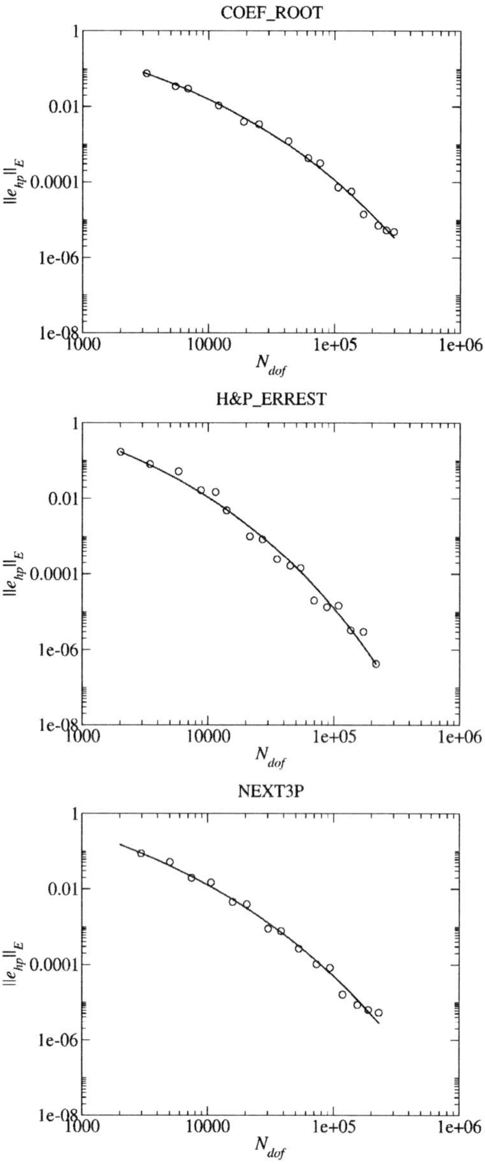

Fig. 10.4 Convergence plots for the wave front problem (continued)

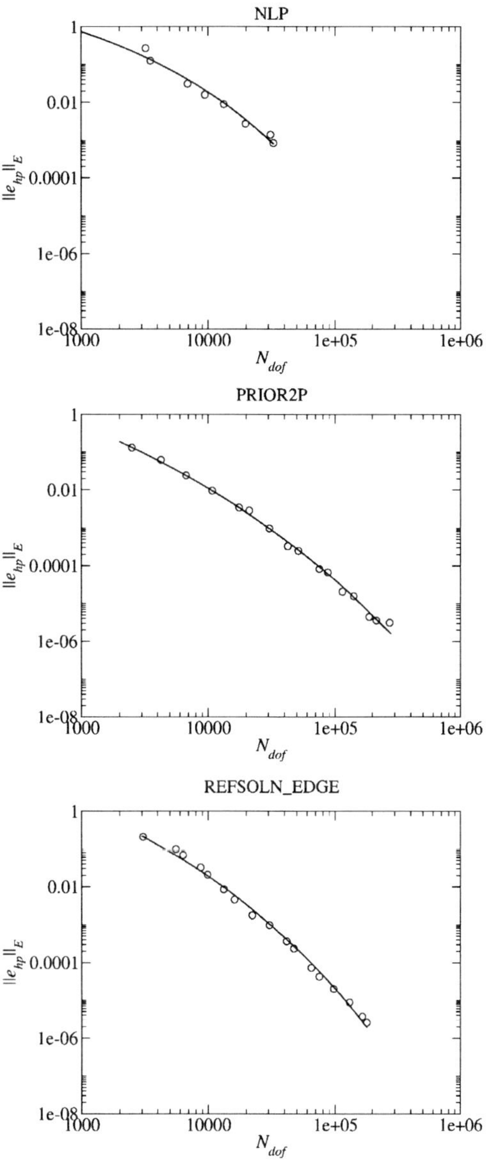

Fig. 10.5 Convergence plots for the wave front problem (continued)

Fig. 10.6 Convergence plots for the wave front problem (continued)

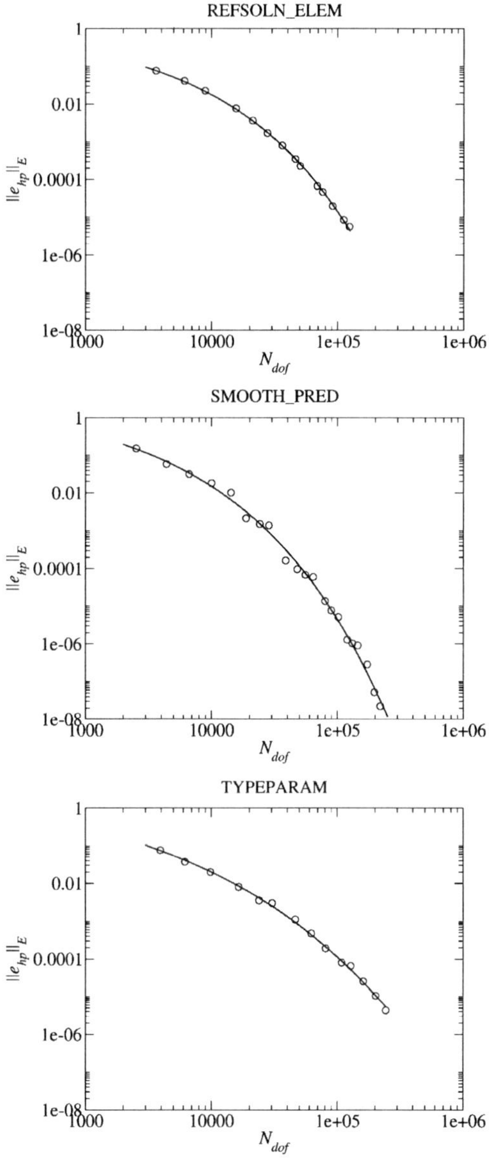

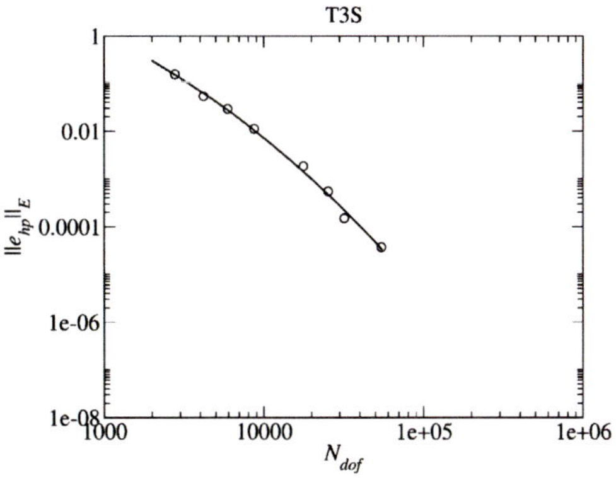

Fig. 10.7 Convergence plots for the wave front problem (continued)

Fig. 10.8 The solution of the L-domain problem

Table 10.1 Exponent on N_{dof} from the exponential least squares fit to the convergence data for the wave front problem

Strategy	Exponent C
ALTERNATE	0.27
APRIORI	0.23
COEF_DECAY	0.14
COEF_ROOT	0.25
H&P_ERREST	0.27
NEXT3P	0.23
NLP	0.30
PRIOR2P	0.16
REFSOLN_EDGE	0.21
REFSOLN_ELEM	0.44
SMOOTH_PRED	0.40
TYPEPARAM	0.28
T3S	0.18

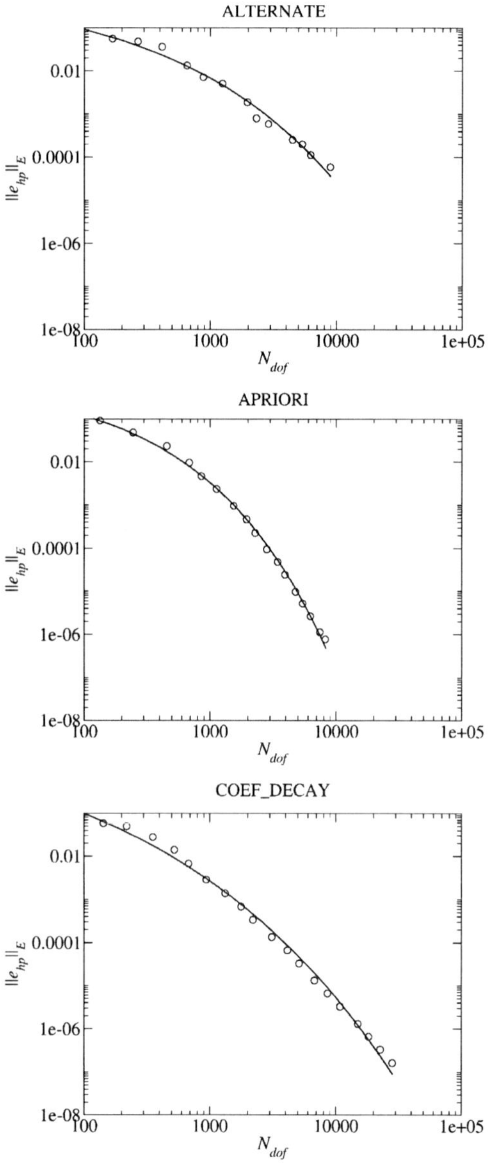

Fig. 10.9 Convergence plots for the L-domain problem

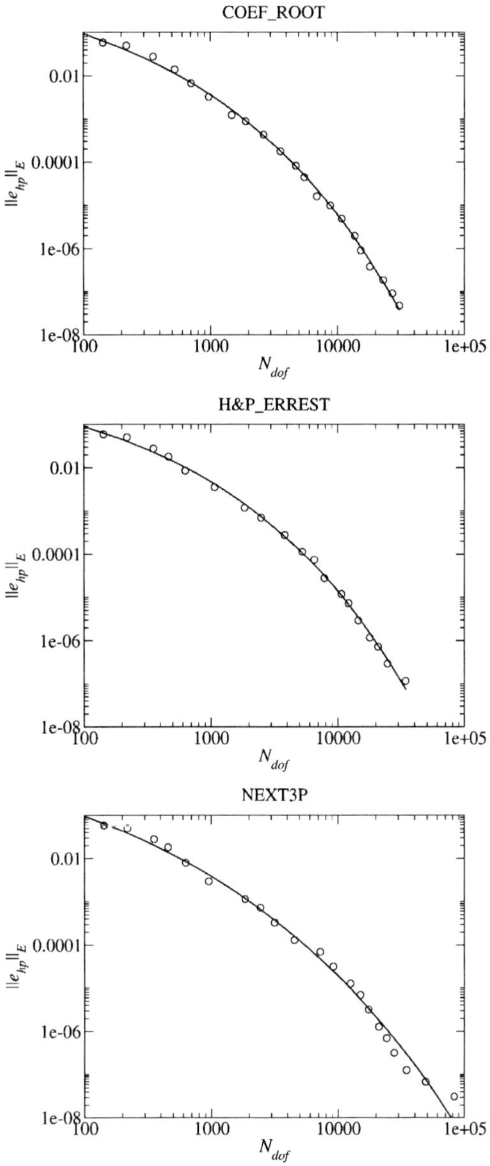

Fig. 10.10 Convergence plots for the L-domain problem (continued)

Fig. 10.11 Convergence plots for the L-domain problem (continued)

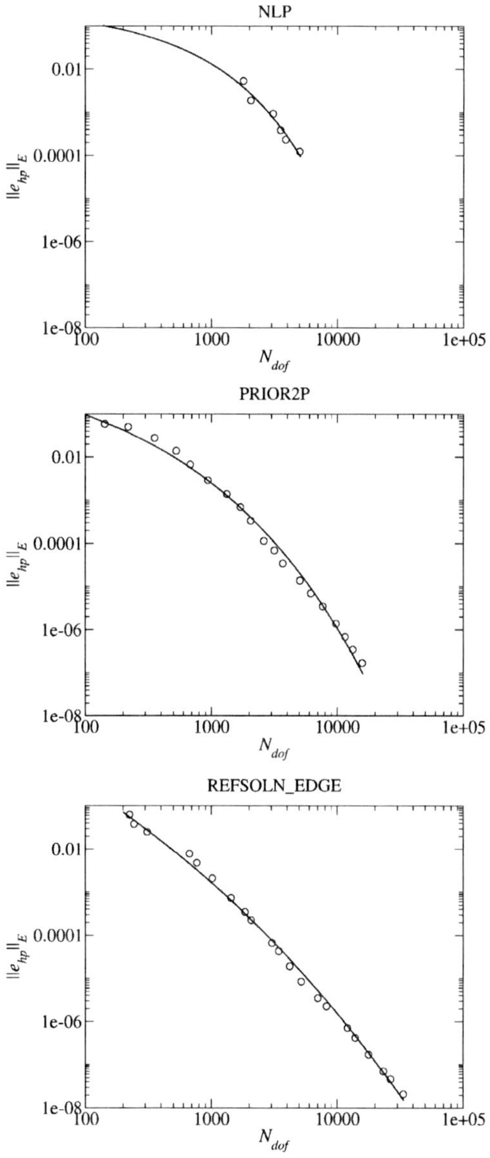

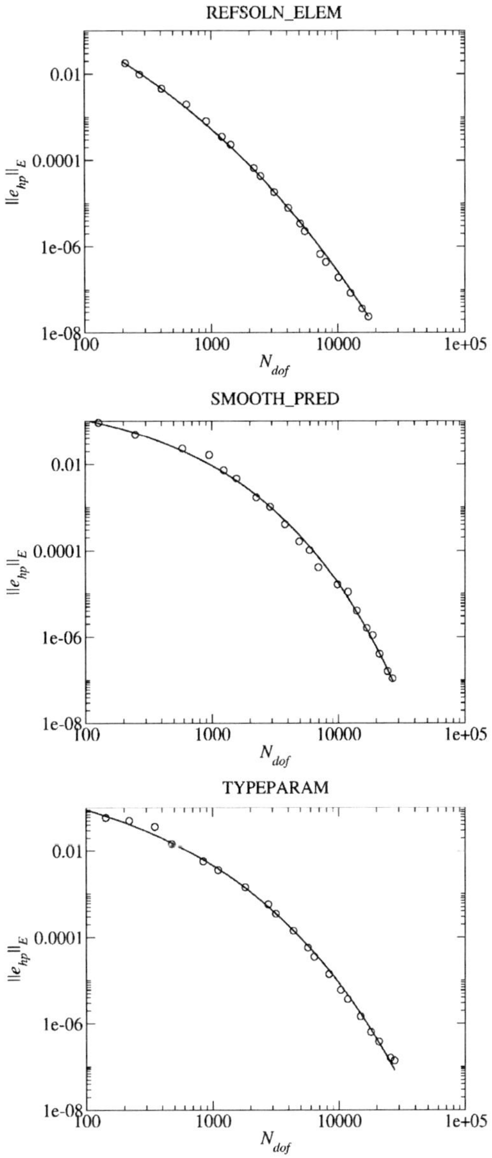

Fig. 10.12 Convergence plots for the L-domain problem (continued)

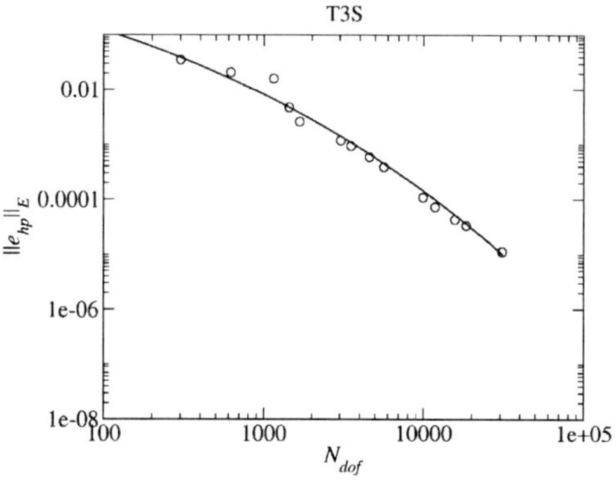

Fig. 10.13 Convergence plots for the L-domain problem (continued)

Table 10.2 Exponent on N_{dof} from the exponential least squares fit to the convergence data for the L-domain problem

Strategy	Exponent C
ALTERNATE	0.34
APRIORI	0.45
COEF_DECAY	0.23
COEF_ROOT	0.30
H&P_ERREST	0.30
NEXT3P	0.22
NLP	0.61
PRIOR2P	0.32
REFSOLN_EDGE	0.13
REFSOLN_ELEM	0.20
SMOOTH_PRED	0.41
TYPEPARAM	0.32
T3S	0.17

other strategies. Toward this end, future research involves a numerical experiment using a large collection of 2D elliptic PDEs that exhibit several types of difficulties, a uniform software base, and a consistent methodology.

References

1. Adjerid, S., Aiffa, M., Flaherty, J.E.: Computational methods for singularly perturbed systems. In: Cronin, J., O'Malley, R.E. (eds.) Singular Perturbation Concepts of Differential Equations. AMS, Providence (1998)
2. Ainsworth, M., Oden, J.T.: A Posteriori Error Estimation in Finite Element Analysis. Wiley, New York (2000)
3. Ainsworth, M., Senior, B.: An adaptive refinement strategy for h-p finite element computations. Appl. Numer. Math. **26**(1–2), 165–178 (1997)

4. Ainsworth, M., Senior, B.: *hp*-finite element procedures on non-uniform geometric meshes: adaptivity and constrained approximation. In: Bern, M.W., Flaherty, J.E., Luskin, M. (eds.) Grid Generation and Adaptive Algorithms. IMA Volumes in Mathematics and Its Applications, vol. 113, pp. 1–28. Springer, New York (1999)
5. Andreani, R., Birgin, E.G., Martnez, J.M., Schuverdt, M.L.: On augmented Lagrangian methods with general lower-level constraints. SIAM J. Optim. **18**, 1286–1309 (2007)
6. Babuška, I., Suri, M.: The h-p version of the finite element method with quasiuniform meshes. RAIRO Modél. Math. Anal. Numér. **21**, 199–238 (1987)
7. Babuška, I., Suri, M.: The p- and h-p versions of the finite element method, an overview. Comput. Methods Appl. Mech. Eng. **80**, 5–26 (1990)
8. Bey, K.S.: An *hp* adaptive discontinuous Galerkin method for hyperbolic conservation laws. Ph.D. thesis, University of Texas at Austin, Austin, TX (1994)
9. Birgin, E.G.: TANGO home page, http://www.ime.usp.br/~egbirgin/tango/
10. Demkowicz, L.: Computing with *hp*-Adaptive Finite Elements, vol. 1. One and Two Dimensional Elliptic and Maxwell Problems. Chapman & Hall/CRC, Boca Raton (2007)
11. Demkowicz, L., Rachowicz, W., Devloo, Ph.: A fully automatic *hp*-adaptivity. J. Sci. Comput. **17**, 127–155 (2002)
12. Gui, W., Babuška, I.: The h, p and h-p versions of the finite element method in 1 dimension. Part 3: The adaptive h-p version. Numer. Math. **49**, 659–683 (1986)
13. Guo, B., Babuška, I.: The h-p version of the finite element method. Part 1: The basic approximation results. Comput. Mech. **1**, 21–41 (1986)
14. Houston, P., Senior, B., Süli, E.: Sobolev regularity estimation for *hp*-adaptive finite element methods. In: Brezzi, F., Buffa, A., Corsaro, S., Murli, A. (eds.) Numerical Mathematics and Advanced Appplications, pp. 619–644. Springer, Berlin (2003)
15. Mavriplis, C.: Adaptive mesh strategies for the spectral element method. Comput. Methods Appl. Mech. Eng. **116**, 77–86 (1994)
16. Melenk, J.M., Wohlmuth, B.I.: On residual-based a-posteriori error estimation in *hp*-FEM. Adv. Comput. Math. **15**, 311–331 (2001)
17. Mitchell, W.F.: PHAML home page, http://math.nist.gov/phaml
18. Mitchell, W.F.: Adaptive refinement for arbitrary finite element spaces with hierarchical bases. J. Comput. Appl. Math. **36**, 65–78 (1991)
19. Novotny, A.A., Pereira, J.T., Fancello, E.A., de Barcellos, C.S.: A fast *hp* adaptive finite element mesh design for 2D elliptic boundary value problems. Comput. Methods Appl. Mech. Eng. **190**, 133–148 (2000)
20. Oden, J.T., Patra, A.: A parallel adaptive strategy for *hp* finite element computations. Comput. Methods Appl. Mech. Eng. **121**, 449–470 (1995)
21. Oden, J.T., Patra, A., Feng, Y.: An *hp* adaptive strategy. In: Noor, A.K. (ed.) Adaptive Multilevel and Hierarchical Computational Strategies, vol. 157, pp. 23–46. ASME Publication (1992)
22. Patra, A.: Private communication
23. Patra, A., Gupta, A.: A systematic strategy for simultaneous adaptive *hp* finite element mesh modification using nonlinear programming. Comput. Methods Appl. Mech. Eng. **190**, 3797–3818 (2001)
24. Rachowicz, W., Oden, J.T., Demkowicz, L.: Toward a universal h-p adaptive finite element strategy, Part 3. Design of h-p meshes. Comput. Methods Appl. Mech. Eng. **77**, 181–212 (1989)
25. Schmidt, A., Siebert, K.G.: A posteriori estimators for the h-p version of the finite element method in 1D. Appl. Numer. Math. **35**, 43–66 (2000)
26. Šolín, P.: Hermes home page, http://hpfem.org
27. Šolín, P.: Private communication
28. Šolín, P., Segeth, K., Doležel, I.: Higher-Order Finite Element Methods. Chapman & Hall/CRC, New York (2004)
29. Šolín, P., Červený, J., Doležel, I.: Arbitrary-level hanging nodes and automatic adaptivity in the *hp*-FEM. Math. Comput. Simul. **77**, 117–132 (2008)

30. Süli, E., Houston, P., Schwab, Ch.: *hp*-finite element methods for hyperbolic problems. In: Whiteman, J.R. (ed.) The Mathematics of Finite Elements and Applications X. MAFELAP, pp. 143–162. Elsevier, Amsterdam (2000)
31. Szabo, B., Babuška, I.: Finite Element Analysis. Wiley, New York (1991)
32. Verfürth, R.: A Review of a Posteriori Error Estimation and Adaptive Mesh-Refinement Techniques. Wiley/Teubner, Chichester/Stuttgart (1996)

Chapter 11
Vectorized Solution of ODEs in MATLAB with Control of Residual and Error

L.F. Shampine

Abstract Vectorization is very important to the efficiency of computation in the popular problem-solving environment MATLAB. Here we develop an explicit Runge–Kutta (7,8) pair of formulas that exploits vectorization. Conventional Runge–Kutta pairs control local error at the end of a step. The new method controls the extended local error at 8 points equally spaced in the span of a step. This is a byproduct of a control of the residual at these points. A new solver based on this pair, `odevr7`, not only has a very much stronger control of error than the recommended MATLAB solver `ode45`, but on standard sets of test problems, it competes well at modest tolerances and is notably more efficient at stringent tolerances.

Keywords MATLAB · Vectorization · Ordinary differential equations · Initial value problems

Mathematics Subject Classification (2000) 65L05 · 65L20

11.1 Introduction

The problem-solving environment (PSE) MATLAB [7] is in very wide use. The costs of certain computations in a PSE are quite different from the costs in general scientific computation. In particular, it is possible to reduce run times in MATLAB very substantially by using the compiled functions that are built into the PSE and by vectorizing the computation so as to exploit fast array operations. Advice about how to vectorize code in MATLAB and pointers to other documents about efficient computation are available at [15]. Vectorization is so important to efficient

L.F. Shampine (✉)
1204 Chesterton Dr., Richardson, TX 75080, USA
e-mail: lfshampine@aol.com

T.E. Simos (ed.), *Recent Advances in Computational and Applied Mathematics*,
DOI 10.1007/978-90-481-9981-5_11,

computation in MATLAB that all the programs for approximating $\int_a^b f(x)\,dx$ *require* that $f(x)$ be coded to accept a vector $[x^1, x^2, \ldots, x^k]$ and return a vector $[f(x^1), f(x^2), \ldots, f(x^k)]$. That is because the cost of evaluating $f(x)$ generally depends weakly on the number of arguments when the computation is vectorized. In [13] we considered how to exploit vectorization when solving numerically a system of n first-order ODEs

$$y' = f(t, y) \tag{11.1.1}$$

on an interval $[t_0, t_f]$ with initial value $y_0 = y(t_0)$. Following the advice of the programs provided by MATLAB for solving stiff initial value problems and boundary value problems, we assume that when the function for evaluating f is given a vector as the first argument with entries $t^1, t^2, \ldots, t^k$ and a matrix as the second argument with columns $y^1, y^2, \ldots, y^k$, it will return a matrix with columns $f(t^1, y^1), f(t^2, y^2), \ldots, f(t^k, y^k)$. With careful coding of this function it is often the case that the cost of evaluating the function with array arguments is not much more than the cost of evaluating it with a single argument. A method was developed in [13] to exploit this kind of vectorization. On standard sets of test problems, the `BV78` solver of that paper competes well with the recommended MATLAB solver `ode45` at all tolerances and is considerably more efficient at stringent tolerances.

In the present investigation we develop a method and a solver `odevr7` that also competes well with `ode45` at all tolerances and is considerably more efficient at stringent tolerances. The new solver has a very strong control of error. Conventional solvers like `ode45` and `BV78` are based on a pair of formulas. They control the size of an estimate of the local error of the lower order formula at the end of a step. They advance the integration with the higher order formula. It is assumed that the error of this formula is less than the error of the lower order formula, hence less than the specified tolerance. The new solver controls the error in the formula used to advance the integration. It controls the size of a residual at 8 points equally spaced in the span of each step. This implies a control of estimates of the extended local error at the 8 points. There is a price to pay for this strong control, but it is a modest one because we have found an analog of the First Same as Last (FSAL) technique for reducing the cost of a successful step. Also, the basic formula is slightly more accurate than the corresponding formula of the BV78 pair. We prove the surprising result that the new pair of formulas has *exactly* the same stability regions as the BV78 pair. Our goal was to develop a solver that has an exceptionally strong error control and is still competitive with a good conventional solver like `ode45`. Certainly we have achieved that with `odevr7` provided that our assumptions about vectorization are valid.

11.2 Block RK Methods

The explicit block one-step methods suggested by Milne [8] are based on implicit Runge-Kutta (RK) formulas that in the course of a step from t_n to $t_n + h = t_{n+1}$, form an accurate approximation not only at t_{n+1}, but also points equally spaced in the span of the step. The implicit formulas are derived in [13] by collocation: A polynomial $P(t)$ with $P(t_n) = y_n$ is to collocate at equally spaced points $t_{n,j} = t_n + jh/r$ for $j = 0, 1, \ldots, r$. With the notation $y_{n,j} = P(t_{n,j})$, $f_{n,j} = f(t_{n,j}, y_{n,j})$ and the definition $y_{n,0} = y_n$, the resulting formulas have the form

$$\begin{aligned}
y_{n,1} &= (y_{n,0} + hA_{1,0} f_{n,0}) + h[A_{1,1} f_{n,1} + \cdots + A_{1,r} f_{n,r}],\\
y_{n,2} &= (y_{n,0} + hA_{2,0} f_{n,0}) + h[A_{2,1} f_{n,1} + \cdots + A_{2,r} f_{n,r}],\\
&\vdots\\
y_{n,r} &= (y_{n,0} + hA_{r,0} f_{n,0}) + h[A_{r,1} f_{n,1} + \cdots + A_{r,r} f_{n,r}].
\end{aligned}$$

It is shown in [17] that the $y_{n,j}$ are all of local order $r + 2$. In particular, this is true of the approximation $y_{n,r} = y_{n+1}$ used to advance the integration, so the implicit RK method is of global order $r + 1$. It is also shown that if r is even, $y_{n,r}$ has a higher local order than at intermediate points, namely $r + 3$, so the method is of global order $r + 2$.

Following Rosser [9], we investigated explicit methods in [13] that are formed by starting with the locally second order approximations $y_{n,j}^{[1]} = y_{n,0} + (jh/r) f_{n,0}$ for $j = 1, \ldots, r$ and then making a *fixed* number of simple iterations in the implicit formulas. An iteration begins by evaluating the $f_{n,j}^{[m]} = f(t_{n,j}, y_{n,j}^{[m]})$ and then computing the $y_{n,j}^{[m+1]}$ from

$$\begin{aligned}
y_{n,1}^{[m+1]} &= (y_{n,0} + hA_{1,0} f_{n,0}) + h\big[A_{1,1} f_{n,1}^{[m]} + \cdots + A_{1,r} f_{n,r}^{[m]}\big],\\
y_{n,2}^{[m+1]} &= (y_{n,0} + hA_{2,0} f_{n,0}) + h\big[A_{2,1} f_{n,1}^{[m]} + \cdots + A_{2,r} f_{n,r}^{[m]}\big],\\
&\vdots\\
y_{n,r}^{[m+1]} &= (y_{n,0} + hA_{r,0} f_{n,0}) + h\big[A_{r,1} f_{n,1}^{[m]} + \cdots + A_{r,r} f_{n,r}^{[m]}\big].
\end{aligned}$$

Each iteration raises the local order of the approximations $y_{n,j}^{[m]}$ by one (up to a maximum order determined by the order of the underlying quadrature formulas). For the BV78 pair of [13] we took $r = 6$, so after 6 iterations, all the approximations have local order 8. When another iteration is done, the local order remains 8 at all the interior points, but the local error at the end of the step is increased to 9, resulting in a formula of global order 8. Along with the previous iterate, this results in an explicit (7,8) pair that we called BV78. As this pair is implemented in the solver `BV78`, the function $f(t, y)$ is evaluated with a single argument to obtain $f_{n,0}$. Each iteration requires evaluation of f at 6 arguments, which can be accomplished with a single array evaluation. Seven iterations are needed to reach the desired order, so each step costs 8 (array) evaluations. This is much better than the 13 evaluations of conventional (7,8) pairs. An attractive aspect of this kind of formula is that the values $y_{n,j}^{[m+1]}$ are just $P^{[m]}(t_{n,j})$ for the polynomial $P^{[m]}(t)$ that interpolates y_0 and $f_{n,j}^{[m]}$ for $j = 0, 1, \ldots, r$. Accordingly, when we reach the m that provides the desired order, we already have available a continuous extension $P^{[m]}(t)$ for "free". This contrasts with the excellent conventional (7,8) pair of Verner [16] that uses three additional evaluations to form a continuous extension of the lower order formula.

In this paper we investigate the (7,8) pair in this family that has $r = 7$. The explicit formulas are evaluated in exactly the same number of array evaluations as the BV78 pair. It is plausible that the extra collocation point would provide a more accurate formula. Indeed, a result in [17] shows that the underlying implicit formula for y_{n+1} with $r = 7$ has a significantly smaller truncation error than the formula with $r = 6$. Unfortunately, this does not appear to be the case for the corresponding explicit formulas. To investigate the matter we substituted the pair with $r = 7$ into `BV78` and compared the resulting code to `BV78` precisely as we compared `BV78` to `ode45` in [13]. The two pairs behaved almost the same, though there were some problems for which the new pair was a little more efficient.

The implicit formulas underlying both pairs are shown in [17] to be A-stable, but, of course, the explicit formulas have finite stability regions. The stability regions for the two formulas of BV78 appear as Fig. 1 in [13]. When the author computed the corresponding stability regions for the new (7,8) pair, he was surprised to find that the regions were the *same* as those of the BV78 pair! This is an interesting consequence of the form of the formulas. The stability region is obtained from the stability polynomial, which results from applying the formula to the test equation $y' = \lambda y$. Like Rosser [9], we use as initial approximation the same polynomial of degree 1 for each choice of r. Each iterate constructs a new polynomial by applying a set of quadrature formulas to the current polynomial. As long as the degree of this polynomial is no greater than the degree of precision of the quadrature formulas, the polynomial is independent of r. In the present circumstances the stability polynomials are the same, so the new explicit (7,8) pair has exactly the same stability regions as those of BV78.

The (7,8) pair with $r = 7$ costs the same number of array evaluations as the BV78 pair with $r = 6$, it has the same stability, and is only a little more efficient. Nevertheless, it has some important advantages. In this section we show how to reduce the cost of a typical step. This improves not only efficiency, but also stability. In the next section we show how to achieve a very much stronger control of error.

The (4,5) Runge–Kutta pair derived by Dormand and Prince [1] that is implemented in the recommended MATLAB ODE solver `ode45` is FSAL (First Same As Last). A pair of this kind first forms the result y_{n+1} that will be used to advance the integration. The function $f(t, y)$ is then evaluated at (t_{n+1}, y_{n+1}), and possibly other arguments. The values are used to compute the other formula in the pair and then the local error of the lower order formula is estimated by comparison. If the step is a success, and most are, the value $f(t_{n+1}, y_{n+1}) = f_{n+1,0}$ is the first value needed in the next step and so is "free" in that step. A pair might be derived so that the integration is advanced with either the lower or the higher order formula. Generally the higher order formula is used, which is known as local extrapolation, because if the error estimate is valid, the higher order formula is the more accurate. However, the BDFs that are so popular for solving stiff problems are not implemented with local extrapolation because the companion formulas have unsatisfactory stability. The Dormand/Prince pair is typical of popular explicit Runge–Kutta pairs in that the higher order formula has a better stability region, so there is quite a good case for local extrapolation. The same is true of the BV78 pair, so we do local extrapolation in the `BV78` solver. It struck the author that something analogous to FSAL was

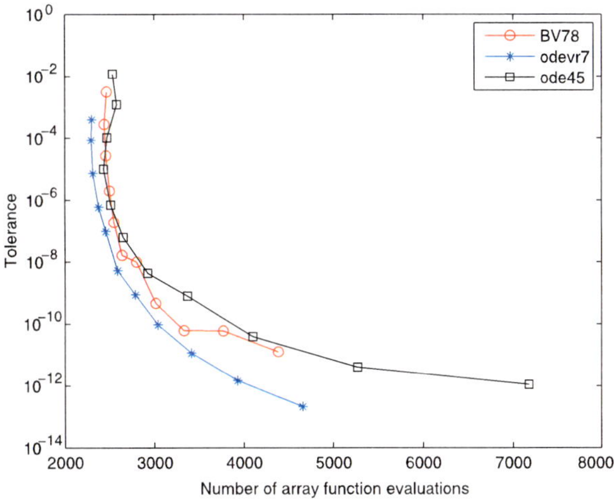

Fig. 11.1 K7 tests stability along the negative real axis

possible with the explicit block RK formulas if local extrapolation is *not* done. To be specific, we discuss this for the (7,8) pair. After forming the results $y_{n,j}$ of global order 7, we form all the $f(t_{n,j}, y_{n,j})$ in a single array evaluation to construct the result of global order 8 used for error estimation. If the step is a success, we advance the integration with $y_{n+1} = y_{n,r}$, so we have available the value $f(t_{n+1}, y_{n+1})$ that we need to start the next step. Just as with an FSAL pair like DOPRI5, this reduces the cost of a successful step by one evaluation.

We had hoped that increasing r would provide more stable explicit formulas. Although that did not prove to be the case, it is shown in [13] that the stability of the BV78 pair is quite satisfactory. Indeed, the stability regions of the DOPRI5 pair used in `ode45`, as seen in Fig. 7.4 of [1], are both uniformly smaller than the region for the block RK formula of order 7. The DOPRI5 pair is FSAL and costs only 6 evaluations per successful step. As we have just seen, something similar is true of the new (7,8) pair so that a successful step costs only 7 evaluations per step. Comparing the regions and the cost per step, it appears that the stability of the new `odevr7` is comparable to `ode45`. It is stated in [13] that the average radius of the stability region for the formula of order 8 in the BV78 pair is about 4.66 and the average radius of the region for the formula of order 7 is about 4.26. Although the `BV78` program advances the integration with the more stable formula of order 8, this costs 8 array evaluations per step. Each successful step in `odevr7` using the formula of order 7 costs only 7 evaluations. Taking into account the sizes of the stability regions and the costs per step, `odevr7` has a small theoretical advantage. In [13] we compared numerically the efficiency of `BV78` and `ode45` when applied to problem K7 of a test set of Krogh [6]. This problem appears in the test set to illuminate the behavior of a solver when stability is an issue. Figure 11.1 shows how much accuracy is achieved for a given number of function calls for all three solvers. In this no distinction is made between calling the function with one, six, or seven arguments. As we expected on theoretical grounds, the codes perform much the same, but `odevr7` is somewhat more efficient.

Our scheme for reducing the cost per step requires that we advance the integration with the lower order formula of the pair. This affects the behavior of the solver with respect to a change of tolerance. In a step from t_n to $t_n + h$, an error per step control (EPS) chooses the step size so that the local error is bounded by a given tolerance τ and an error per unit step control (EPUS) chooses it so that the local error is bounded by $h\tau$. The paper [10] works out the behavior of the global error of a one-step method with each of the four possibilities of error control and local extrapolation. An error per step control with local extrapolation results in a global error that is asymptotically proportional to the tolerance τ. This combination is found in `ode45` and `BV78`. The popular BDF codes use EPS and do not do local extrapolation. A one-step method of global order p implemented in this way has a global error proportional to $\tau^{p/(p+1)}$. Accordingly, the implementation of the new (7,8) pair in `odevr7` has a global error proportional to $\tau^{7/8}$. As with the popular BDF codes, at orders this high, the global error is sufficiently close to being proportional to the tolerance that users do not seem to be troubled by the fact that it is not quite proportional. The experiments reported in Sect. 11.4 consider the behavior of the global error as the tolerance is successively reduced by a factor of 10. At stringent tolerances, it can be seen in plots like Fig. 11.2, where observed accuracy is plotted against the number of array function evaluations, that the reduction when using `odevr7` is not as close to being an order of magnitude as it is with `BV78`.

11.3 Error Control

In this section we exploit an important difference between the two (7,8) block RK formulas to control error in a much more robust way. In Sect. 11.2 we saw that for the BV78 pair, all the approximations $y_{n,j}$ corresponding to the lower order formula have local order 8, but for the higher order formula, only the approximation at the end of the step is of local order 9. For the new pair with block size $r = 7$, all the approximations $y_{n,j}$ corresponding to the higher order formula are of local order 9. Accordingly, we can compare these approximations to those of local error 8 to estimate the local error in *all* of the intermediate approximations, not merely the one used to advance the integration. This is quite remarkable. At each step we compute 7 new approximations evenly spread throughout the step and we can estimate the local error of each in exactly the same way that the local error at the end of the step is ordinarily estimated. Not only do we get an extraordinarily robust assessment of the local error, but we also account for the fact that a solution might change substantially over the course of a step.

A more formal statement of this control of local error will be needed for subsequent developments. The local solution $u(t)$ through (t_n, y_n) is defined by

$$u'(t) = f(t, u(t)), \qquad u(t_n) = y_n. \tag{11.3.1}$$

Typical methods approximate the local solution at the end of a step to $t_n + h$, but we consider here methods that also produce a polynomial $P(t)$ that approximates $u(t)$ throughout the span of the step. How well it does this is measured by

$$\max_{t_n \le t \le t_n+h} \|P(t) - u(t)\| = \|\|P - u\|\|. \tag{11.3.2}$$

The usual local error is the quantity $P(t_n + h) - u(t_n + h)$, so at other t in the span of the step, the quantity $P(t) - u(t)$ is sometimes called the *extended local error*. When the meaning is clear, we shorten this to *local error*. The local errors of the approximate solutions $y_{n,j}$,

$$\max_{j=0,\dots,r} \|P(t_{n,j}) - u(t_{n,j})\| = \|\|P - u\|\|_r \tag{11.3.3}$$

are of particular interest for the methods we study. For our investigation of the new formula implemented in odevr7, it will be convenient to write $y_{n,j} = y_{n,j}^{[6]}$ for the locally eighth order results that are returned as approximations to $y(t_{n,j})$ and $P(t)$ for the polynomial with $P(t_{n,j}) = y_{n,j}$, $j = 0, \dots, 7$. We further write $y_{n,j}^* = y_{n,j}^{[7]}$ for the locally ninth order results that are used to estimate the local errors of the $y_{n,j}$. In this notation a conventional estimate of the local error of $y_{n,j}$ is

$$y_{n,j}^* - y_{n,j} = u(t_{n,j}) - P(t_{n,j}) + O(h^9) \quad \text{for } j = 0, \dots, 7.$$

With this we have an estimate of the maximum local error at 8 points equally spaced in the interval of interest,

$$\max_{j=0,\dots,r} \|y_{n,j}^* - y_{n,j}\| = \|\|P - u\|\|_r + O(h^9).$$

Rather than control this quantity directly, we have preferred in odevr7 to control a scaled residual. In the rest of this section we study how the two kinds of controls are related.

The residual of a smooth approximate solution $Q(t)$ is

$$R(t) = Q'(t) - f(t, Q(t)).$$

When we recognize that

$$f_{n,j}^{[6]} = f(t_{n,j}, y_{n,j}^{[6]}) = f(t_{n,j}, P(t_{n,j}))$$

and

$$f_{n,j}^{[5]} = \frac{dP^{[5]}}{dt}(t_{n,j}) = P'(t_{n,j})$$

we find that the residual of $P(t)$ at the nodes $t_{n,j}$ is

$$R(t_{n,j}) = P'(t_{n,j}) - f(t_{n,j}, P(t_{n,j})) = f_{n,j}^{[5]} - f_{n,j}^{[6]}. \tag{11.3.4}$$

From the definition of the iterates we find that for given m,

$$\begin{aligned}
y_1^{[m+1]} - y_1^{[m]} &= h\big[A_{1,1}\big(f_1^{[m]} - f_1^{[m-1]}\big) + \cdots + A_{1,r}\big(f_r^{[m]} - f_r^{[m-1]}\big)\big],\\
y_2^{[m+1]} - y_2^{[m]} &= h\big[A_{2,1}\big(f_1^{[m]} - f_1^{[m-1]}\big) + \cdots + A_{2,r}\big(f_r^{[m]} - f_r^{[m-1]}\big)\big],\\
&\vdots\\
y_r^{[m+1]} - y_r^{[m]} &= h\big[A_{r,1}\big(f_1^{[m]} - f_1^{[m-1]}\big) + \cdots + A_{r,r}\big(f_r^{[m]} - f_r^{[m-1]}\big)\big].
\end{aligned}$$

Using the new notation and (11.3.4) in these equations for $m = 6$ leads to

$$\begin{aligned} y_1^* - y_1 &= -h[A_{1,1}R(t_1) + \cdots + A_{1,r}R(t_r)], \\ y_2^* - y_2 &= -h[A_{2,1}R(t_1) + \cdots + A_{2,r}R(t_r)], \\ &\vdots \\ y_r^* - y_r &= -h[A_{r,1}R(t_1) + \cdots + A_{r,r}R(t_r)]. \end{aligned}$$

It is then straightforward to show that

$$\max_{j=0,\ldots,r} \|y_{n,j}^* - y_{n,j}\| \le h\|A\|_\infty \|\|R(t)\|\|_r. \tag{11.3.5}$$

At each step the `odevr7` program controls the size of the scaled residual at the nodes of the step, which is to say that it controls $h\|\|R(t)\|\|_r$. For the new formula, $\|A\|_\infty \approx 0.96$, so this also controls the extended local error at 8 points evenly spread throughout $[t_n, t_n + h]$.

To better understand this error control, we derive a general result relating control of (extended) local error to control of a scaled residual. This aspect of the present investigation is closely related to the work of [12, 14]. Suppose now that $P(t)$ is an approximate solution that has the correct value at the beginning of the step and satisfies the ODEs with residual $R(t)$,

$$P'(t) = f(t, P(t)) + R(t), \qquad P(t_n) = y_n.$$

Subtracting (11.3.1) satisfied by the local solution $u(t)$ from this equation for $P(t)$, integrating the difference from t_n to t, and accounting for the values at the beginning of the step leads first to

$$P(t) - u(t) = \int_{t_n}^{t} [f(x, P(x)) - f(x, u(x))]dx + \int_{t_n}^{t} R(x)dx$$

and then to

$$\|P(t) - u(t)\| \le \int_{t_n}^{t} \|f(x, P(x)) - f(x, u(x))\|dx + \int_{t_n}^{t} \|R(x)\|dx.$$

As usual in the study of numerical methods for ODEs, we suppose that $f(t, y)$ satisfies a Lipschitz condition with constant L. It then follows easily that

$$\|\|P - u\|\| \le hL\|\|P - u\|\| + h\|\|R\|\|.$$

When solving non-stiff problems, it is generally assumed that hL is rather smaller than one. To be concrete, if we assume that $hL \le 1/2$, then

$$\|\|P - u\|\| \le \frac{h}{1 - hL}\|\|R\|\| \le 2h\|\|R\|\|. \tag{11.3.6}$$

With this modest assumption on the step size we find that a control of the scaled residual $h\|\|R\|\|$ provides a control of the extended local error. We favor a residual control because it is meaningful even when the asymptotic results about accuracy that justify estimates of local error are of dubious validity. In principle we can always compute a good estimate of the size of the residual because we can evaluate $R(t)$ wherever we like.

It is useful to regard the inequality (11.3.5) for explicit block RK formulas as a discrete analog of the general result (11.3.6), but we must recognize that there are some important differences. Obviously one takes into account only $r + 1$ points in the span of the step and the other, the whole interval $[t_n, t_n + h]$. The quantity on the left side of (11.3.5) is only an (asymptotically correct) *estimate* of the extended local error at the points of interest. Unfortunately, we have no theoretical results that say the discrete norms of (11.3.5) approximate well the continuous norms of (11.3.6). Certainly it is plausible that sampling the residual at 8 points equally spaced throughout the step would provide a good estimate of the maximum value of the residual, but we have not shown that for the scheme of `odevr7`. There are methods with continuous extensions for which there are asymptotic results that provide the locations of extrema of the residual, see for example [3, 11]. With this information an asymptotically correct estimate of the maximum residual can be readily computed. Alternatively, a quadrature formula is used to obtain an asymptotically correct estimate of an integral norm of the residual for the method of [5]. Enright and Li [2] discuss estimation of the size of the residual and in particular, how an improved estimate can lead to a better performance. Schemes with asymptotically correct estimates of the size of the residual are used in [12, 14] to provide a control not only of the size of a scaled residual, but also a control of the extended local error as in (11.3.6).

In `odevr7` we control the size of a scaled residual at 8 points equally spaced in the span of each step. We have shown that this bounds asymptotically correct estimates of the extended local error at those 8 points. It is conventional to control an estimate of the local error only at the end of a step. Also, it is conventional to do local extrapolation with the consequence that the error actually incurred is smaller than the quantity being controlled only if the expected asymptotic behavior is evident. Certainly we have a far stronger control of error in `odevr7`. We have *not* proved that the 8 equally spaced samples provide an asymptotically correct estimate of the residual for the method of `odevr7`. Indeed, we do not believe that to be true. Still, with this many samples spread throughout the interval we think it reasonable to expect that `odevr7` will enjoy some of the robustness of a full control of the residual.

11.4 Illustrative Computations

In [13] we compared `BV78` and `ode45` using two standard sets of test problems [4, 6]. Here we compare these solvers to `odevr7` in the same way. Full details of the tests are reported in [13], so here we provide only an overview. We have compared the three solvers on both sets of test problems, but the results are consistent, so for brevity we report here only what happened with Krogh's set. The programs for both sets and `odevr7` itself are available at http://faculty.smu.edu/shampine/current.html. There are 10 problems in the set, but we must exclude K5 because it is a pair of second order ODEs and none of the solvers can treat such problems directly. For this reason Krogh includes in his set

Table 11.1 Run times for test set of Krogh [6]

Problem	$K1$	$K2$	$K3$	$K4\_5$	$K6$	$K7$	$K8$	$K9$	$K10$
BV78	0.2	0.5	0.5	0.6	0.6	0.9	0.8	0.9	0.5
odevr7	0.2	0.5	0.5	0.6	0.6	1.0	0.9	1.0	0.5
ode45	0.5	1.8	1.7	2.0	1.9	1.2	2.1	2.7	1.7

the problem K4, which is K5 written as a first order system. We display results for this problem in Table 11.1 under the heading "K4_5". The problems are to be solved with either pure relative or pure absolute error. The solvers were applied to each of the 9 problems of the test set with tolerances $10^{-2}, 10^{-3}, \ldots, 10^{-12}$. Tolerances were excluded from the comparison when one of the solvers had an error bigger than 1. Using reference values obtained as in [13], we compared the accuracies at 200 points equally spaced in the interval of integration. In this way we test not only the accuracy of the method, but also the accuracy of the continuous extension. Moreover, the cost of evaluating the continuous extension is included in the run times reported. As emphasized in [13], it is difficult to obtain consistent run times in MATLAB, especially when the run times are small in absolute terms, *as they are for these test problems*. To reduce the effects of this, we report run times summed over the whole range of tolerances. For each test problem, the entries of Table 11.1 were obtained in a single run of a program preceded by a "`clear all`" command. We have done this repeatedly and found the times to be consistent, though the last digit displayed might vary. Nevertheless, *the run times displayed in the table should be considered only a rough guide as to relative costs*. We summarize these results by comparing the total time required to solve this wide range of problems over a wide range of tolerances. As we found already in [13], the total run time of `ode45` is about 2.8 times that of `BV78` and we now see that it is about 2.7 times that of `odevr7`.

For both sets of test problems it was possible to vectorize the evaluation of the ODEs so that the cost depends weakly on the number of arguments. With the exception of the C5 problem of the test set [4], this was easy. In the case of the five body problem of C5, we actually tried several ways of coding the equations. Array operations and vectorization are so important to efficient computation in MATLAB that the experienced user routinely takes this into account. The programs for solving stiff initial value problems for ODEs and those for solving boundary value problems for ODEs have already demonstrated the advantages of accounting for vectorization when developing numerical algorithms. It is not always easy and it may not be possible to code evaluation of a system of ODEs so that the cost is a weak function of the number of arguments. Whether or not the ODEs can be vectorized well, the new `odevr7` solver is effective and provides an exceptionally strong control of error.

Each of our programs plots efficiency in terms of the accuracy achieved versus the number of array evaluations. The performance differs from problem to problem, but Fig. 11.2 shows what might be described as a typical plot. The problem K9 is a two body problem in elliptic motion with eccentricity 0.6. `BV78` is somewhat more

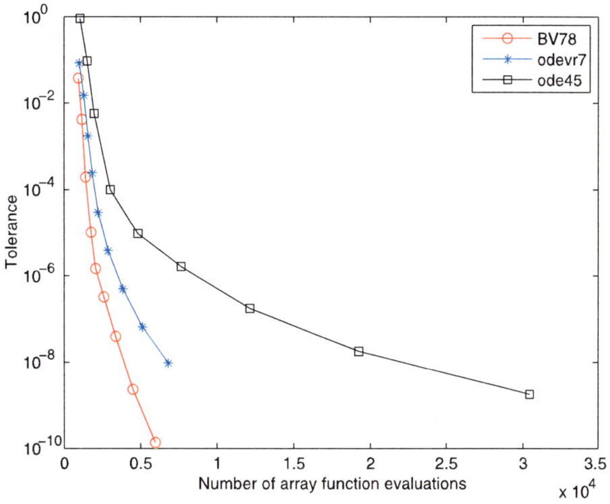

Fig. 11.2 K9 is a two body problem with eccentricity 0.6

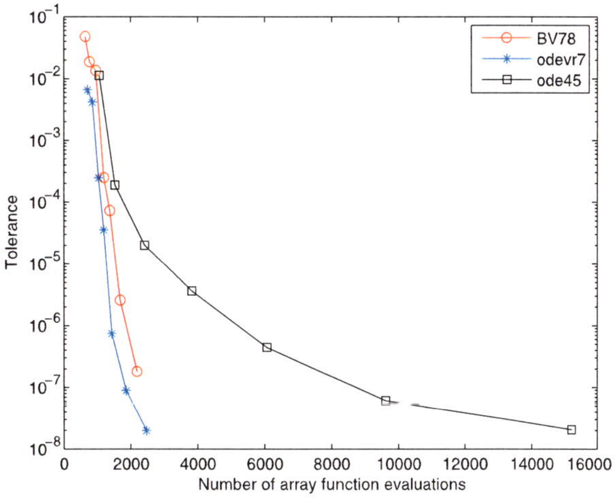

Fig. 11.3 K10 is a restricted three body problem

efficient than odevr7, especially at the most stringent tolerances where the fact that it integrates at order 8 is most important. Though BV78 is typically more efficient in this sense for Krogh's test set, odevr7 is comparably efficient for some problems and more efficient for a few. As we saw in Fig. 11.1, odevr7 is somewhat more efficient than BV78 when stiffness is an issue. The performance of the solvers on the restricted three body problem K10 is interesting. Figure 11.3 shows that odevr7 solves K10 a little more efficiently than BV78. The fact that it performs better at crude tolerances is not unusual, but the fact that it performs better at the most stringent tolerances is.

The new odevr7 has a very much stronger control of error than BV78. It is less efficient than BV78, but it is competitive because each step is cheaper and the formulas are a little more accurate. That, however, is not the important conclusion from these tests and analysis: odevr7 has a very much stronger control of error than ode45 and if the cost of evaluating the ODEs depends weakly on the number of arguments, it is comparably efficient at modest tolerances and notably more efficient at stringent tolerances.

11.5 Conclusions

We assume that evaluation of $f(t, y)$ in the ODEs (11.1.1) is vectorized and that the cost of evaluating this function with several arguments is not much greater than the cost of evaluating it with a single argument. This is a good assumption for the standard sets of test problems [4, 6]. The solver BV78 developed in [13] then competes well with the recommended MATLAB solver ode45 at all tolerances and is considerably more efficient at stringent tolerances. The same is true of the solver odevr7 developed here. Indeed, for a wide range of standard test problems solved for a wide range of tolerances, ode45 had a run time that was about 2.7 times the run time of odevr7. This is gratifying when it is appreciated that the new solver has a remarkably strong control of error. The ode45 solver has a conventional control of the size of an estimate of the local error of the formula of order 4. However, it advances the integration with the formula of order 5 (local extrapolation). This assumes that its error is smaller than the error of the formula of order 4, which it certainly will be if the expected asymptotic behavior is evident, hence the local error of the step will be smaller than the specified tolerance. The odevr7 solver advances the integration with a formula of order 7. Not only does it control the size of an estimate of the local error of this formula at the end of the step, but even the size of estimates of the extended local error at 7 other points equally spaced in the span of the step. In odevr7 this control of the extended local error is a byproduct of the control of the size of the residual of a continuous extension at 8 points equally spaced throughout the span of the step.

References

1. Dormand, J.R.: Numerical Methods for Differential Equations a Computational Approach. CRC Press, Boca Raton (1996)
2. Enright, W.H., Li, Y.: The reliability/cost trade-off for a class of ODE solvers. Numer. Alg. **53**, 239–260 (2010)
3. Higham, D.J.: Robust defect control with Runge–Kutta schemes. SIAM J. Numer. Anal. **26**, 1175–1183 (1989)
4. Hull, T.E., Enright, W.H., Fellen, B.M., Sedgwick, A.E.: Comparing numerical methods for ordinary differential equations. SIAM J. Numer. Anal. **9**, 603–637 (1972)
5. Kierzenka, J., Shampine, L.F.: A BVP solver based on residual control and the MATLAB PSE. ACM Trans. Math. Softw. **27**, 299–316 (2001)
6. Krogh, F.T.: On testing a subroutine for the numerical integration of ordinary differential equations. J. ACM **20**, 545–562 (1973)
7. Matlab: The MathWorks, Inc., 3 Apple Hill Dr., Natick, MA 01760
8. Milne, W.E.: Numerical Solution of Differential Equations. Dover, Mineola (1970)

9. Rosser, J.B.: A Runge-Kutta for all seasons. SIAM Rev. **9**, 417–452 (1967)
10. Shampine, L.F.: Local error control in codes for ordinary differential equations. Appl. Math. Comput. **3**, 189–210 (1977)
11. Shampine, L.F.: Interpolation for Runge–Kutta methods. SIAM J. Numer. Anal. **22**, 1014–1027 (1985)
12. Shampine, L.F.: Solving ODEs and DDEs with residual control. Appl. Numer. Math. **52**, 113–127 (2005)
13. Shampine, L.F.: Vectorized solution of ODEs in MATLAB. Scalable Comput.: Pract. Experience **10**, 337–345 (2010). A preprint is available at http://faculty.smu.edu/shampine/current.html
14. Shampine, L.F., Kierzenka, J.: A BVP solver that controls residual and error. J. Numer. Anal. Ind. Appl. Math. **3**, 27–41 (2008)
15. The MathWorks, Code Vectorization Guide, available at http://www.mathworks.com/support/tech-notes/1100/1109.html
16. Verner, J.H.: A 'most efficient' Runge-Kutta (13:7,8) pair. Available at http://www.math.sfu.ca/~jverner/
17. Watts, H.A., Shampine, L.F.: A-stable block implicit one step methods. BIT **12**, 252–256 (1972)

Chapter 12
Forecasting Equations in Complex-Quaternionic Setting

W. Sprössig

Abstract We consider classes of fluid flow problems under given initial value and boundary value conditions on the sphere and on ball shells in $\mathbb{R}^3$. Our attention is focused to the forecasting equations and the deduction of a suitable quaternionic operator calculus.

Keywords Forecasting equations · Quaternionic operator calculus · Toroidal flows

Mathematics Subject Classification (2000) Primary 30G35 · Secondary 35G15

12.1 Introduction

Hamilton's discovery of the skew-field of quaternions was made in 1843. He found the first division ring and opened the possibility of its algebraic and geometric use. Almost 90 years later, the Romanian mathematicians G.C. Moisil and N. Teodorescu [17] from Cluj University as well the group around the Swiss mathematician R. Fueter from the Federal Institute of Technology Zurich started with a quaternionic analysis and Clifford analysis, which is an analogue of complex function theory in higher dimensions. The breakthrough came with the book "Clifford analysis" by the Belgian research group around R. Delanghe at Ghent University. In the following years, right and left analogues of complex analysis were developed. In this connection, independent work of H.R. Malonek and F. Sommen are significant. They both discussed possible equivalence of the three different approaches to real

W. Sprössig (✉)
Fakultaet fuer Mathematik und Informatik, TU Bergakademie Freiberg, Prueferstraße 9, 09596 Freiberg, Germany
e-mail: sproessig@math.tu-freiberg.de

T.E. Simos (ed.), *Recent Advances in Computational and Applied Mathematics*,
DOI 10.1007/978-90-481-9981-5_12, © Springer Science+Business Media B.V. 2011

Clifford analysis (series expansions, differentiability, generalised Cauchy-Riemann equations). In particular it means that this type of "hypercomplex" analysis has the same power like the classical complex analysis in the plane.

Complex methods play an important role for the treatment of boundary value problems. In our books [9, 10] and [12], we developed a new strategy for solving linear and non-linear boundary and initial-value boundary problems of partial differential equations of mathematical physics using quaternionic analysis. It was found that stationary problems are related to the algebra of real quaternions and initial-boundary value problems are associated to the algebra of complex quaternions, which is isomorphic (in the sense of associative algebras) with the famous Pauli algebra. It was necessary to develop a special quaternionic operator calculus with three basic operators: Dirac type operator or Cauchy-Riemann operator, Teodorescu transform and Cauchy-Fueter operator. In the complex plane, these operators correspond to the Cauchy-Riemann equations, the T-operator and the Cauchy operator. Null solutions of the Dirac type operator are called holomorphic or monogenic. Boundary conditions are handled using Plemelj type formulae relative to the Cauchy-Fueter operator. A good understanding of initial value problems requires a change in the basic operator trinity. In this way, the Dirac operator with zero mass is replaced by the Dirac operator with mass. The kernel of the Teodorescu transform changes from the Cauchy kernel to a kernel generated by the MacDonald function. The Cauchy-Fueter operator has the same kernel. As a result, a new operator calculus satisfying again a formula of Borel–Pompeiu type was developed.

In 1989, we introduced a Hodge-Bergman decomposition for the quaternionic Hilbert space in our first book [9]. Such a decomposition separates the kernel of the Dirac type operator. We were able to describe explicitly its orthogonal complement space. The space is just the image of the operator adjoint to the Dirac type operator on the quaternionic Sobolev space $\overset{\circ}{W}{}_1^2$ with zero trace on the boundary. The corresponding orthogonal projections onto these subspaces are called the Bergman projection and the Pompeiu projection, respectively. This basic principle has been generalised for forms on differentiable manifolds (cf. [20], 1995). Both orthogonal projections can be described with help of the generating operator trinity and an isomorphism between two subspaces of suitable Sobolev-Slobodetzkij spaces over the boundary of the domain. The first one is the space of all functions holomorphically extentable to the domain, the other is the space of all functions holomorphically extentable to the exterior of the domain, vanishing at infinity.

Finally, the quaternionic factorisation of the Klein-Gordon operator into two Dirac operators with masses plays the key role. The relation $\Delta = -D^2$ was found already by Paul Dirac.

In this paper, we consider classes of fluid flow problems on the sphere and in ball shells with given initial value and boundary value conditions. We focus our attention to the corresponding Navier-Stokes equations and its linearisations—the so called forecasting equations. Shallow water equations are rather similar to these set of equations, we shall discuss them as well. The physical background of such type of equations is described in the book "Turbulence in fluids" by M. Lesnieur [14]. For a better understanding, we will give a brief introduction to the corresponding physical problems. The main aim of the article is to construct quaternionic operator calculus tailored for the above mentioned applications.

12.2 Forecasting Equations—A Physical Description

Taking into account the Earth's rotation, Newton's second law reads as follows:

Let ν be the *dynamic viscosity* and a the *angular velocity* of the rotating frame of reference, we then have

$$\frac{Du}{Dt} = -\frac{1}{\rho}\nabla p + \nu\Delta u + f - 2a \wedge u + \phi \quad \text{(Newton's second law)}, \tag{12.2.1}$$

where the left hand side describes the *inertial acceleration* of a fluid element with respect to all forces acting on it. The notation Du/Dt expresses the Lagrangian rate of change of the radial velocity u of an infinitesimal box of fluid. The *flow vector* u is considered relatively to the center of the rotating reference frame. Furthermore, f stands for the outer forces and the term $2a \wedge u$ is the *Coriolis force*. We set

$$\phi := g - a \times (a \times x), \tag{12.2.2}$$

where g is *Newtonian gravity* and $a \times (a \times x)$ the *centrifugal force* (cf. [18]). We denote by x the position vector in a frame which rotates with the Earth, its origin is in the center of the Earth.

There are the following relations between temperature T, pressure p and density ρ. At first we have

$$c_R = \frac{p}{T\rho}, \tag{12.2.3}$$

where c_R denotes the *universal gas constant per unit mass*. Further the total time derivative of the temperature is the sum of the local rate of change of the temperature and the advection term of T.

$$\frac{DT}{Dt} = \partial_t T + (u \cdot \nabla)T. \tag{12.2.4}$$

The left hand side depends on Q, p, ρ, c_V. Here Q is the *a quantity of heat* and c_V the *specific heat at constant volume* V. In a fixed volume the change of mass is measured by inflow and outflow. Gauss' law yields

$$\partial_t \rho = -\nabla \cdot (u\rho) \tag{12.2.5}$$

or in another formulation

$$\frac{D\rho}{Dt} + \rho\nabla \cdot u = 0. \tag{12.2.6}$$

Hence

$$\frac{D\rho}{Dt} = 0 \quad \Leftrightarrow \quad \nabla \cdot u = 0 \quad \text{(incompressibility condition).} \tag{12.2.7}$$

Summing up we obtain by substitution of Lagrange's derivatives

$$\partial_t u = \nu \Delta u - (u \cdot \nabla)u - 2a \wedge u - \frac{1}{\rho}\nabla p - \phi + f, \tag{12.2.8}$$

$$c_V \partial_t T = -c_V (u \cdot \nabla)T - \frac{p}{\rho}(\nabla \cdot u) + Q, \tag{12.2.9}$$

$$\partial_t \rho = -(u \cdot \nabla)\rho - \rho(\nabla \cdot u), \tag{12.2.10}$$

$$p = \rho c_R T \tag{12.2.11}$$

and initial value and boundary value conditions. It should be noted that the quantities Q, c_V, c_R consider also the water particles in the air.

If density and heating changes can be neglected, i.e.,

$$\frac{D\rho}{Dt} = 0, \tag{12.2.12}$$

we get the following simplified model of the simplified forecasting equations (SFE) in a ball-shell.

Let G_{12} be a ball-shell with an inner boundary Γ_1 and an outer boundary Γ_2. Then we have

$$\partial_t u = \nu \Delta u - (v(u) \cdot \nabla)u - \frac{1}{\rho}\nabla p - 2a \wedge u + F \quad \text{in } G_{12}, \tag{12.2.13}$$

$$\nabla \cdot u = 0 \quad \text{in } G_{12}, \tag{12.2.14}$$

$$\rho = \text{const}, \tag{12.2.15}$$

$$u(t,x) = g(t,x) \quad \text{on } \Gamma_1 \cup \Gamma_2, \tag{12.2.16}$$

$$u(0,x) = u_0(x) \quad \text{in } G_{12}. \tag{12.2.17}$$

We have fixed $F := f - \phi$, where ϕ is the *apparent gravity*. A suitable reference for such physical interpretation of the system is the book [14]. In particular, it is also explained there, why the assumption of incompressibility is physical realistic. The argument goes as follows: One has to compare the anticipated velocity with the sound speed. In case of atmospherical flows, we have, as a rule, velocities smaller than the speed of sound and so the condition of incompressibility makes sense.

Remark 12.1 The key role plays the *advection term* $(v(u) \cdot \nabla)u$. For $v = \text{const}$ we have the (linear) Oseen-type of the SFE. For $v = 0$ we obtain the linear Stokes-type of the SFE) and for $v = u$ the non-linear Navier-Stokes type SFE arise.

12.3 Toroidal Flows on the Sphere

12.3.1 Tangential Flows on the Sphere

The height of the Earth's atmosphere is only about ten kilometers. When we compare this with the diameter of a low pressure area of thousand kilometers and more, it is justified to reduce the ball shell to a sphere. Then we allow only surface curl divergence-free flows in tangential directions. This is a strong restriction, only horicontal movements are allowed. Unfortunately, hurricans are excluded.

Let be Ω a domain on the sphere with the sufficient smooth boundary C. We then have the equation (cf. [7])

$$\partial_t u + (v(u)\cdot\nabla_S)u = \nu\Delta_S u - \frac{1}{\rho}\nabla_S p - 2a\wedge u + F \quad \text{in } \Omega \tag{12.3.1}$$

with the *vector derivative* ∇_S, the *Beltrami operator* Δ_S, the *surface gradient* $\nabla_S p$ and the *surface divergence* $\nabla_S \cdot u$. The vector of outer forces F also includes the so-called *apparent gravity*, which means that gravity is reduced by the centrifugal force. Moreover it is assumed that

$$\nabla_S \cdot u = 0 \tag{12.3.2}$$

with Dirichlet boundary conditions on $\partial\Omega =: C$.

Remark 12.2 We note that a velocity field is called *toroidal*, if it is tangential and surface-divergence-free

Note There is a similarity to shallow water (or Barré de Saint-Vainant) equations, which can be described in a normalised form as follows:

$$\partial_t u + (v(u)\cdot\nabla_S)u = -2a\wedge u + \nu\Delta_S, u - g\nabla_S h, \tag{12.3.3}$$

$$\partial_t H = -(u\cdot\nabla_S)H + H\nabla_S\cdot u, \tag{12.3.4}$$

$$H := h(t,x) - h_G(t,x), \tag{12.3.5}$$

$$u(0,x) = u_0, \tag{12.3.6}$$

$$h(0,x) = h_0, \tag{12.3.7}$$

where H is *total depth* of the fluid, h_G describes the *ground profil* and h is the *surface function* of the fluid. The reader may compare it with [15].

12.3.2 Quaternionic Algebras and Functional Spaces

Let $\mathbb{H}$ be the algebra of real quaternions and $a \in \mathbb{H}$, then $a = \sum_{k=0}^{3}\alpha_k e_k$. α_0 is called the *scalar part* and denoted by $\mathrm{Sc}\,a$. Further let $e_k^2 = -e_0 = -1$; $e_1 e_2 =$

$-e_2e_1 = e_3$, $e_2e_3 = -e_3e_2 = e_1$, $e_3e_1 = -e_1e_3 = e_2$. Natural operations of addition and multiplication in $\mathbb{H}$ turn $\mathbb{H}$ into a skew-field. Quaternionic conjugation is given by

$$\overline{e_0} = e_0, \quad \overline{e_k} = -e_k \quad (k = 1, 2, 3),$$
$$\overline{a} = a_0 - \sum_{k=1}^{3} \alpha_k e_k =: \alpha_0 - \mathbf{a}. \tag{12.3.8}$$

Further relations are

$$\overline{a}a = a\overline{a} = |a|^2_{\mathbb{R}^4} =: |a|^2_{\mathbb{H}}, \tag{12.3.9}$$

$$a^{-1} := \frac{1}{|a|^2}\overline{a}, \quad \overline{ab} = \overline{b}\,\overline{a}. \tag{12.3.10}$$

Remark 12.3 Quaternions as structure were discovered by Sir R.W. Hamilton in 1843 [13]. Already 100 years earlier L. Euler used such units in his theory of kinematics [1].

We denote by $\mathbb{H}(\mathbb{C})$ the set of quaternions with complex coefficients, i.e.

$$a = \sum_{k=0}^{3} \alpha_k e_k \quad (\alpha_k \in \mathbb{C}). \tag{12.3.11}$$

For $k = 0, 1, 2, 3$ we have the commutator relation $ie_k = e_k i$. Any complex quaternion a has the decomposition $a = a^1 + ia^2$ ($a^j \in \mathbb{H}$), leading to notation $\mathbb{C}\mathbb{H}$. We have three possible conjugations:

1. $\overline{a}^{\mathbb{C}} := a^1 - ia^2$,
2. $\overline{a}^{\mathbb{H}} := \overline{a}^1 + i\overline{a}^2$,
3. $\overline{a}^{\mathbb{C}\mathbb{H}} := \overline{a}^1 - i\overline{a}^2$.

We assume that we have a sufficiently smooth bounded domain G with the boundary Γ in $\mathbb{R}^3$ or in a domain Ω with the smooth boundary curve C on the sphere S^2. Function spaces of quaternionic valued functions are defined componentwise. We will use in our paper Hölder spaces, C^∞, subspaces of some kind of quaternionic holomorphic (monogenic) functions and quaternionic Sobolev spaces as well as their trace spaces on the corresponding boundary.

12.3.3 Tangential Derivatives

We have to work with Sobolev spaces on smooth 2-dimensional manifolds M in $\mathbb{R}^3$, so-called *hypersurfaces*. We denote by $B_r(x)$ the ball with the radius r around x. Let h be a function defined on $M \cap B_r(x)$ and let H be a smooth extension into $B_r(x) \subset \mathbb{R}^3$. Furthermore, let P_x be the orthogonal projection onto the tangent space

T_xM of the manifold M at a point x. Then the vector derivative ∇_M is defined at the point x by

$$(\nabla_M h)(x) = \sum_{i=1}^{3} P_x e_i \partial_i H. \qquad (12.3.12)$$

If $y \in \mathbb{R}^3$, then P_x can be computed as $P_x y = y - n_x(n_x \cdot y) \in T_xM$, because of $n_x \cdot P_x y = 0$. Using $n_x \times n_y = n_x \wedge n_y$ and the double cross product we have

$$P_x y = -n_x \wedge (n_x \wedge y). \qquad (12.3.13)$$

A good reference is the book by J. Cnops [4]. Now let $D = \sum_{i=1}^{3} e_i \partial_i$ be the massless Dirac operator, then the vector derivative is given by

$$\begin{aligned} \nabla_M &= P_x(D) = D - n_x(n_x \cdot D) \\ &= \sum_{j=1}^{3} [\partial_j - n_j \partial_j] e_j = \sum_{j=1}^{3} \mathcal{D}_j e_j. \end{aligned} \qquad (12.3.14)$$

In this case the vector derivative ∇_M is called *Günter's gradient* and $\mathcal{D}_j$ *Günter's partial derivatives.*

Note Günter's derivatives were studied for the first time by N. Günter [8] in 1953.

There are close relations to Stokes derivatives and the Gaussian curvature. Indeed, we have

$$\nabla_M = -n_x(n_x \times D) = -n_x \sum_{i<j} e_{ij}(n_i \partial_j - n_j \partial_i) = n_x \sum_{i<j} e_{ij} \mathcal{M}_{ij}. \qquad (12.3.15)$$

The derivatives $\mathcal{M}_{ij}$ are called *Stokes derivatives*. Günter's partial derivatives can be expressed by Stokes partial derivatives

$$\mathcal{D}_j = \sum_{k=1}^{3} n_k \mathcal{M}_{kj}, \qquad \sum_{k=1}^{3} n_k \mathbf{D}_k = 0. \qquad (12.3.16)$$

Moreover, we have the relation

$$\sum_{k=1}^{3} \mathcal{D}_k n_k = G, \qquad (12.3.17)$$

where G is the *Gaussian curvature* (cf. [6, 16]).

Now we return to the sphere. The vector derivative on the sphere is given by

$$P_x(D) = \nabla_S = -\omega(\omega \wedge D) = -\omega \sum_{i<j} e_i e_j (\omega_i \partial_j - \omega_j \partial_i), \qquad (12.3.18)$$

with ($\omega^2 = -1$). The operators $\omega_i \partial_j - \omega_j \partial_i$ are called *angular momentum operators*. The *spherical Dirac operator* is given by

$$\Gamma_S = \sum_{i<j} e_i e_j (\omega_i \partial_j - \omega_j \partial_i) \tag{12.3.19}$$

and connected to the vector derivative as follows:

$$\nabla_S = -\omega \Gamma_S \quad \text{and} \quad \omega \nabla_S = \Gamma_S. \tag{12.3.20}$$

$\Gamma_S \cdot u$ is called *surface-curl-divergence*. It is well-known that the Laplacian permits the factorisation

$$\Delta = \partial_r^2 + \frac{2}{r}\partial_r + \frac{1}{r^2}\Delta_S = \nabla \cdot \nabla = \sum_{i=1}^{3} \partial_i^2. \tag{12.3.21}$$

R. Duduchava [6] proved that a similar decomposition is also valid for the Beltrami operator using Günter's derivatives. We have

$$\Delta_S = \Gamma_S(-2 - \Gamma_S) = \nabla_S \cdot \nabla_S = \sum_{i=1}^{3} \mathcal{D}_i^2. \tag{12.3.22}$$

12.4 Oseen's Problem on the Sphere

In this section we deal with the Oseen linearisation of the simplified forecasting equations. We intent to study these equations on the sphere and in a ball-shell. For this reason a corresponding quaternionic operator calculus will be introduced. Corresponding versions of a Dirac operator, a Teodorescu transform and a Cauchy-Fueter type operator will play a key role again. But also formulae of Plemelj type and a Bergman-Hodge decomposition are necessary. The best introduction to these topics can be found in the PhD thesis of P. Van Lancker (Ghent) [23].

12.4.1 Discretized Oseen's Problem on the Sphere

At least there are three different methods to involve time-derivatives in a quaternionic operator system:

(i) The quaternionic basis is extended by addition of two further formal algebraic elements. Then we get a so called *Witt basis*. In this basis a quaternionic operator trinity can be introduced. A corresponding Borel-Pompeiu formula keeps valid. This approach is proposed in the papers [3] and [2].

(ii) A time-discretisation leads to a new quaternionic operator trinity. We now study a Dirac operator with a mass and integral operators with more general kernels. This approach was used for the first time in our paper [11].

(iii) A quaternionic method of harmonic extension is worked out. This is based on the use of operator exponentials of the Dirac operator and the Laplacian. First results in this direction can be found in [21].

Here we follow the second method and we construct a time-discretisation. Consider $T > 0, \subset S$. Oseen's problem is defined in $[0, T] \times \overline{\Omega}$ with the known vector function v. In the sequel we shall work with the quaternionic formulation, i.e. we have shall identify the 3-dimensional vector u with the quaternion $(0, u)$. In our sense this is completely unspectacular.

$$\partial_t u - \nu \Delta_S u - \frac{1}{\rho} \nabla_S p = \mathcal{F} \quad \text{in } \Omega, \tag{12.4.1}$$

$$\nabla_S \cdot u = 0 \quad \text{in } \Omega, \tag{12.4.2}$$

$$u(0, \cdot) = u_0 \quad \text{in } \Omega, \tag{12.4.3}$$

$$u = g \quad \text{on } \partial\Omega = C, \tag{12.4.4}$$

where $\mathcal{F} = \mathcal{F}(u, a, v) := -2a \wedge u + (v \cdot \nabla_S)u + F$.

Now let us consider $T = n\tau$, with the meshwidth τ. We set $u_k := u(k\tau, \cdot)$, $p_k := p(k\tau, \cdot) g_k = g(k\tau, \cdot)$ for $0 \leq k \leq m$.

We take the forward differences: $\partial_t u \approx (u_{k+1} - u_k)/\tau$ in $(k\tau, x)$, introducing the so-called *kinematic viscosity* $\eta := \nu\rho$. A division by ν leads to the quaternionic formulation of Oseen's equation

$$\frac{1}{\nu\tau} u_{k+1} - \Gamma_S(-\Gamma_S - 2) u_{k+1} + \frac{1}{\eta} \nabla_S p_{k+1} = \mathcal{F}_k = \frac{1}{\nu} \mathcal{F}(u_k, \ldots) + \frac{u_k}{\nu}. \tag{12.4.5}$$

By symmetric factorisation we obtain the decomposition

$$(\Gamma_S + \alpha_+)(\Gamma_S + \alpha_-) u_{k+1} + \frac{1}{\eta} \nabla_S p_{k+1} = \mathcal{F}_k. \tag{12.4.6}$$

Using Günters gradient and taking into account $\omega^2 = -1$ (quaternionic multiplication!) we find

$$\omega(\Gamma_S - 2 - \alpha_+)\omega(\Gamma + \alpha_+) u_{k+1} + \frac{1}{\eta} \nabla_S p_{k+1} = \mathcal{F}_k =: \mathcal{F}_k(u_k), \tag{12.4.7}$$

hence

$$D_{-2-\alpha_+} D_{\alpha_+} u_{k+1} + \frac{1}{\eta} D_0 p_{k+1}, = \mathcal{F}_k, \tag{12.4.8}$$

where

$$\alpha_- = -2 - \alpha_+ = \beta \quad \text{and} \quad \alpha_+ = \alpha = -1 + i\sqrt{-1 + \frac{1}{\nu\tau}},$$

$$\beta := -2 - \alpha, \tag{12.4.9}$$

$$D_\beta D_{\overline{\alpha}} u_{k+1} + \frac{1}{\eta} D_0 p_{k+1} = \mathcal{F}_k. \tag{12.4.10}$$

12.4.2 Quaternionic Operator Calculus on the Sphere

We introduce the following operators (cf. [23]): $\Gamma_S + \alpha$, $\alpha \in \mathbb{C} \setminus (\mathbb{N} \cup -\mathbb{N})$, $\alpha \neq 0$.

$$D_\alpha := \omega(\Gamma_S + \alpha) \quad \text{(Günter's gradient)}, \tag{12.4.11}$$

$$T_\alpha := -\int_\Omega E_\alpha(\omega, \xi) \cdot dS(\omega) \quad \text{(Teodorescu transform)}, \tag{12.4.12}$$

$$F_{C,\alpha} := -\int_{-C} E_\alpha(\omega, \xi) n(\omega) \cdot dC(\omega) \quad \text{(Cauchy-Fueter type operator)}. \tag{12.4.13}$$

A corresponding Borel-Pompeiu formula is given by

$$F_{C,\alpha} u + T_\alpha D_\alpha u = \begin{cases} u & \text{in } \Omega, \\ 0 & \text{in } S \setminus \overline{\Omega}. \end{cases} \tag{12.4.14}$$

The result can be found in [23]. Let $\alpha \in \mathbb{C} \setminus \{\mathbb{N} \cup \{-2 - \mathbb{N}\}\}$. Then

$$E_\alpha(\omega, \xi) = \frac{\pi}{\sigma_3 \sin \pi \alpha} K_\alpha(-\xi, \omega)\omega, \tag{12.4.15}$$

where σ_3 is the surface area of the unit sphere. Further we have

$$\omega \cdot \xi = -\sum_{i=1}^{3} \xi_i \omega_i, \tag{12.4.16}$$

$$K_\alpha(-\xi, \omega) = C_\alpha^{3/2}(\omega \cdot \xi) - \xi \omega C_{\alpha-1}^{3/2}(\omega \cdot \xi), \tag{12.4.17}$$

with the *Gegenbauer polynomials* $C_\alpha^\mu(t)$.

Using *Kummer's function* ${}_2F_1(a, b; c; z)$ we get the representation:

$$C_\alpha^{3/2}(z) = \frac{\Gamma(\alpha+3)}{\Gamma(\alpha+1)} \frac{1}{4} \, {}_2F_1\left(-\alpha, \alpha+3; 2; \frac{1-z}{2}\right),$$
$$z \in \mathbb{C} \setminus \{-\infty, 1\}. \tag{12.4.18}$$

Kummer's function is for $|z| < 1$ defined by:

$${}_2F_1(a, b; c; z) := \sum_{k=0}^{\infty} \frac{(a)_k (b)_k}{(c)_k} \frac{z^k}{k!}, \quad (a)_k = \frac{\Gamma(a+k)}{\Gamma(a)}. \tag{12.4.19}$$

Solutions of $D_\alpha u = 0$ in Ω are called *inner spherical holomorphic (monogenic) functions of order* α in Ω. We have

$$D_\alpha E_\alpha(\omega, \xi) = \delta(\omega - \xi). \tag{12.4.20}$$

Good references to this topic are the works [23] and [5].

12.4.3 Plemelj Decompositions on the Boundary of Spherical Domains

Now we consider the Hilbert module $L_2(\partial\Omega) =: L_2(C)$ of all square integrable complex-quaternionic valued functions defined on C, which is a C^∞-Liapunov curve. There are two possibilities to introduce an inner product

$$(u, v)_C = \int_C \overline{u} v dS \in \mathbb{CH}. \tag{12.4.21}$$

Here dS denotes the Lebesgue measure on C. Another complex valued inner product can be obtained by putting

$$[u, v]_C = \mathrm{Sc}(u, v)_C. \tag{12.4.22}$$

The latter definition leads to a norm and therefore to a quaternionic Hilbert space.

Let u be a quaternion valued C^∞-function on C. We already know that $F_{C,\alpha}u$ belongs to the kernel of the operator $D_\alpha = \omega(\Gamma_S + \alpha)$. We introduce a singular integral operator of Fueter-Bitzadse type

$$(S_{C,\alpha}u)(\xi) := 2 \lim_{\varepsilon \to 0} \int_{C \setminus B_\varepsilon(\xi)} E_\alpha(\omega, \xi) n(\omega) u(\omega) dS(\omega) \tag{12.4.23}$$

$$= 2 v.p. \int_C E_\alpha(\omega, \xi) n(\omega) u(\omega) dS(\omega). \tag{12.4.24}$$

The vector-valued quaternion $n(\omega)$ is orthogonal to the curve C in the point $\omega \in S^2$ and belongs to the 2-dimensional tangent space $T_\omega S^2$ as well as to the 1-dimensional tangent space $T_\omega C$.

Using ideas in [19] and [23] one gets $S^2_{C,\alpha} = I$. Let $\Omega^+ := \Omega$, $\Omega^- := \mathrm{co}\,\Omega$. Applying the general trace operator as non-tangential limit on the sphere towards the boundary C we get Plemelj-type formulae

$$n.t. - \lim_{\substack{t \to \xi \\ t \in \Omega^\pm}} (F_{C,\alpha}u)(t) = \frac{1}{2}[\pm I + S_{C,\alpha}]u(\xi)$$

$$=: \begin{cases} P_{C,\alpha}u(\xi), & t \in \Omega^+, \\ -Q_{C,\alpha}u(\xi), & t \in \Omega^-. \end{cases} \tag{12.4.25}$$

The operators

$$Q_{C,\alpha} := \frac{1}{2}[I - S_{C,\alpha}], \qquad P_{C,\alpha} := \frac{1}{2}[I + S_{C,\alpha}] \tag{12.4.26}$$

are called *Plemelj projections*. Let $M^{\alpha,\infty}$ be the submodule of $C^\infty(\Omega)$ of $\alpha + \Gamma_S$-holomorphic functions (cf. [22]). Further we denote by $M^{\alpha,\infty}(C)$ the subspace of traces of $M^{\alpha,\infty}(\Omega)$. The space $M^{\alpha,\infty}$ is isomorphic to $M^{\alpha,\infty}$ considered as subspace of $L_2(C)$. The *Hardy space* $HS^\alpha(\Omega)$ is defined as the closure of $M^{\alpha,\infty}$ in $L_2(C)$. Such a decomposition is now given by

$$L_2(C) = HS^\alpha(\Omega^+) \oplus HS^\alpha(\Omega^-) \tag{12.4.27}$$

with the Plemelj-type projection $P_{C,\alpha}$, $Q_{C,\alpha}$, on the first and the second Hardy space, respectively.

The Stokes theorem on the sphere is proved in [23]. It reads as follows

$$\int_C \overline{v} n g ds = \int_\Omega [-(\overline{D_\beta v})g + \overline{v}(D_\alpha g)] dS(\omega) \tag{12.4.28}$$

with $\beta + \alpha = -2$ and with respect to the inner product

$$(u, v) = \int_\Omega \overline{u} v d\Omega. \tag{12.4.29}$$

Using $E_\alpha(\xi, \omega) = -E_\beta(\omega, \xi)$, $\overline{E}_\alpha(\xi, \omega) = -\xi E_\beta(\xi, \omega)\omega$ we obtain the following Bergman-Hodge decomposition of the quaternionic Hilbert space

$$L_2(\Omega) = \ker D_\alpha \cap L_2(\Omega) \oplus D_\beta \overset{\circ}{W}{}_2^1(\Omega). \tag{12.4.30}$$

The operators $\mathbb{P}_{\alpha,\beta}$, $\mathbb{Q}_{\alpha,\beta} = I - \mathbb{P}_{\alpha,\beta}$ are called Bergman type projection, Pompeiu type operator, respectively. The proof is analogous to that in the book [9]. In a similar way to the space case one can show that the *Bergman type projection* $\mathbb{P}_{\alpha,\beta}$ permits the explicit representation:

$$\mathbb{P}_{\alpha,\beta} := F_{C,\alpha}(\mathrm{tr}_C T_\beta F_{C,\alpha})^{-1} \mathrm{tr}_C T_\beta. \tag{12.4.31}$$

12.4.4 Time-Discrete Representation of Oseen's Problem

Set

$$\mathcal{H}_{k+1} := T_\beta F_{C,\alpha}(\mathrm{tr}_C T_\beta F_{C,\alpha})^{-1} Q_{C,\beta} g_{k+1} + F_{C,\beta} g_{k+1}, \tag{12.4.32}$$

then follows that

$$u_{k+1} = -\frac{1}{\eta} T_\beta \mathbb{Q}_{\alpha,\beta} \tilde{p}_{k+1} + T_\beta \mathbb{Q}_{\alpha,\beta} T_\alpha \mathcal{F}_k + \mathcal{H}_{k+1}, \tag{12.4.33}$$

where $\tilde{p}_{k+1} := p_{k+1} - \alpha T_\alpha p_{k+1}$. Note that

$$D_\alpha \tilde{p}_{k+1} = D_0 p_{k+1}. \tag{12.4.34}$$

The approximation and stability can be proved in a similar way as in [11], because α tends to finite a complex value for τ, which are going to zero and do not lie on the axes. We should remark here, that we have the identity

$$K_\alpha(-\xi,\omega) = \frac{\sin\pi\alpha}{\pi}\sum_{k=0}^{\infty}\left[\frac{K_k(\xi,\omega)}{\alpha-k} - \frac{\xi K_k(\xi,\omega)\omega}{\alpha+k+2}\right] \tag{12.4.35}$$

at least for all non-real α (cf. [23], p. 120). The convergence of the expansion has to be understood in the distributional sense. We emphasize that all information on the boundary values is included in the terms $\mathcal{H}_{k+1}$.

12.4.5 Forecasting Equations in the Ball Shell

We would like to come back to the forecasting equations on the ball shell, which were described as

$$\frac{1}{\nu}\partial_t u + DDu + \frac{1}{\eta}Dp + \frac{1}{\eta}(u\cdot D)u = \frac{1}{\eta}F - 2\frac{1}{\eta}a\wedge u =: \mathcal{F}(\cdot,u), \tag{12.4.36}$$

$$\operatorname{Sc} Du = 0 \quad \text{in } G_{12}, \tag{12.4.37}$$

$$\rho(t,x) = 0, \tag{12.4.38}$$

$$u(0,x) = u_0(x) \quad \text{in } G_{12}, \tag{12.4.39}$$

$$u(t,x) = g(t,x) \quad \text{on } \Gamma_1\cup\Gamma_2 =: \Gamma. \tag{12.4.40}$$

12.4.6 Quaternionic Operator Formulation of Forecasting Equations

Again we use forward differences in order to approximate the time derivative. With the same notations as before and with $a := \sqrt{\frac{\rho}{\tau\eta}}$, we obtain

$$(D+ia)(D-ia)u_{k+1} + \frac{1}{\eta}Dp_{k+1}$$
$$= \mathcal{F}(\cdot,u_k) - M^*(u_k) + a^2u_k =: M(u_k) \quad (k=0,\dots,n-1) \tag{12.4.41}$$

and

$$u_{k+1} = -\frac{1}{\eta}T_{-ia}\mathbb{Q}_{ia}T_{ia}Dp_{k+1} - T_{-ia}\mathbb{Q}_{ia}T_{ia}M(u_k) \tag{12.4.42}$$

$$+ \underbrace{T_{-ia}F_{ia}(\operatorname{tr}_\Gamma T_{-ia}F_{ia})^{-1}Q_{\Gamma,-ia}g_{k+1} + F_{-ia}g_{k+1}}_{H_{k+1}}. \tag{12.4.43}$$

Then we get the following statement.

Corollary 12.1 *Set*

$$u_{k+1}^{(n)} - H_{k+1} = -T_{-ia}\mathbb{Q}_{ia}T_{ia}M(u_k^{(n-1)}) - \frac{1}{\eta}T_{-ia}\mathbb{Q}_{ia}p_{k+1}^{(n)} \tag{12.4.44}$$

with

$$\frac{1}{\eta}\mathrm{Sc}\,\mathbb{Q}_{ia}p_k^{(n)} = -\mathrm{Sc}\,\mathbb{Q}_{ia}T_{ia}M(u_k^{(n-1)}). \tag{12.4.45}$$

Under suitable "smallness" conditions (cf. [9, 10]) *the sequence* $u_{k+1}^{(n)}$ *converges for* $(n \to \infty)$ *in* $W_2^1(G_{12})$ *to* u_{k+1}.

Proof The proof is similar to the proof in ([10], p. 178). □

Remark 12.4 We have in a neighbourhood of the "quaternion" $\mathcal{H}_k$ which reflects for each k the boundary value information, for any time in the interval $[0, T]$ a sequence, which is strongly coverging to the solution of the forecasting problem.

The author thanks Mrs. Le Thu Hoai (TU Hanoi) for carefully reading, improving this manuscript and many fruitful discussions. Furthermore, the author wishes also to thank the reviewers for their valuable hints and improvements.

References

1. Blaschke, W.: Anwendung dualer Quaternionen auf Kinematik. Ann. Acad. Sci. Fennicae, Ser. A I, Math. 250/3 (1958)
2. Cerejeiras, P., Vieira, N.: Factorization of the Non-stationary Schrödinger Operator. Adv. Appl. Clifford Algebras **17**, 331–341 (2007)
3. Cerejeiras, P., Kähler, U., Sommen, F.: Parabolic Dirac operators and the Navier-Stokes equations over time-varying domains. Math. Methods Appl. Sci. **28**, 1715–1724 (2005)
4. Cnops, J.: An Introduction of Dirac Operators on Manifolds. Birkhäuser, Basel (2002)
5. Delanghe, R., Sommen, F., Soucek, V.: Clifford Algebra and Spinor Valued Functions. Kluwer, Dordrecht (1992)
6. Duduchava, R.: Boundary value problems on a smooth surface with smooth boundary. Universität Stuttgart, Preprint 2002-5,1-19 (2002)
7. Fengler, M.J.: Vector spherical harmonic and vector wavelet based non-linear Galerkin schemes for solving the incompressible Navier-Stokes equation on the sphere. Shaker, D386 (2005)
8. Günter, N.: Potential Theory and Its Application to the Basic Problems of Mathematical Physics. Fizmatgiz, Moscow (1953). (Russian translation in Frenc Gauthier-Villars, Paris (1994))
9. Gürlebeck, K., Sprössig, W.: Quaternionic Analysis and Elliptic Boundary Value Problems. Birkhäuser, Basel (1989)
10. Gürlebeck, K., Sprößig, W.: Quaternionic and Clifford Calculus for Physcists and Engineers. Mathematical Methods in Practice. Wiley, Chichester (1997)
11. Gürlebeck, K., Sprössig, W.: Representation theory for classes of initial value problems with quaternion analysis. Math. Methods Appl. Sci. **25**, 1371–1382 (2002)
12. Gürlebeck, K., Habetha, K., Sprößig, W.: Holomorphic Functions in the Plane and n-Dimensional Space. Birkhäuser, Basel (2007)

13. Hamilton, W.R.: Elements of Quaternions. Longmans Green, London (1866). Reprinted by Chelsea, New York (1969)
14. Lesieur, M.: Turbulence in Fluids. Third Revised and Enlarged Edition. Kluwer, Boston (1997)
15. Majda, A.: Introduction to PDE's and waves for atmosphere and ocean. New York, Courant-Lecture Notes, vol. 9 (2003)
16. Mitrea, M.: Boundary value problems for Dirac operators and Maxwell's equations in non-smooth domains. Math. Methods Appl. Sci. **25**, 1355–1369 (2002)
17. Moisil, G., Teodorescu, N.: Fonctions holomorphes dans l'espace. Mat. Cluj **5**, 142–150 (1931)
18. Norburg, J., Roulstone, I.: Large-Scale Atmosphere-Ocean Dynamics I. Cambridge University Press, Cambridge (2002)
19. Prößdorf, S.: Einige Klassen singulärer Gleichungen. Birkhäuser, Basel (1974)
20. Schwarz, G.: Hodge Decompositions—A Method for Solving Boundary Value Problems. LMN, vol. 1607. Springer, Berlin (1995)
21. Sprössig, W.: Exponentials of the Dirac operator and an application (2009)
22. Sprössig, W., Le, T.H.: On a new notion of holomorphy and its applications. Cubo, Math. J. **11**(1), 145–162 (2008)
23. Van Lancker, P.: Clifford analysis on the unit sphere. Thesis, University of Ghent (1997)

Chapter 13
Symplectic Exponentially-Fitted Modified Runge-Kutta Methods of the Gauss Type: Revisited

G. Vanden Berghe and M. Van Daele

Abstract The construction of symmetric and symplectic exponentially-fitted Runge-Kutta methods for the numerical integration of Hamiltonian systems with oscillatory solutions is reconsidered. In previous papers fourth-order and sixth-order symplectic exponentially-fitted integrators of Gauss type, either with fixed or variable nodes, have been derived. In this paper new such integrators are constructed by making use of the six-step procedure of Ixaru and Vanden Berghe (Exponential Fitting, Kluwer Academic, Dordrecht, 2004). Numerical experiments for some oscillatory problems are presented and compared to the results obtained by previous methods.

Keywords Exponential fitting · Symplecticness · RK-methods · Oscillatory Hamiltonian systems

Mathematics Subject Classification (2000) 65L05 · 65L06

13.1 Introduction

The construction of Runge-Kutta (RK) methods for the numerical solution of ODEs, which have periodic or oscillating solutions has been considered extensively in the literature [1–5, 9, 12–17]. In this approach the available information on the solutions is used in order to derive more accurate and/or efficient algorithms than the general purpose algorithms for such type of problems. In [8] a particular six-step flow chart

G. Vanden Berghe (✉) · M. Van Daele
Vakgroep Toegepaste Wiskunde en Informatica, Universiteit Gent, Krijgslaan 281-S9, 9000 Gent, Belgium
e-mail: guido.vandenberghe@ugent.be

T.E. Simos (ed.), *Recent Advances in Computational and Applied Mathematics*,
DOI 10.1007/978-90-481-9981-5_13,

is proposed by which specific exponentially-fitted algorithms can be constructed. Up to now this procedure has not yet been applied in all its aspects for the construction of symplectic RK methods of Gauss type.

In principle the derivation of exponentially-fitted (EF) RK methods consists in selecting the coefficients of the method such that it integrates exactly all functions of a particular given linear space, i.e. the set of functions

$$\{1, t, \ldots, t^K, \exp(\pm\lambda t), t\exp(\pm\lambda t), \ldots, t^P\exp(\pm\lambda t)\}, \qquad (13.1.1)$$

where $\lambda \in \mathbb{C}$ is a prescribed frequency. In particular when $\lambda = i\omega$, $\omega \in \mathbb{R}$ the couple $\exp(\pm\lambda t)$ is replaced by $\sin(\omega t)$, $\cos(\omega t)$. In all previous papers other set of functions have been introduced.

On the other hand, oscillatory problems arise in different fields of applied sciences such as celestial mechanics, astrophysics, chemistry, molecular dynamics and in many cases the modelling gives rise to Hamiltonian systems. It has been widely recognized by several authors [7, 10, 11, 13, 14] that symplectic integrators have some advantages for the preservation of qualitative properties of the flow over the standard integrators when they are applied to Hamiltonian systems. In this sense it may be appropriate to consider symplectic EFRK methods that preserve the structure of the original flow. In [14] the well-known theory of symplectic RK methods is extended to modified (i.e. by introducing additional parameters) EFRK methods, where the set of functions $\{\exp(\pm\lambda t)\}$ has been introduced, giving sufficient conditions on the coefficients of the method so that symplecticness for general Hamiltonian systems is preserved. Van de Vyver [14] was able to derive a two-stage fourth-order symplectic modified EFRK method of Gauss type. Calvo et al. [2–4] have studied two-stage as well as three-stage methods. In their applications they consider pure EFRK methods as well as modified EFRK methods. Their set of functions is the trigonometric polynomial one consisting essentially of the functions $\exp(\pm\lambda t)$ combined with $\exp(\pm 2\lambda t)$ and sometimes $\exp(\pm 3\lambda t)$ or a kind of mixed set type where $\exp(\pm\lambda t)$ is combined with $1, t$ and t^2. In all cases they constructed fourth-order (two-stage case) and sixth-order (three-stage case) methods of Gauss type with fixed or frequency dependent knot points. On the other hand Vanden Berghe et al. have constructed a two-stage EFRK method of fourth-order integrating the set of functions (13.1.1) with ($K = 2$, $P = 0$) and ($K = 0$, $P = 1$), but unfortunately these methods are not symplectic. In addition it has been pointed out in [7] that symmetric methods show a better long time behavior than non-symmetric ones when applied to reversible differential systems.

In this paper we investigate the construction of two-stage (fourth-order) and three-stage (sixth-order) symmetric and symplectic modified EFRK methods which integrate exactly first-order differential systems whose solutions can be expressed as linear combinations of functions present in the set (13.1.1). Our purpose consists in deriving accurate and efficient modified EF geometric integrators based on the combination of the EF approach, followed from the six-step flow chart by Ixaru and Vanden Berghe [8], and symmetry and symplecticness conditions. A sketch of this six-step flow is given in Sect. 13.2. The paper is organized as follows. In Sect. 13.2 we present the notations and definitions used in the rest of the paper as well as some properties of symplectic and symmetric methods also described in [4]. In Sect. 13.3 we derive a class of new two-stage symplectic modified EFRK integrators with frequency dependent nodes and in Sect. 13.4 we consider the analogous class of new three-stages method. In Sect. 13.5 we present some numerical experiments for sixth-order methods with oscillatory Hamiltonian systems and we compare them with the results obtained by other symplectic (EF)RK Gauss integrators given in [4, 7].

13.2 Notations and Definitions

We consider initial value problems for first-order differential systems

$$y'(t) = f(t, y(t)), \qquad y(t_0) = y_0 \in \mathbb{R}^m. \tag{13.2.1}$$

In case of Hamiltonian systems $m = 2d$ and there exits a scalar Hamiltonian function $H = H(t, y)$, so that $f(y) = -J\nabla_y H(t, y)$, where J is the $2d$-dimensional skew symmetric matrix

$$J = \begin{pmatrix} 0_d & I_d \\ -I_d & 0_d \end{pmatrix}, \qquad J^{-1} = -J,$$

and where $\nabla_y H(t, y)$ is the column vector of the derivatives of $H(t, y)$ with respect to the components of $y = (y_1, y_2, \dots, y_{2d})^T$. The Hamiltonian system can then be written as

$$y'(t) = -J\nabla_y H(t, y(t)), \qquad y(t_0) = y_0 \in \mathbb{R}^{2d}. \tag{13.2.2}$$

For each fixed t_0 the flow map of (13.2.1) will be denoted by $\phi_h : \mathbb{R}^m \to \mathbb{R}^m$ so that $\phi_h(y_0) = y(t_0 + h; t_0, y_0)$. In particular, in the case of Hamiltonian systems, ϕ_h is a symplectic map for all h in its domain of definition, i.e. the Jacobian matrix of $\phi_h(y_0)$ satisfies

$$\phi_h'(y_0) J \phi_h'(y_0)^T = J.$$

A desirable property of a numerical method ψ_h for the numerical integration of a Hamiltonian system is to preserve qualitative properties of the original flow ϕ_h such as the symplecticness, in addition to provide an accurate approximation of the exact ϕ_h.

Definition 13.1 A numerical method defined by the flow map ψ_h is called symplectic if for all Hamiltonian systems (13.2.2) it satisfies the condition

$$\psi_h'(y_0) J \psi_h'(y_0)^T = J. \tag{13.2.3}$$

One of the well-known examples of symplectic numerical methods is the s-stage RK Gauss methods which possess order $2s$. In this paper we shall deal with so-called modified implicit RK-methods, introduced for the first time to obtain explicit EFRK methods [15] and re-used by Van de Vyver [14] for the construction of two-stage symplectic RK methods.

Definition 13.2 A s-stage modified RK method for solving the initial value problems (13.1.1) is a one step method defined by

$$y_1 = \psi_h(y_0) = y_0 + h \sum_{i=1}^{s} b_i f(t_0 + c_i h, Y_i), \tag{13.2.4}$$

$$Y_i = \gamma_i y_0 + h \sum_{i=1}^{s} a_{ij} f(t_0 + c_j h, Y_j), \quad i = 1, \ldots, s, \tag{13.2.5}$$

where the real parameters c_i and b_i are respectively the nodes and the weights of the method. The parameters γ_i make the method modified with respect to the classical RK method, where $\gamma_i = 1, i = 1, \ldots, s$. The s-stage modified RK-method (13.2.4)–(13.2.5) can also be represented by means of its Butcher's tableau

$$\begin{array}{c|c|ccc} c_1 & \gamma_1 & a_{11} & \ldots & a_{1s} \\ c_2 & \gamma_2 & a_{21} & \ldots & a_{2s} \\ \vdots & \ldots & \vdots & \ddots & \vdots \\ c_s & \gamma_s & a_{s1} & \ldots & a_{ss} \\ \hline & & b_1 & \ldots & b_s \end{array} \tag{13.2.6}$$

or equivalently by the quartet (c, γ, A, b).

The conditions for a modified RK method to be symplectic have been obtained by Van de Vyver [14] and they are given in the following theorem.

Theorem 13.1 *A modified RK-method* (13.2.4)–(13.2.5) *for solving the Hamiltonian system* (13.2.2) *is symplectic if the following conditions are satisfied*

$$m_{ij} \equiv b_i b_j - \frac{b_i}{\gamma_i} a_{ij} - \frac{b_j}{\gamma_j} a_{ji} = 0, \quad 1 \leq i, j \leq s. \tag{13.2.7}$$

In [2] it is shown that a modified RK-method not only preserves the linear invariants but also quadratic invariants if its coefficients satisfy conditions (13.2.7).

Definition 13.3 The adjoint method ψ_h^* of a numerical method ψ_h is the inverse map of the original method with reverse time step $-h$, i.e. $\psi_h^* := \psi_{-h}^{-1}$. In other words, $y_1 = \psi_h^*(y_0)$ is implicitly defined by $\psi_{-h}(y_1) = y_0$. A method for which $\psi_h^* = \psi_h$ is called symmetric.

One of the properties of a symmetric method $\psi_h^* = \psi_h$ is that its accuracy order is even. For s-stage modified RK methods whose coefficients are h-dependent, as it is the case of EF methods, it is easy to see that the coefficients of ψ_h and ψ_h^* are related by

$$c(h) = e - Sc^*(-h), \qquad b(h) = Sb^*(-h),$$
$$\gamma(h) = S\gamma^*(-h), \qquad A(h) = S\gamma^*(-h)b^T(h) - SA(-h)S,$$

where

$$e=(1,\dots,1)^T\in\mathbb{R}^s \quad\text{and}\quad S=(s_{ij})\in\mathbb{R}^{s\times s} \quad\text{with } s_{ij}=\begin{cases}1, & \text{if } i+j=s+1,\\ 0, & \text{if } i+j\neq s+1.\end{cases}$$

It has been remarked by Hairer et al. [7] that symmetric numerical methods show a better long time behavior than nonsymmetric ones when applied to reversible differential equations, as it is the case of conservative mechanical systems. In [3] it is observed that for modified RK methods whose coefficients are even functions of h the symmetry conditions are given by

$$\begin{aligned} &c(h)+Sc(h)=e, &&b(h)=Sb(h),\\ &\gamma(h)=S\gamma(h), &&SA(h)+A(h)S=\gamma(h)b^T(h). \end{aligned} \tag{13.2.8}$$

Since for symmetric EFRK methods the coefficients contain only even powers of h, the symmetry conditions can be written in a more convenient form by putting [3]

$$\begin{aligned} &c(h)=\frac{1}{2}e+\theta(h), \qquad A(h)=\frac{1}{2}\gamma(h)b^T(h)+\Lambda(h),\\ &\theta(h)=(\theta_1,\dots,\theta_s)^T\in\mathbb{R}^s \quad\text{and}\quad \Lambda=(\lambda_{ij})\in\mathbb{R}^{s\times s}. \end{aligned} \tag{13.2.9}$$

Therefore, for a symmetric EFRK method whose coefficients a_{ij} are defined by

$$a_{ij}=\frac{1}{2}\gamma_i b_j+\lambda_{ij}, \quad 1\le i,j\le s,$$

the symplecticness conditions (13.2.7) reduce to

$$\mu_{ij}\equiv\frac{b_i}{\gamma_i}\lambda_{ij}+\frac{b_j}{\gamma_j}\lambda_{ji}=0, \quad 1\le i,j,\le s. \tag{13.2.10}$$

The idea of constructing symplectic EFRK taking into account the six-step procedure [8] is new. We briefly shall survey this procedure and suggest some adaptation in order to make the comparison with previous work more easy.

In step (i) we define the appropriate form of an operator related to the discussed problem. Each of the s internal stages (13.2.5) and the final stage (13.2.4) can be regarded as being a generalized linear multistep method on a non-uniform grid; we can associated with each of them a linear functional, i.e.

$$\begin{aligned} &\mathcal{L}_i[h,\mathbf{a}]y(t)=y(t+c_ih)-\gamma_i y(t)-h\sum_{j=1}^{s}a_{ij}y'(t+c_jh),\\ &i=1,2,\dots s \end{aligned} \tag{13.2.11}$$

and

$$\mathcal{L}[h,\mathbf{b}]y(t)=y(t+h)-y(t)-h\sum_{i=1}^{s}b_i y'(t+c_ih). \tag{13.2.12}$$

We further construct the so-called moments which are for Gauss methods the expressions for $L_{i,j}(h,\mathbf{a})=\mathcal{L}_i[h,\mathbf{a}]t^j$, $j=0,\dots,s-1$ and $L_i(h,\mathbf{b})=\mathcal{L}[h,\mathbf{b}]t^j$, $j=0,\dots,2s-1$ at $t=0$, respectively.

In step (ii) the linear systems

$$L_{ij}(h, \mathbf{a}) = 0, \quad i = 1, \ldots, s, \; j = 0, 1, \ldots, s-1,$$
$$L_i(h, \mathbf{b}) = 0, \quad i = 0, 1, \ldots, 2s-1$$

are solved to reproduce the classical Gauss RK collocation methods, showing the maximum number of functions which can be annihilated by each of the operators.

The steps (iii) and (iv) can be combined in the present context. First of all we have to define all reference sets of s and $2s$ functions which are appropriate for the internal and final stages respectively. These sets are in general hybrid sets of the following form

$$1, \quad t, \quad t^2, \quad \ldots, \quad t^K \quad \text{or} \quad t^{K'}$$
$$\exp(\pm\lambda t), \quad t\exp(\pm\lambda t), \quad \ldots, \quad t^P\exp(\pm\lambda t) \quad \text{or} \quad t^{P'}\exp(\pm\lambda t),$$

where for the internal stages $K + 2P = s - 3$ and for the final stage $K' + 2P' = 2s - 3$. The set in which there is no classical component is identified by $K = -1$ and $K' = -1$, while the set in which there is no exponential fitting component is identified by $P = -1$ or $P' = -1$. It is important to note that such reference sets should contain all successive functions inbetween. Lacunary sets are in principle not allowed.

Once the sets chosen the operators (13.2.11)–(13.2.12) are applied to the members of the sets, in this particular case by taking into account the symmetry and the symplecticness conditions described above. The obtained independent expressions are put to zero and in step (v) the available linear systems are solved. Detailed examples of these technique follow in Sects. 13.3 and 13.4. The numerical values for $\lambda_{ij}(h)$, $b_i(h)$, $\gamma_i(h)$ and $\theta_i(h)$ are expressed for real values of λ (the pure exponential case) or for pure imaginary $\lambda = i\omega$ (oscillatory case). In order to make the comparison with previous work transparable we have opted to denote the results for real λ-values.

After the coefficients in the Butcher tableau have been filled in, the principal term of the local truncation error can be written down (step (vi)). Essentially, we know [17] that the algebraic order of the EFRK methods remains the same as the one of the classical Gauss method when this six-step procedure is followed, in other words the algebraic order is $\mathcal{O}(h^{2s})$, while the stage order is $\mathcal{O}(h^s)$. Explicit expressions for this local truncation error will not be discussed here.

13.3 Two-Stage Methods

In this section we analyze the construction of symmetric and symplectic EFRK Gauss methods with $s = 2$ stages whose coefficients are even functions of h. These EFRK methods have stage order 2 and algebraic order 4. From the symmetry conditions (13.2.8), taking into account (13.2.9) it follows that the nodes $c_j = c_j(h)$ and weights $b_j = b_j(h)$ satisfy

$$c_1 = \frac{1}{2} - \theta, \qquad c_2 = \frac{1}{2} + \theta, \qquad b_1 = b_2,$$

θ being a real parameter, and the coefficients $a_{ij} = a_{ij}(h)$ and $\gamma_i(h)$ satisfy:

$$a_{11} + a_{22} = \gamma_1 b_1, \qquad a_{21} + a_{12} = \gamma_2 b_1.$$

The symplecticness conditions (13.2.7) or (13.2.10) are equivalent to

$$a_{11} = \gamma_1 b_1/2, \qquad \frac{a_{12}}{\gamma_1} + \frac{a_{21}}{\gamma_2} = b_1, \qquad a_{22} = \gamma_2 b_2/2,$$

which results in

$$\gamma_1 = \gamma_2, \qquad \lambda_{21} = -\lambda_{12}.$$

Taking into account the above relations the Butcher tableau can be expressed in terms of the unknowns θ, γ_1, λ_{12} and b_1:

$$\begin{array}{c|c|cc} \frac{1}{2}-\theta & \gamma_1 & \frac{\gamma_1 b_1}{2} & \frac{\gamma_1 b_1}{2}+\lambda_{12} \\ \frac{1}{2}+\theta & \gamma_1 & \frac{\gamma_1 b_1}{2}-\lambda_{12} & \frac{\gamma_1 b_1}{2} \\ \hline & & b_1 & b_1 \end{array} \tag{13.3.1}$$

For the internal stages, the relation $K + 2P = -1$ results in the respective (K, P)-values:

- $(K = 1, P = -1)$ (the classical polynomial case with set $\{1, t\}$), and
- $(K = -1, P = 0)$ (the full exponential case with set $\{\exp(\lambda t), \exp(-\lambda t)\}$).

For the outer stage, we have $K' + 2P' = 1$, resulting in the respective (K', P')-values:

- $(K' = 3, P' = -1)$ (the classical polynomial case with set $\{1, t, t^2, t^3\}$),
- $(K' = 1, P' = 0)$ (mixed case with hybrid set $\{1, t, \exp(\pm\lambda t)\}$), and
- $(K' = -1, P' = 1)$ (the full exponential case with set $\{\exp(\pm\lambda t), t\exp(\pm\lambda t)\}$.

The hybrid sets $(K = 1, P = -1)$ and $(K' = 3, P' = -1)$ are related to the polynomial case, giving rise to the well-known RK order conditions and to the fourth order Gauss method [6]

$$\begin{array}{c|c|cc} \frac{1}{2}-\frac{\sqrt{3}}{6} & 1 & \frac{1}{4} & \frac{1}{4}-\frac{\sqrt{3}}{6} \\ \frac{1}{2}+\frac{\sqrt{3}}{6} & 1 & \frac{1}{4}+\frac{\sqrt{3}}{6} & \frac{1}{4} \\ \hline & & \frac{1}{2} & \frac{1}{2} \end{array}$$

Let us remark that considering the $(K = 1, P = -1)$ set for the internal stages gives rise to $\gamma_1 = 1$, a value which is not compatible with the additional symmetry, symplecticity and order conditions imposed. Therefore in what follows we combine the $(K = -1, P = 0)$ case with either $(K' = 1, P' = 0)$ or $(K' = -1, P' = 1)$.

Case $(K' = 1, P' = 0)$

The operators (13.2.11) and (13.2.12) are applied to the functions present in the occurring hybrid sets, taking into account the structure of the Butcher tableau (13.3.1). Following equations arise with $z = \lambda h$:

$$2b_1 = 1, \tag{13.3.2}$$

$$2b_1 \cosh(z/2)\cosh(\theta z) = \frac{\sinh(z)}{z}, \tag{13.3.3}$$

$$\lambda_{12}\cosh(\theta z) = -\frac{\sinh(\theta z)}{z}, \tag{13.3.4}$$

$$\lambda_{12}\sinh(\theta z) - \frac{\cosh(\theta z)}{z} = -\frac{\gamma_1}{z}\cosh(z/2), \tag{13.3.5}$$

resulting in

$$b_1 = 1/2, \qquad \theta = \frac{\operatorname{arccosh}(\frac{2\sinh(z/2)}{z})}{z}, \qquad \lambda_{12} = -\frac{\sinh(\theta z)}{z\cosh(\theta z)},$$

$$\gamma_1 = \frac{(\frac{\sinh(\theta z)^2}{z\cosh(\theta z)} + \frac{\cosh(\theta z)}{z})z}{\cosh(z/2)}.$$

The series expansions for these coefficients for small values of z are given by

$$\theta = \sqrt{3}\bigg(\frac{1}{6} + \frac{1}{2160}z^2 - \frac{1}{403200}z^4 + \frac{1}{145152000}z^6 + \frac{533}{9656672256000}z^8 - \frac{2599}{2789705318400000}z^{10} + \cdots\bigg),$$

$$\lambda_{12} = \sqrt{3}\bigg(-\frac{1}{6} + \frac{1}{240}z^2 - \frac{137}{1209600}z^4 + \frac{143}{48384000}z^6 - \frac{81029}{1072963584000}z^8 + \frac{16036667}{8369115955200000}z^{10} + \cdots\bigg),$$

$$\gamma_1 = 1 - \frac{1}{360}z^4 + \frac{11}{30240}z^6 - \frac{71}{1814400}z^8 + \frac{241}{59875200}z^{10} + \cdots,$$

showing that for $z \to 0$ the classical values are retrieved.

Case ($K' = -1, P' = 1$)

In this approach equations (13.3.3)–(13.3.5) remain unchanged and they deliver expressions for b_1, γ_1 and λ_{12} in terms of θ. Only (13.3.2) is replaced by

$$b_1(\cosh(\theta z)(2\cosh(z/2) + z\sinh(z/2)) + 2\theta z\cosh(z/2)\sinh(\theta z)) = \cosh(z) \tag{13.3.6}$$

By combining (13.3.3) and (13.3.6) one obtains an equation in θ and z, i.e.:

$$\theta\sinh(z)\sinh(\theta z) = \cosh(\theta z)\bigg(\cosh(z) - \frac{\sinh(z)}{z} - \sinh^2(z/2)\bigg).$$

It is not anymore possible to write down an analytical solution for θ, but iteratively a series expansion can be derived. We give here those series expansions as obtained for the four unknowns

$$\theta = \sqrt{3}\bigg(\frac{1}{6} + \frac{1}{1080}z^2 + \frac{13}{2721600}z^4 - \frac{1}{7776000}z^6 - \frac{1481}{1810626048000}z^8 + \frac{573509}{63552974284800000}z^{10} + \cdots\bigg),$$

$$b_1 = \frac{1}{2} - \frac{1}{8640}z^4 + \frac{1}{1088640}z^6 + \frac{1}{44789760}z^8 - \frac{149}{775982592000}z^{10} + \cdots,$$

$$\lambda_{12} = \sqrt{3}\bigg(-\frac{1}{6} + \frac{1}{270}z^2 - \frac{223}{2721600}z^4 + \frac{17}{9072000}z^6 - \frac{259513}{5431878144000}z^8 + \frac{9791387}{7944121785600000}z^{10} + \cdots\bigg),$$

$$\gamma_1 = 1 - \frac{1}{480}z^4 + \frac{17}{60480}z^6 - \frac{2629}{87091200}z^8 + \frac{133603}{43110144000}z^{10} + \cdots.$$

13.4 Three-Stage Methods

The Gauss methods with $s = 3$ stages have been analyzed in detail by Calvo et al. [3, 4]. We just shall report here the final results they have obtained by taking into account the symmetry and symplecticity conditions:

$$c_1 = \frac{1}{2} - \theta, \qquad c_2 = \frac{1}{2}, \qquad c_3 = \frac{1}{2} + \theta,$$

$$b_3 = b_1, \qquad \gamma_3 = \gamma_1$$

$$\Lambda = \begin{pmatrix} 0 & -\alpha_2 & -\alpha_3 \\ -\alpha_4 & 0 & \alpha_4 \\ \alpha_3 & \alpha_2 & 0 \end{pmatrix}$$

and

$$\frac{b_1}{\gamma_1}\alpha_2 + \frac{b_2}{\gamma_2}\alpha_4 = 0. \tag{13.4.1}$$

The three-stage modified RK-methods are given by the following tableau in terms of the unknowns θ, γ_1, γ_2, α_2, α_3, α_4, b_1 and b_2:

$$\begin{array}{c|c|ccc} \frac{1}{2} - \theta & \gamma_1 & \frac{\gamma_1 b_1}{2} & \frac{\gamma_1 b_2}{2} - \alpha_2 & \frac{\gamma_1 b_1}{2} - \alpha_3 \\ \frac{1}{2} & \gamma_2 & \frac{\gamma_2 b_1}{2} - \alpha_4 & \frac{\gamma_2 b_2}{2} & \frac{\gamma_2 b_1}{2} + \alpha_4 \\ \frac{1}{2} + \theta & \gamma_1 & \frac{\gamma_1 b_1}{2} + \alpha_3 & \frac{\gamma_1 b_2}{2} + \alpha_2 & \frac{\gamma_1 b_1}{2} \\ \hline & & b_1 & b_2 & b_1 \end{array}$$

For the internal stages the relation $K + 2P = 0$ results in the respective (K, P)-values:

- $(K = 2, P = -1)$ (the classical polynomial case with set $\{1, t, t^2\}$), and
- $(K = 0, P = 0)$ (with hybrid set $\{1, \exp(\pm\lambda t)\}$).

For the final state we have $K' + 2P' = 3$, resulting in the respective (K', P')-values:

- $(K' = 5, P' = -1)$ (the classical polynomial case with set $\{1, t, t^2, t^3, t^4, t^5\}$),
- $(K' = 3, P' = 0)$ (with hybrid set $\{1, t, t^2, t^3, \exp(\pm\lambda t)\}$),
- $(K' = 1, P' = 1)$ (with hybrid set $\{1, t, \exp(\pm\lambda t), t\exp(\pm\lambda t)\}$), and
- $(K' = -1, P' = 2)$ (the full exponential case with set $\{\exp(\pm\lambda t), t\exp(\pm\lambda t), t^2\exp(\pm\lambda t)\}$).

The sets $(K = 2, P = -1)$ and $(K' = 5, P' = -1)$ related to the polynomial case gives rise to the order conditions for the three-stage Gauss method of order six [6]

$$
\begin{array}{c|c|ccc}
\frac{1}{2} - \frac{\sqrt{15}}{10} & 1 & \frac{5}{36} & \frac{2}{9} - \frac{\sqrt{15}}{15} & \frac{5}{36} - \frac{\sqrt{15}}{30} \\
\frac{1}{2} & 1 & \frac{5}{36} + \frac{\sqrt{15}}{24} & \frac{2}{9} & \frac{5}{36} - \frac{\sqrt{15}}{24} \\
\frac{1}{2} + \frac{\sqrt{15}}{10} & 1 & \frac{5}{36} + \frac{\sqrt{15}}{30} & \frac{2}{9} + \frac{\sqrt{15}}{15} & \frac{5}{36} \\
\hline
 & & \frac{5}{18} & \frac{4}{9} & \frac{5}{18}
\end{array}
$$

Following the ideas developed in this paper it should be obvious that we combine the $(K = 0, P = 0)$ case with the three non-polynomial cases for the final stage. However keeping the 1 in the hybrid set for $(K = 0, P = 0)$ delivers in $\gamma_1 = \gamma_2 = 1$, a result which is not compatible with the symplecticity condition (13.4.1). Therefore we choose for the internal stages the hybrid set $\{\exp(\pm\lambda t)\}$, omitting the constant 1; in other words we accept exceptionally a lacunary set, what is principally not allowed by the six-step procedure [8]. Under these conditions, and taking into account the symmetry conditions the α_i $(i = 2, 3, 4)$ parameters are the solutions in terms of θ, γ_1 and γ_2 of the following three equations [4]:

$$
\begin{aligned}
&1 - \gamma_2\cosh(z/2) - 2z\alpha_4\sinh(\theta z) = 0,\\
&\cosh(\theta z) - \gamma_1\cosh(z/2) + z\alpha_3\sinh(\theta z) = 0,\\
&\sinh(\theta z) - z\alpha_3\cosh(\theta z) - z\alpha_2 = 0,
\end{aligned}
\tag{13.4.2}
$$

thus giving:

$$
\begin{aligned}
\alpha_2 &= \frac{\cosh(2\theta z) - \gamma_1\cosh(z/2)\cosh(\theta z)}{z\sinh(\theta z)},\\
\alpha_3 &= \frac{\gamma_1\cosh(z/2) - \cosh(\theta z)}{z\sinh(\theta z)}, \qquad \alpha_4 = \frac{1 - \gamma_2\cosh(z/2)}{2z\sinh(\theta z)}.
\end{aligned}
\tag{13.4.3}
$$

For small values of z series expansions are introduced for these expressions (see also next paragraphs). The solution for the other parameters depends essentially on the chosen values of K' and P'.

Case $(K' = 3, P' = 0)$

The operators (13.2.11) and (13.2.12) are applied to the functions present in the occurring hybrid set, taking into account the symmetry conditions; we derive three independent equations in b_1, b_2 and θ, i.e.

$$2b_1 + b_2 = 1, \tag{13.4.4}$$

$$b_1\theta^2 = \frac{1}{24}, \tag{13.4.5}$$

$$b_2 + 2b_1 \cosh(\theta z) = \frac{2\sinh(z/2)}{z}. \tag{13.4.6}$$

Taking into account (13.4.4) and (13.4.6) b_1 and b_2 can be expressed in terms of θ:

$$b_1 = \frac{z - 2\sinh(z/2)}{2z(1-\cosh(\theta z))}, \qquad b_2 = \frac{2\sinh(z/2) - z\cosh(\theta z)}{z(1-\cosh(\theta z))}.$$

These expressions combined with (13.4.5) results in the following equation for θ:

$$\theta^2 - \frac{z(1-\cosh(\theta z))}{12(z - 2\sinh(z/2))} = 0.$$

If now the symplecticness condition (13.4.1) is imposed, the parameter γ_1 is determined by

$$\gamma_1 = \frac{\gamma_2(2\sinh(z/2) - z)\cosh(2\theta z)}{2\sinh(z/2) - \gamma_2\sinh(z) + (\gamma_2\sinh(z) - z)\cosh(\theta z)}.$$

The obtained parameters define a family of EFRK methods which are symmetric and symplectic for all $\gamma_2 \in \mathbb{R}$. Following [4] we choose from now on $\gamma_2 = 1$.

Now it is easy to give the series expansions for all the coefficients for small values of z:

$$\begin{aligned}
\theta &= \sqrt{15}\bigg(\frac{1}{10} + \frac{1}{21000}z^2 - \frac{131}{1058400000}z^4 + \frac{13487}{48898080000000}z^6 \\
&\quad - \frac{1175117}{3203802201600000000}z^8 - \frac{505147}{915372057600000000000}z^{10} + \cdots\bigg), \\
\gamma_1 &= 1 - \frac{3}{70000}z^6 + \frac{13651}{1176000000}z^8 - \frac{2452531}{862400000000}z^{10} + \cdots, \\
b_1 &= \frac{5}{18} - \frac{1}{3780}z^2 + \frac{167}{190512000}z^4 - \frac{23189}{8801654400000}z^6 \\
&\quad + \frac{7508803}{1153368792576000000}z^8 - \frac{87474851}{8073581548032000000000}z^{10} + \cdots, \\
b_2 &= \frac{4}{9} + \frac{1}{1890}z^2 - \frac{167}{95256000}z^4 + \frac{23189}{4400827200000}z^6 \\
&\quad - \frac{7508803}{576684396288000000}z^8 + \frac{87474851}{4036790774016000000000}z^{10} + \cdots, \\
\alpha_2 &= \sqrt{15}\bigg(\frac{1}{15} - \frac{1}{6000}z^2 + \frac{11623}{3175200000}z^4 - \frac{213648613}{73347120000000}z^6 \\
&\quad + \frac{1669816359863}{2135868134400000000}z^8 - \frac{409429160306437}{2135868134400000000000}z^{10} + \cdots\bigg),
\end{aligned}$$

$$\alpha_3 = \sqrt{15}\left(\frac{1}{30} + \frac{3}{14000}z^2 - \frac{24739}{793800000}z^4 + \frac{14753813}{2993760000000}z^6 - \frac{7187933379103}{6407604403200000000}z^8 + \frac{48242846122937}{177989011200000000000}z^{10} + \cdots\right),$$

$$\alpha_4 = \sqrt{15}\left(-\frac{1}{24} + \frac{13}{67200}z^2 - \frac{37}{12700800}z^4 + \frac{19922401}{469421568000000}z^6 - \frac{733072729}{12204960768000000000}z^8 + \frac{1539941201}{183074411520000000000}z^{10} + \cdots\right).$$

Case $(K' = 1, P' = 1)$

Equations (13.4.4) and (13.4.6) remain unchanged. Equation (13.4.5) is replaced by the equation obtained by applying the operator (13.2.12) with $s = 3$ on $t\exp(\pm\lambda t)$ resulting in:

$$2b_1 z^2\theta \sinh(\theta z) = z\cosh(z/2) - 2\sinh(z/2). \tag{13.4.7}$$

Taking into account (13.4.6) and (13.4.7) b_1 and b_2 can be expressed in terms of θ:

$$b_1 = \frac{z\cosh(z/2) - 2\sinh(z/2)}{2z^2\theta\sinh(\theta z)}, \tag{13.4.8}$$

$$b_2 = \frac{-\cosh(\theta z)z\cosh(z/2) + 2\cosh(\theta z)\sinh(z/2) + 2\sinh(z/2)z\theta\sinh(\theta z)}{z^2\theta\sinh(\theta z)}. \tag{13.4.9}$$

Introducing these results for b_1 and b_2 into (13.4.4) provides an equation for θ:

$$\frac{(1-\cosh(\theta z))(z\cosh(z/2) - 2\sinh(z/2)) + z\theta\sinh(\theta z)(2\sinh(z/2) - z)}{z^2\theta\sinh(\theta z)} = 0.$$

From the symplecticness condition (13.4.1) an expression for γ_1 follows:

$$\gamma_1 = \frac{\gamma_2\cosh(2\theta z)(z\cosh(z/2) - 2\sinh(z/2))}{\cosh(\theta z)(z\cosh(z/2) - 2\sinh(z/2)) - 2\sinh(z/2)z\theta\sinh(\theta z)(1 - \gamma_2\cosh(z/2))}. \tag{13.4.10}$$

Again we choose $\gamma_2 = 1$. The series expansions for the different parameters now follow immediately:

$$\theta = \sqrt{15}\left(\frac{1}{10} + \frac{1}{10500}z^2 - \frac{31}{117600000}z^4 + \frac{2869}{5433120000000}z^6 - \frac{332933}{355978022400000000}z^8 + \frac{1792783}{711956044800000000000}z^{10} + \cdots\right),$$

$$\gamma_1 = 1 - \frac{9}{280000}z^6 + \frac{6861}{784000000}z^8 - \frac{3685091}{1724800000000}z^{10} + \cdots,$$

$$b_1 = \frac{5}{18} - \frac{1}{1890}z^2 - \frac{23}{21168000}z^4 + \frac{3383}{244490400000}z^6 - \frac{6186473}{128152088064000000}z^8 + \frac{6259951}{448532308224000000000}z^{10} + \cdots,$$

$$b_2 = \frac{4}{9} + \frac{1}{945}z^2 + \frac{23}{10584000}z^4 - \frac{3383}{122245200000}z^6 + \frac{6186473}{64076044032000000}z^8 - \frac{6259951}{224266154112000000000}z^{10} + \cdots,$$

$$\alpha_2 = \sqrt{15}\left(\frac{1}{15} - \frac{1}{18000}z^2 + \frac{1063}{352800000}z^4 - \frac{4445759}{2037420000000}z^6 + \frac{1250913246151}{2135868134400000000}z^8 - \frac{305480839860709}{2135868134400000000000}z^{10} + \cdots\right),$$

$$\alpha_3 = \sqrt{15}\left(\frac{1}{30} + \frac{19}{126000}z^2 - \frac{2179}{88200000}z^4 + \frac{8735197}{2328480000000}z^6 - \frac{1798803442789}{2135868134400000000}z^8 + \frac{216068604952379}{1067934067200000000000}z^{10} + \cdots\right),$$

$$\alpha_4 = \sqrt{15}\left(-\frac{1}{24} + \frac{43}{201600}z^2 - \frac{59}{28224000}z^4 + \frac{1419377}{52157952000000}z^6 - \frac{431537179}{1220496076800000000}z^8 + \frac{237023071}{533967033600000000000}z^{10} + \cdots\right).$$

Case $(K' = -1, P' = 2)$

Equations (13.4.6) and (13.4.7) remain unchanged. A third equation is added which follows from the application of the operator (13.2.12) with $s = 3$ on $t^2 \exp(\pm\lambda t)$, i.e.:

$$\begin{aligned} &b_1 \cosh(z\theta)\left(2\cosh(z/2) + \frac{1}{2}z\sinh(z/2) + 2z\theta^2\sinh(z/2)\right) - \cosh(z) \\ &+ 2b_1\sinh(z\theta)(2\theta\sinh(z/2) + z\theta\cosh(z/2)) \\ &+ b_2\left(\cosh(z/2) + \frac{1}{4}z\sinh(z/2)\right) = 0. \end{aligned} \tag{13.4.11}$$

The formal expression for b_1 and b_2 remain respectively (13.4.8) and (13.4.9). Introducing these expression for b_1 and b_2 into (13.4.11) gives us an equation for θ. From the symplecticness condition (13.4.1) again the expression (13.4.10) for γ_1 follows. Again by choosing $\gamma_2 = 1$, the series expansion of the different parameters follow:

$$\theta = \sqrt{15}\left(\frac{1}{10} + \frac{1}{7000}z^2 - \frac{37}{88200000}z^4 - \frac{2323}{679140000000}z^6 + \frac{466717}{889945056000000}z^8 - \frac{15014807}{667458792000000000}z^{10} + \cdots\right),$$

$$\gamma_1 = 1 - \frac{27}{1120000}z^6 + \frac{41379}{6272000000}z^8 - \frac{22149861}{13798400000000}z^{10} + \cdots,$$

$$b_1 = \frac{5}{18} - \frac{1}{1260}z^2 - \frac{187}{31752000}z^4 + \frac{11887}{91683900000}z^6 - \frac{14932867}{1601901100800000}z^8 - \frac{16262011}{28033269264000000}z^{10} + \cdots,$$

$$b_2 = \frac{4}{9} + \frac{1}{630}z^2 + \frac{187}{15876000}z^4 + \frac{173633}{733471200000}z^6 - \frac{52835987}{32038022016000000}z^8 + \frac{817009801}{224266154112000000000}z^{10} + \cdots,$$

$$\alpha_2 = \sqrt{15}\left(\frac{1}{15} + \frac{1}{18000}z^2 + \frac{719}{176400000}z^4 - \frac{157253603}{97796160000000}z^6 + \frac{468408965117}{1067934067200000000}z^8 - \frac{76002203332597}{711956044800000000000}z^{10} + \cdots\right),$$

$$\alpha_3 = \sqrt{15}\left(\frac{1}{30} + \frac{11}{126000}z^2 - \frac{3523}{176400000}z^4 + \frac{40063763}{13970880000000}z^6 - \frac{675385487507}{1067934067200000000}z^8 + \frac{322656693230117}{2135868134400000000000}z^{10} + \cdots\right),$$

$$\alpha_4 = \sqrt{15}\left(-\frac{1}{24} + \frac{47}{201600}z^2 - \frac{73}{56448000}z^4 + \frac{1520789}{156473856000000}z^6 - \frac{220181869}{1220496076800000000}z^8 + \frac{47152907}{140633292800000000000}z^{10} + \cdots\right).$$

Remark Sixth-order symmetric and symplectic modified Runge-Kutta methods of Gauss type have been constructed by others. In [3] the authors constructed such methods by making use of a basic set consisting of $\{\exp(\pm\lambda t), \exp(\pm 2\lambda t), \exp(\pm 3\lambda t)\}$ with fixed θ-values and frequency dependent θ-values. In [4] analogous constructions are discussed based on a reference set $\{t, t^2, \exp(\pm\lambda t)\}$, again with fixed and frequency dependent θ-values. In both cases the results are in a sense comparable with ours and in the numerical experiments we shall compare the results of [4] with the ones we have obtained.

13.5 Numerical Experiments

In this section we report on some numerical experiments where we test the effectiveness of the new and the previous [4] modified Runge-Kutta methods when they are applied to the numerical solution of several differential systems. All the considered codes have the same qualitative properties for the Hamiltonian systems. In the figures we show the decimal logarithm of the maximum global error versus the number of steps required by each code in logarithmic scale. All computations were carried out in double precision and series expansions are used for the coefficients when $|z| < 0.1$. In all further displayed figures following results are shown: the

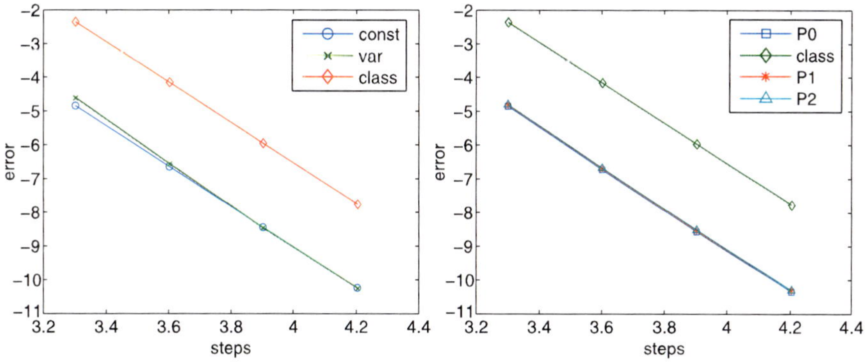

Fig. 13.1 Maximum global error in the solution of Problem 1. In the *left figure* the results obtained by the methods of Calvo et al. [4] are displayed. In the *right figure* the results obtained with the methods of order six derived in this paper are shown

method of Calvo et al. with constant nodes (const) and with variable nodes (var), the classical Gauss results (class) and the results obtained with the new methods with $P' = 0$ (P0), $P' = 1$ (P1) and $P' = 2$ (P2).

Problem 1 Kepler's plane problem defined by the Hamiltonian function

$$H(p,q) = \frac{1}{2}(p_1^2 + p_2^2) - (q_1^2 + q_2^2)^{-1/2},$$

with the initial conditions $q_1(0) = 1 - e$, $q_2(0) = 0$, $p_1(0) = 0$, $p_2(0) = ((1+e)/(1-e))^{\frac{1}{2}}$, where e $(0 \le e < 1)$ represents the eccentricity of the elliptic orbit. The exact solution of this IVP is a 2π-periodic elliptic orbit in the (q_1, q_2)-plane with semimajor axis 1, corresponding the starting point to the pericenter of this orbit. In the numerical experiments presented here we have chosen the same values as in [4], i.e. $e = 0.001$, $\lambda = i\omega$ with $\omega = (q_1^2 + q_2^2)^{-\frac{3}{2}}$ and the integration is carried out on the interval [0, 1000] with the steps $h = 1/2^m$, $m = 1, \ldots, 4$. The numerical behavior of the global error in the solution is presented in Fig. 13.1. The results obtained by the three new constructed methods are falling together. One cannot distinguish the results. They are comparable to the ones obtained by Calvo and more accurate than the results of the classical Gauss method of the same order. Remark that e has been kept small as it was the case in previous papers. We have however observed that increasing e does not changed the conclusions reached.

Problem 2 A perturbed Kepler's problem defined by the Hamiltonian function

$$H(p,q) = \frac{1}{2}(p_1^2 + p_2^2) - \frac{1}{(q_1^2 + q_2^2)^{1/2}} - \frac{2\epsilon + \epsilon^2}{3(q_1^2 + q_2^2)^{3/2}},$$

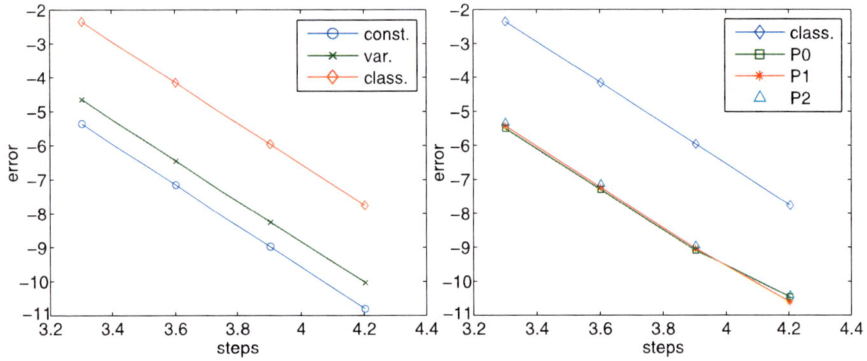

Fig. 13.2 Maximum global error in the solution of Problem 2. In the *left figure* the results obtained by the methods of Calvo et al. [4] are displayed. In the *right figure* the results obtained with the methods of order six derived in this paper are shown

with the initial conditions $q_1(0) = 1$, $q_2(0) = 0$, $p_1(0) = 0$, $p_2(0) = 1 + \epsilon$, where ϵ is a small positive parameter. The exact solution of this IVP is given by

$$q_1(t) = \cos(t + \epsilon t), \qquad q_2(t) = \sin(t + \epsilon t), \qquad p_i(t) = q_i'(t), \quad i = 1, 2.$$

As in [4] the numerical results are computed with the integration steps $h = 1/2^m$, $m = 1, \ldots, 4$. We take the parameter $\epsilon = 10^{-3}$, $\lambda = i\omega$ with $\omega = 1$ and the problem is integrated up to $t_{end} = 1000$. The global error in the solution is presented in Fig. 13.2. For our methods we have the same conclusions as for the Problem 1. On the contrary for the results of Calvo the results obtained with fixed θ-values are more accurate than the ones obtained by variable θ-values.

Problem 3 Euler's equations that describe the motion of a rigid body under no forces

$$\dot{q} = f(q) = ((\alpha - \beta)q_2q_3, (1-\alpha)q_3q_1, (\beta - 1)q_1q_2)^T,$$

with the initial values $q(0) = (0, 1, 1)^T$, and the parameter values $\alpha = 1 + \frac{1}{\sqrt{1.51}}$ and $\beta = 1 - \frac{0.51}{\sqrt{1.51}}$. The exact solution of this IVP is given by

$$q(t) = \left(\sqrt{1.51}\,\mathrm{sn}(t, 0.51), \mathrm{cn}(t, 0.51), \mathrm{dn}(t, 0.51)\right)^T,$$

it is periodic with period $T = 7.45056320933095$, and sn, cn, dn stand for the elliptic Jacobi functions. Figure 13.3 shows the numerical results obtained for the global error computed with the iteration steps $h = 1/2^m$, $m = 1, \ldots, 4$, on the interval $[0, 1000]$, and $\lambda = i2\pi/T$. The results of Calvo et al. are all of the same accuracy while in our approach the EF methods are still more accurate than the classical one. In this problem the choice of the frequency is not so obvious and therefore the differentiation between the classical and the EF methods is not so pronounced.

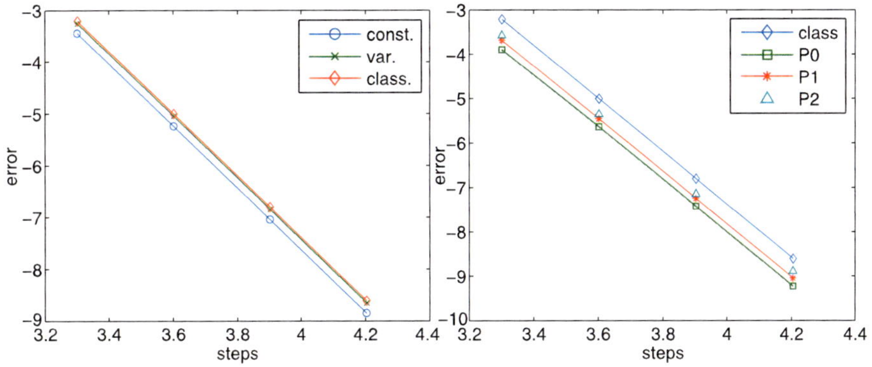

Fig. 13.3 Maximum global error in the solution of Problem 3. In the *left figure* the results obtained by the methods of Calvo et al. [4] are displayed. In the *right figure* the results obtained with the methods of order six derived in this paper are shown

13.6 Conclusions

In this paper another approach for constructing symmetric symplectic modified EFRK methods based upon the sixth-step procedure of [8] is presented. Two-stage fourth-order and three-stage sixth-order integrators of Gauss type which are symmetric and symplectic and which preserve linear and quadratic invariants have been derived. When the frequency used in the exponential fitting process is put to zero all considered integrators reduce to the classical Gauss integrator of the same order. Some numerical experiments show the utility of these new integrators for some oscillatory problems. The results obtained here are quite similar to the ones obtained in [4], but they differ in some of the details. The introduced method can be extended to EFRK with larger algebraic order.

References

1. Bettis, D.G.: Runge-Kutta algorithms for oscillatory problems. J. Appl. Math. Phys. **30**, 699–704 (1979)
2. Calvo, M., Franco, J.M., Montijano, J.I., Rández, L.: Structure preservation of exponentially fitted Runge-Kutta methods. J. Comput. Appl. Math. **218**, 421–434 (2008)
3. Calvo, M., Franco, J.M., Montijano, J.I., Rández, L.: Sixth-order symmetric and symplectic exponentially fitted Runge-Kutta methods of the Gauss type. Comput. Phys. Commun. **178**, 732–744 (2008)
4. Calvo, M., Franco, J.M., Montijano, J.I., Rández, L.: Sixth-order symmetric and symplectic exponentially fitted modified Runge-Kutta methods of the Gauss type. J. Comput. Appl. Math. **223**, 387–398 (2009)
5. Franco, J.M.: Runge-Kutta methods adapted to the numerical integration of oscillatory problems. Appl. Numer. Math. **50**, 427–443 (2004)
6. Hairer, E., Nørsett, S.P., Wanner, G.: Solving Ordinary Differential Equations I, Nonstiff Problems. Springer, Berlin/Heidelberg (1993)

7. Hairer, E., Lubich, C., Wanner, G.: Geometric Numerical Integration: Structure Preserving Algorithms for Ordinary Differential Equations. Springer, Berlin (2002)
8. Ixaru, L.Gr., Vanden Berghe, G.: Exponential Fitting. Mathematics and Its Applications, vol. 568. Kluwer Academic, Dordrecht (2004)
9. Ozawa, K.: A functional fitting Runge-Kutta method with variable coefficients. Jpn. J. Indust. Appl. Math. **18**, 107–130 (2001)
10. Sanz-Serna, J.M.: Symplectic integrators for Hamiltonian problems: An overview. Acta Numer. **1**, 243–286 (1992)
11. Sanz-Serna, J.M., Calvo, M.P.: Numerical Hamiltonian Problems. Chapman and Hall, London (1994)
12. Simos, T.E.: An exponentially-fitted Runge-Kutta method for the numerical integration of initial-value problems with periodic or oscillating solutions. Comput. Phys. Commun. **115**, 1–8 (1998)
13. Simos, T.E., Vigo-Aguiar, J.: Exponentially-fitted symplectic integrator. Phys. Rev. E **67**, 1–7 (2003)
14. Van de Vyver, H.: A fourth order symplectic exponentially fitted integrator. Comput. Phys. Commun. **174**, 255–262 (2006)
15. Vanden Berghe, G., De Meyer, H., Van Daele, M., Van Hecke, T.: Exponentially-fitted explicit Runge-Kutta methods. Comput. Phys. Commun. **123**, 7–15 (1999)
16. Vanden Berghe, G., De Meyer, H., Van Daele, M., Van Hecke, T.: Exponentially-fitted Runge-Kutta methods. J. Comput. Appl. Math. **125**, 107–115 (2000)
17. Vanden Berghe, G., Van Daele, M., Van de Vyver, H.: Exponentially-fitted Runge-Kutta methods of collocation type: Fixed or variable knot points? J. Comput. Appl. Math. **159**, 217–239 (2003)

Index

T.E. Simos (ed.), *Recent Advances in Computational and Applied Mathematics*,
DOI 10.1007/978-90-481-9981-5, © Springer Science+Business Media B.V. 2011